Springer Series in Electrophysics
Volume 13

Edited by Günter Ecker

Springer Series in Electrophysics

Editors: Günter Ecker Walter Engl Leopold B. Felsen

Volume 1 **Structural Pattern Recognition** By T. Pavlidis

Volume 2 **Noise in Physical Systems** Editor: D. Wolf

Volume 3 **The Boundary-Layer Method in Diffraction Problems**
By V. M. Babič, N. Y. Kirpičnikova

Volume 4 **Cavitation and Inhomogeneities** in Underwater Acoustics
Editor: W. Lauterborn

Volume 5 **Very Large Scale Integration (VLSI)**
Fundamentals and Applications
Editor: D. F. Barbe 2nd Edition

Volume 6 **Parametric Electronics** An Introduction
By K.-H. Löcherer, C. D. Brandt

Volume 7 **Insulating Films on Semiconductors**
Editors: M. Schulz, G. Pensl

Volume 8 **Theoretical Fundamentals of the Ocean Acoustics**
By L. Brekhovskikh, Y. P. Lysanov

Volume 9 **Principles of Plasma Electrodynamics**
By A. F. Alexandrov, L. S. Bogdankevich, A. A. Rukhadze

Volume 10 **Ion Implantation Techniques**
Editors: H. Ryssel, H. Glawischnig

Volume 11 **Ion Implantation: Equipment and Techniques**
Editors: H. Ryssel, H. Glawischnig

Volume 12 **Fundamentals of VLSI Technology**
Editor: Y. Tarui

Volume 13 **Physics of Highly Charged Ions**
By R. K. Janev, L. P. Presnyakov, V. P. Shevelko

Volume 14 **Plasma Physics for Physicists**
By I. A. Artsimovich, R. Z. Sagdeev

Volume 15 **Relativity and Engineering**
By J. Van Bladel

Volume 16 **Elements of Random Process Theory**
By S. M. Rytov, Y. A. Kravtsov, V. I. Tatarsky

Volume 17 **Concert Hall Acoustics**
By Y. Ando

Volume 18 **Planar Circuits** for Microwaves and Lightwaves
by T. Okoshi

Volume 19 **Physics of Shock Waves in Gases and Plasmas**
By M. Liberman, A. Velikovich

Volume 20 **Kinetic Theory of Particles and Photons**
By J. Oxenius

R.K. Janev L.P. Presnyakov
V.P. Shevelko

Physics of Highly Charged Ions

With 93 Figures

Springer-Verlag
Berlin Heidelberg New York Tokyo

Professor Dr. Ratko K. Janev

Institute of Physics, P.O. Box 57, 11001 Belgrade, Yugoslavia

Professor Dr. Leonid P. Presnyakov
Dr. Vjatcheslav P. Shevelko

Academy of Sciences of the USSR, P.N. Lebedev Physical Institute, Leninsky Prospect, 53, SU-117924 Moscow, USSR

Series Editors:

Professor Dr. Günter Ecker

Ruhr-Universität Bochum, Theoretische Physik, Lehrstuhl I, Universitätsstraße 150, D-4630 Bochum-Querenburg, Fed. Rep. of Germany

Professor Dr. Walter Engl

Institut für Theoretische Elektrotechnik, Rhein.-Westf. Technische Hochschule, Templergraben 55, D-5100 Aachen, Fed. Rep. of Germany

Professor Leopold B. Felsen Ph.D.

Polytechnic Institute of New York, 333 Jay Street, Brooklyn, NY 11201, USA

ISBN-13:978-3-642-69197-3 e-ISBN-13:978-3-642-69195-9
DOI: 10.1007/978-3-642-69195-9

Library of Congress Cataloging in Publication Data. Janev, R. K. (Ratko K.), 1939– . Physics of highly charged ions. (Springer series in electrophysics; v. 13). 1. Ions. I. Presnyakov, L. P. II. Shevelko, V. P. III. Title. IV. Title: Highly charged ions. V. Series. QC702.J28 1985 539 84-25471

2153/3020-543210

Preface

The physics of highly charged ions has become an essential ingredient of many modern research fields, such as x-ray astronomy and astrophysics, controlled thermonuclear fusion, heavy ion nuclear physics, charged particle accelerator physics, beam-foil spectroscopy, creation of xuv and x-ray lasers, etc. A broad spectrum of phenomena in high-temperature laboratory and astrophysical plasmas, as well as many aspects of their global physical state and behaviour, are directly influenced, and often fully determined, by the structure and collision properties of multiply charged ions. The growth of interest in the physics of highly charged ions, experienced especially in the last ten to fifteen years, has stimulated a dramatic increase in research activity in this field and resulted in numerous significant achievements of both fundamental and practical importance.

This book is devoted to the basic aspects of the physics of highly charged ions. Its principal aim is to provide a basis for understanding the structure and spectra of these ions, as well as their interactions with other atomic particles (electrons, ions, atoms and molecules). Particular attention is paid to the presentation of theoretical methods for the description of different radiative and collision phenomena involving multiply charged ions. The experimental material is included only to illustrate the validity of theoretical methods or to demonstrate those physical phenomena for which adequate theoretical descriptions are still absent. The general principles of atomic spectroscopy are included to the extent to which they are pertinent to the subject matter.

In the course of the work on this book we had valuable and useful discussions with our colleagues Drs. T. P. Grozdanov, B. N. Chichkov and D. B. Uskov. To all of them we express our sincere gratitude.

Belgrade-Moscow
December, 1984

R. K. Janev
L. P. Presnyakov *V. P. Shevelko*

Contents

1. Introduction

The physics of highly ionized atoms is not only a field of extraordinary richness in specific physical phenomena, but also a part of atomic physics which cross-correlates with many other scientific disciplines, such as astrophysics, plasma physics, heavy ion physics, etc. Apart from its pure scientific importance, the physics of highly charged ions also has a strong impact on research in several applied physics and technological areas (fusion research, material sciences, etc.), which provides a further stimulus for the development of this field.

In this introductory chapter, we shall give a brief account of the interdisciplinary and applied aspects of the physics of highly charged ions, and define the scope of the book.

1.1 General Overview

Pioneering work in the study of multiply charged ions started as early as the 1920s when *Millikan* and his coworkers investigated the spectra of the Na I isoelectronic sequence up to Cl VII, created in a condensed vacuum spark [1.1]. For many years, spectroscopic studies were a central topic in the investigations of multiply charged ions, motivated by the challenge of testing various theoretical predictions within quantum-mechanical concepts and bridging the gap between optical and x-ray spectra. A good account of these early investigations of the structure and spectra of highly charged ions can be found in the reviews by *Edlén* [1.2, 3]. The accumulated information on the energy levels of these ions, with references to the line classification work, is contained in the well-known compilation of *Moore* [1.4].

A new era in the studies of the structure and spectra of highly ionized atoms began with the discovery by *Edlén* and *Tyren* [1.5] of so-called satellite lines in the spectra of multiply charged ions, one of the essentially new elements in these spectra with respect to those of neutral atoms. This discovery and its explanation is fundamental to the analysis of the radiation from highly charged ions and its use for diagnostic purposes.

The recent experimental progress in highly charged ion physics has been connected with the important developments in the construction of laboratory sources of multiply charged ions (such as magnetic fusion devices, laser-produced plasmas, beam-foil sources, newly developed sources for charged-particle accelerators), parallelled with advances in the experimental tools and methods used for studying highly charged ions and their collision processes (high-resolution spectrometers, detectors, intersecting beam techniques, ion traps, etc.). Spacecraft-based studies, initiated in the 1960s (rockets, satellites), have provided considerable information on highly charged ions in hot astrophysical plasmas. The experimental progress has been complemented by significant theoretical developments in the description of the structure and collision physics of highly charged ions. Both the degree of sophistication of theoretical models and the availability of powerful computing facilities have considerably increased the accuracy of theoretical predictions.

Despite these remarkable achievements, most of which have been obtained in the last ten to fifteen years, many challenging problems in the physics of highly charged ions still wait to be solved. This is particularly true for the collision physics of highly charged ions, where even the mechanisms for some processes are not properly understood. Many new developments and interesting results are still to be expected in this field of research.

1.2 Basic Features of Highly Charged Ions and Their Interactions

The specific properties distinguishing the highly charged ion from the neutral atom are mainly due to the unshielded Coulomb field of the nucleus of the ion. Its effects show not only in the structure of the ion, but also in its interactions with other atomic particles and with the electromagnetic field. The structure of few-electron highly charged ions is considerably affected by relativistic effects in the electron motion and the absence of any "pure" angular momentum coupling scheme. The formation of the excited states of multiply charged ions is frequently characterized by the high angular momenta of these states, and their decay takes place not only by allowed processes but also through severe violation of the normal selection rules. The typical radiation emitted by highly ionized atoms falls in the vacuum ultraviolet (vuv) and x-ray wavelength regions.

The long-range Coulomb field of multicharged ions decisively influences the collision dynamics and is reflected in all collision interactions of these ions. One important feature of the interaction processes is the strong dependence of their probabilities or cross sections on the ionic charge Z. In most cases, the Z-dependence of the cross sections can be (approximately) represented in the form of a scaling factor Z^{α}. For the electron-ion and ion-ion processes α is negative, while for the radiative and ion-atom (molecule) processes α is a positive number, which for some processes and in certain energy regions may even be considerably larger than one. This feature of the collision processes of highly charged ions, resulting in extremely large (or small, for $\alpha \leq -1$) cross sections, represents the physical basis of their pronounced role in high-temperature astrophysical and laboratory plasmas and in different practical applications.

Another important characteristic of the collisions of highly charged ions with atoms is the great variety of processes which may occur. A highly charged ion brings into the collision a significant amount of additional potential energy which can be disseminated through a large number of inelastic reaction channels: creation of excited states, autoionizing states, and (even copious) production of secondary electrons are the most prominent examples of the results of a collision event.

1.3 Highly Charged Ions in Astrophysics

Analysis of the spectral information in the uv and x-ray regions, obtained from different astrophysical objects, provides a unique possibility for the study of physical processes taking place in these objects. The most abundant emission in the uv and x-ray region comes from the stars, supernova remnants, so-called compact x-ray sources, galactic nuclei and quasi-stellar objects, and it signifies the occurrence of highly energetic phenomena in them. A proper interpretation of the uv and x-ray spectra from these objects can give information about their structure, composition, energy balance, dynamics of mass flow, density and temperature distributions, and other relevant physical parameters.

While galactic and extra-galactic x-ray astronomy is still in its beginning stage, solar uv and x-ray spectroscopy have already given invaluable contributions to our understanding of physical processes and conditions in the hot solar plasma. The spacecraft-based studies were and are an important impetus for these investigations. The analysis of the uv and x-ray spectra from the active solar regions and in particular, from solar flares [1.6], has revealed the existence of highly charged ions (with a considerable abundance) of almost all the elements up to Ni^{26+}. Especially rich are the uv and x-ray spectra of flare emission, originating from transitions in ions of both intermediate and heavier elements (up to Ni and Cu) in almost all stages of ionization. These spectra have provided information not only about the composition and physical conditions in the solar chromosphere and corona, but also, through their correct interpretation, about the elementary collision processes in the solar plasma. On the other hand, precise data about the cross sections of radiative and collision processes of multiply charged ions are a prerequisite both for the diagnostics and modelling of the solar plasma. The physics of highly charged ions is, thus, an essential ingredient of solar physics [1.7, 8].

1.4 Highly Charged Ions in Controlled Thermonuclear Fusion Research

Multiply charged ions are present and play an essential role in magnetically and inertially confined high-temperature fusion plasmas. In magnetic fusion devices, the ions of heavier elements originate from the material released from the walls (and the limiter, in the case of tokamaks) due to the plasma – solid surface interaction processes. In the inertial fusion plasmas, they originate from the high-Z outer layer material of the pellet. In contrast to the basic fusion plasma constituents (electrons and hydrogenic isotopes), the ions of heavier elements are qualified as "plasma impurities" and have many deleterious effects. In tokamak plasmas, small amounts of highly charged impurity ions lead to an enormous increase in the radiation and particle losses, a decrease in the Ohmic heating, the creation of local instabilities, etc. [1.9, 10]. Taken together, these effects strongly modify the conditions of thermonuclear fusion burning (the Lawson criterion) and may even prevent the tokamak reactor ignition. The reason for the strong effects of impurity ions on the plasma energy balance and other properties is the strong charge dependence of their radiative and collision processes. The latter may also impair the efficiency of the neutral beam method for supplementary heating of magnetically confined fusion plasmas [1.11].

However, the presence of highly charged ions and their processes in a fusion plasma may also have positive effects since they can be used for diagnostics and burning control purposes. In the case of inertial confinement fusion, utilizing either laser or charged-particle (electrons, ions) beams, the uv and x-ray radiation from the highly charged ions (or products of their collision processes) may be used for studies of the dynamics of pellet implosion, the coupling of the beam-driven energy with the pellet material, and other parameters defining the physical state of the hot plasma. Controlled thermonuclear fusion research has greatly stimulated studies of the structure and collision processes of highly charged ions by providing intense laboratory sources of such ions (tokamaks, laser-produced plasmas) and imposing stringent data requirements [1.12] on the atomic physics research in this field [1.13, 14].

1.5 Highly Charged Ions in Accelerator-Based Physics and Other Research Fields

Modern accelerator-based ion physics includes many areas of research, such as heavy ion nuclear physics, beam-foil interaction physics, high-energy atomic collision physics, ion-solid surface collision phenomena, etc. The new generations of charged-particle accelerators use ion sources, able to produce ions of heavy elements in high charge states. Among these, the ECR (electron cyclotron resonance) source and EBIS (electron beam ion source) have proved to be the most efficient in producing high-intensity, low-energy ion beams with charges up to $Z \approx 30$ (at present). Construction of such ion sources needs knowledge of the collision processes involving multiply charged ions. Using standard stripping techniques, accelerated heavy ion beams may further be stripped to much higher charge states (up to $Z \approx 70$). Highly charged energetic ions may be used to probe the inner-shell structure of atoms and to study inner-shell collision phenomena (vacancy production and associated radiative and Auger processes), to make ion penetration and channelling studies in condensed media (stopping power of matter, coherent radiation and blocking of channelled ions), and to simulate neutron radiation damage and other radiobiological effects, etc. [1.15–18]. Violent energetic ion collisions provide a possibility for testing quantum electrodynamics in supercritical electric fields and studying the structure and radiative properties of superheavy atoms [1.19, 20].

Another field of research where multiply charged ions are involved is the creation of powerful lasers in the xuv and x-ray wavelength regions [1.21, 22]. Although there are serious technical difficulties, the search for adequate inversion schemes continues [1.23].

1.6 Scope of the Book

It is obvious that it would be impossible to reflect all the aspects of the physics of highly charged ions pertaining to the research areas discussed above. Therefore, we have adopted the philosophy of presenting in this book only the basic features of the structure and spectra of highly charged ions and their most important radiative and collision processes with other particles (electrons, ions, atoms, molecules). We have put the emphasis on general theoretical methods for the description of the structure and collision properties of multicharged ions, while the experimental results are included only to illustrate the validity of a particular theoretical method or the physical effect considered in cases when adequate theoretical descriptions do not exist. However, without aiming to be complete, we have made an attempt to mention also the most essential experimental achievements by referring to original papers.

The structure and spectra of highly charged ions are discussed in Chap. 2, while their radiative processes involving continuous states are considered in Chap. 3. Electron-ion collision processes are discussed in Chaps. 4 and 5. The inelastic collision processes of highly charged ions with heavy atomic particles (atoms, low-charged ions and molecules) are discussed in Chaps. 6–9, of which Chap. 6 is devoted to the general theoretical methods for the description of these processes. Finally, Chap. 10 deals with the rate constants for the collision processes of highly charged ions.

In some of the chapters (Chaps. 4–9), we found it more convenient to use the atomic system of units: $m_e = \hbar = e^2 = 1$. In these units, the characteristic atomic length $a_0 = \hbar^2/m e^2$ (the Bohr radius) and atomic velocity $v_0 = e^2/\hbar$ (the Bohr velocity) are also equal to one. The exceptions to this convention will be stated explicitly.

2. Structure and Spectra of Highly Charged Ions

The line spectra of highly charged ions are mostly studied in high-temperature laboratory sources (tokamak discharges, low-inductance sparks, exploding wires, laser-produced plasmas) and astrophysical plasmas (solar flares and active regions). Such spectra are widely used for plasma diagnostics, providing important information on plasma densities and temperatures, chemical impurities and other parameters. An increase in plasma temperature leads to an increase of the ionization stages of heavy elements, which radiate spectral lines mostly in the vacuum ultra-violet (vuv) and x-ray regions. Along with the usual, traditional methods for plasma diagnostics in the visible and uv regions, diagnostic methods which are new in principle and are closely connected with features of the spectra of multiply charged ions are possible in the vuv and x-ray regions.

The spectroscopic symbol of the ion X^{q+} is Z, where

$$Z = Z_n - N + 1 = q + 1 , \tag{2.1}$$

Z_n being the nuclear charge, N the total number of electrons and q the net ionic charge. Sometimes the notation X_Z is used for the ion and sometimes spectroscopic notation, where Roman numerals denote the value of Z (e.g., the ion Ne^{8+} becomes Ne IX and the neutral Ne atom becomes Ne I). Throughout this chapter, we shall call an ion X^{q+} multiply (or highly) charged if $Z \geq 5$.

2.1 General Properties of the Spectra

The general spectral structure of a multiply charged ion resembles the spectrum of a neutral atom in the same isoelectronic sequence. An important difference is, however, the fact that the ion spectrum is shifted to the shorter-wavelength region. The energy levels and the spectra of neutral and low-ionized atoms have been studied elsewhere, see, for example [2.1, 2]. This section is mainly concerned with the study of the basic properties of the spectra caused by the strong Coulomb field of multiply charged ions. Some of these properties have easily understandable origins. For instance, the wavelengths of the resonance lines of the neutral H and He atoms are equal to 1216 and 593 Å, respectively, i.e. their ratio is close to 2. With increasing nuclear charge, the ion spectrum excited from K shells (configurations $1\,s$ and $1\,s^2$ in the ground state) looks increasingly similar to the hydrogen one. For example, the resonance wavelengths in Fe XXVI hydrogen-like, and Fe XXV helium-like, ions are equal to 1.78 Å and 1.85 Å, i.e. they differ by 5% only. The same behaviour can also be seen for ions along other isoelectronic sequences. In other words, with increasing nuclear charge the wavelengths of the lines of multiply charged ions with similar ion charges are very close to each other.

Another difference between the atomic and ionic spectra is connected with the clear spectral resolution of ionic lines which are not resolved in the corresponding atomic case. First of all, this concerns the fine structure components. The spacing between the fine structure components in the neutral hydrogen atom (transitions $2\,^2P_{3/2} - 1\,^2S_{1/2}$ and $2\,^2P_{1/2} - 1\,^2S_{1/2}$) is $0.36\ \mathrm{cm}^{-1}$. This doublet is unresolved in the emission spectra of low-temperature plasmas, because the Doppler width of the resonance line is two orders of magnitude larger at temperatures $T \gtrsim 1$ eV. In the spectra of multiply charged ions the distance between fine structure components is much larger than the Doppler width even at temperatures $T \gtrsim 10^6$ K. In H-like ions, such as Ca XX for example, the wavelengths of the resonance doublet are equal to 3.018 and 3.024 Å, respectively, and these lines are well resolved in the emission spectra of laboratory and astrophysical sources.

In the case of multiply charged ions, the relativistic corrections to the characteristics of orbital electrons increase strongly when the ionic charge increases. This leads to a significant increase in those transition probabilities which are usually small in the neutral atom in the same isoelectronic sequence. As an example, one can refer to the intercombination line $2\,^3P_1 - 1\,^1S_0$ in the spectra of He-like ions, the intensity of which is comparable with the intensity of the resonance line if $Z \gtrsim 10$. In the neutral helium atom, the magnetic dipole transition $2\,^3S_1 - 1\,^1S_0$ is considered to be a forbidden transition, while in the spectra of multiply charged ions this line is well registered and widely used for plasma diagnostics purposes.

Finally, the spectra of multiply charged ions are characterized by the presence of so-called satellite lines, along with the lines associated with the usual radiative transitions. The satellite lines arise due to radiative decay of autoionizing states, i.e. of the electron configurations with two (or more) excited electrons. Such states are located above the first ionization threshold and exist only in non-hydrogenic ions. For ions with $Z \lesssim 5$ these states decay predominantly by autoionization. The probability of radiative decay in highly charged ions is comparable or even greater than the autoionization probability, which leads to the appearance of a large number of satellite lines. Therefore, any one-electron parent line of an ion X_Z is accompanied by satellites arising from the ions X_{Z-1}, X_{Z-2}, etc. Moreover, for highly charged ions with $Z \gtrsim 10$, the satellite lines have intensities comparable with the intensity of the resonance line.

As a consequence of the physical effects mentioned above, there are many lines with comparable intensities in any limited spectral interval. Some of these lines belong to the ion with a given spectroscopic symbol Z and arise due to the resolution of fine structure components, the change of selection rules due to relativistic effects, etc. The satellite lines are emitted by ions of the preceding ionization stages. At first sight it seems that the large number of additional lines makes the identification and the analysis of the ion spectra difficult. However, with modern high-resolution instrumentation, this situation turns out to be one of the main advantages of x-ray spectroscopy. Each limited interval of the spectrum contains quantitative information not only about the ion structure, but also about the parameters of the plasma in which these ions exist. In this respect, the satellite lines play an essential role. To a certain extent, x-ray spectroscopy may be considered as a spectroscopy of the autoionizing states.

2.2 Satellites of the Spectral Lines

Let γ_0 and γ_1 denote the sets of quantum numbers of the ground and excited states of the ion X_Z, respectively. Then the satellite to the parent line, $\gamma_1 - \gamma_0$, corresponds to the transition $\gamma_1\, n\, l - \gamma_0\, n\, l$ of the ion X_{Z-1}, where n and l are the principal and orbital quantum numbers of the additional electron. If the ion charge is large enough, the presence of the additional electron does not change much the wavelength of the transition $\gamma_1 - \gamma_0$. Figure 2.1 shows the level diagrams of the H-, He- and Li-like ions with the same nuclear charge Z_n. The resonance doublet $2\,{}^2P_{3/2,1/2} - 1\,{}^2S_{1/2}$ in H-like ions has a group of satellites arising from the transitions between the configurations $2p\,n\,l - 1s\,n\,l$ and $2s\,n\,l - 1s\,n\,l$ in He-like ions. The resonance line of the He-like ion, $1s\,2p\,({}^1P) - 1s^2\,({}^1S)$, has satellites from Li-like ions (transitions $1s\,2p\,n\,l - 1s^2\,n\,l$, $1s\,2s\,n\,l - 1s^2\,n\,l$). The pattern is followed for Be-like ions, B-like ions, and so on.

In Fig. 2.1 the so-called doubly excited states are explicitly shown. The number of excited electrons may even be more than two. For example, in Be-like ions (4 electrons), autoionizing states $1s\,2s^2\,2p$, $1s\,2s\,2p^2$, $1s\,2p^3$, or generally, $1s\,n\,l\,n'\,l'\,n''\,l''$ are possible.

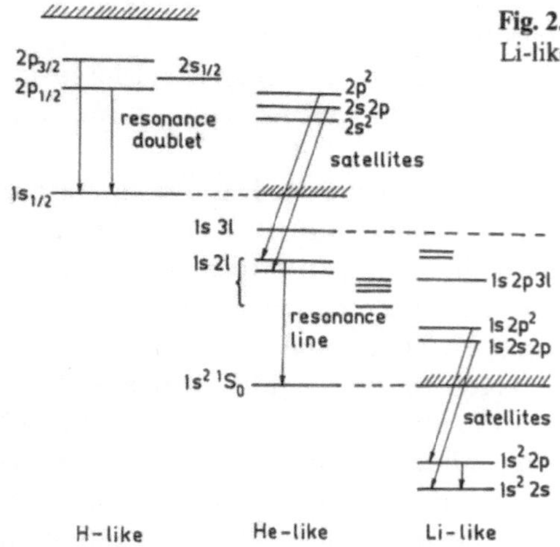

Fig. 2.1. Energy level scheme of H-, He- and Li-like ions

It is essential for all satellite lines that the upper level is always an autoionizing one, i.e. that it lies above the first ionization limit, while the lower level may be associated either with the ground state or with one of the excited states (Fig. 2.1). The intensity of the satellites increases as the ion charge increases. Actually, the autoionizing state can spontaneously decay via two channels,

$$\text{radiation:} \quad X_Z^{**}(\gamma_1\, n\, l) \rightarrow X_Z(\gamma_0\, n\, l) + \hbar\,\omega\,, \quad \text{or} \tag{2.2}$$

$$\text{autoionization:} \quad X_Z^{**}(\gamma_1\, n\, l) \rightarrow X_{Z+1}(\gamma_0) + e\,. \tag{2.3}$$

It is known that the autoionization probability, A_a, has a weak dependence on Z, whereas the radiative probability, A_r, increases strongly with Z (for electric dipole transitions, $A_r \sim Z^4$). Therefore, for ions with a small Z the autoionization channel (2.3) is the dominant one, for $Z = 10-20$ one has $A_r \approx A_a$, and for $Z > 30$ the radiative channel (2.2) dominates. These statements are valid only if none of the channels is forbidden due to symmetry constraints. Let us note that the satellites with $n = 2$ are well separated from the parent line and have a red shift; the $n = 3$ satellites are much closer to the parent line and may be located on both (the red and the blue) sides of it. The satellites with $n \geq 4$ are usually located within the Doppler width of the parent line.

The optical transitions of the type

$$1s\,2p\,n\,l - 1s^2\,n\,l\,, \tag{2.4}$$

i.e. satellites to the lines $1s\,2p - 1s^2$ of He-like ions were first observed and identified in 1939 by *Edlén* and *Tyren* [2.3] in the spectra of light elements using a vacuum spark source. These "unusual" transitions were afterwards studied in the spectra of multiply charged ions in both high-temperature laboratory sources and astrophysical plasmas (active regions and solar flares). At present, there exist many hundreds of original papers devoted to satellite spectra. Rather complete information on the main research directions in this field and the results obtained during the last 10–12 years is contained in several review articles, see, e.g. [2.4–11].

Among the laboratory sources used for investigation of the structure and the spectra of highly charged ions, an essential role belongs to laser-produced plasmas. A detailed review of this activity up to 1981 is given in the monograph by *Boiko* et al. [2.7]; other aspects can be found in [2.6, 9, 11]. A lot of results on x-ray spectra, obtained in solar flares, are given in [2.4, 5, 10, 12–20]. It is also necessary to mention the experiments with low-inductance sparks [2.21–24] and exploding wires [2.25, 26]. Recently, spectra of multiply charged ions with high spectroscopic resolution were obtained using fusion plasma devices of the tokamak type [2.27, 28]. Besides using plasma sources, it is also possible to investigate the spectra and energy levels of highly ionized atoms using accelerated ion beams passing through a very thin foil (beam-foil spectroscopy) [2.29–31]. In all of the high-temperature sources mentioned above, the satellite lines have intensities comparable with the intensities of "usual" lines (resonance and other lines).

At present, the spectral resolution of the wavelengths measured in laboratory and astrophysical plasmas is comparable with the accuracy of theoretical calculations and lies in the range $\Delta\lambda/\lambda = 10^{-4} - 10^{-5}$ for $\lambda = 1 - 10$ Å.

2.3 Classification of the States and Spectral Lines of Multiply Charged Ions

The classification and calculation of the energy levels of multiply charged ions require a choice of the correct coupling scheme for the angular and spin momenta of the electrons. Let us recall that LS-coupling is used when the relativistic effects (spin-orbit, spin-spin and others) are smaller than the electrostatic ones. The jj-coupling corresponds to the opposite case, when relativistic interactions are much stronger than electrostatic ones. The situation when the relativistic and the electrostatic effects are comparable and, consequently, neither the LS- nor the jj-coupling schemes can be used, is called intermediate coupling.

Another important factor, which should be taken into account in many-electron systems, is the effect of the correlation interaction between the electrons. This effect leads to deviations of the many-electron wave function from the usual antisymmetric combination of one-electron wave functions. The correlation effect can be described by the interaction of the configurations with the same principal quantum number.

The wave functions and energy levels of highly ionized atoms are calculated in the following way. The Hamiltonian of an N-electron ion with nuclear charge Z_n has the form

$$H = H_0 + H' , \tag{2.5}$$

where H_0 is the non-relativistic Hamiltonian

$$H_0 = \sum_{i=1}^{N} \left(-\frac{\hbar^2}{2m} \nabla_i^2 - \frac{Z_n e^2}{r_i} \right) + \sum_{i>j} \frac{e^2}{|\boldsymbol{r}_i - \boldsymbol{r}_j|} . \tag{2.6}$$

Here $\boldsymbol{r}_i$ is the coordinate of the ith electron, $\hbar$ is Planck's constant, and m and e are the mass and the charge of the electron. The relativistic part H' for a system of electrons in the nuclear field can be written as an expansion in powers of the fine structure constant $\alpha = e^2/\hbar c$, where c is the velocity of light in vacuum. The interaction between atomic particles is described by terms of order α^2 (without taking into account the radiative corrections) in the framework of the Breit-Pauli Hamiltonian, which will be considered in Sect. 2.4.

The solution of the eigenvalue problem with the total Hamiltonian (2.5) and, subsequently, the systematic calculation of the spectra of multiply charged ions (including both usual lines and satellites) can be performed as follows. First, the Schrödinger equation is solved with the non-relativistic Hamiltonian (2.6) as a zero-order approximation, using either the Hartree-Fock method [2.17–20] or an expansion of the many-electron wave functions in terms of one-electron Coulomb functions [2.32–36]. In the initial stage of the calculations, a pure type of angular momentum coupling is used, usually LS-coupling. The inclusion of relativistic effects and the transformation to the intermediate coupling scheme can be made by the diagonalization of the energy matrix.

Let a denote a set of quantum numbers SLJ, where S is the spin, L is the orbital angular momentum and J is the total angular momentum. Then in a zero-order approximation, one has

$$H_0 \, \varphi_n(a) = \varepsilon_n(a) \, \varphi_n(a) , \quad a = \{SLJ\} , \tag{2.7}$$

$$\varphi_n(a) = \sum_q \prod_{i=1}^{N} C_{iq}^n \, \varphi_q(\boldsymbol{r}_i) . \tag{2.8}$$

Here all the $\varphi_q(\boldsymbol{r}_i)$ are one-electron wave functions and the coefficients C^n_{iq} should be written taking into account antisymmetrization in the LS-coupling scheme. The wave functions $\psi_n(\gamma)$ of the Hamiltonian H can also be written in the form

$$\psi_n(\gamma) = \sum_{a'} V(a' \mid \gamma)\, \varphi_n(a') \tag{2.9}$$

where the coefficients $V(a \mid \gamma)$ are eigenvectors of the transformation to intermediate coupling. The matrix elements of the total Hamiltonian (2.5) are equal to

$$H_{\gamma\gamma'} = \langle \gamma \mid H \mid \gamma' \rangle = \sum_{a,a'} V(a \mid \gamma)\, H_{aa'}\, V(a' \mid \gamma'), \tag{2.10}$$

$$H_{aa'} = \langle a \mid H \mid a' \rangle . \tag{2.11}$$

Table 2.1. Key letter system used for designating the individual lines of He- and Li-like ions [2.17]

Array	Multiplet	Line $J - J'$	Key letter
$1s\,2p^2 - 1s^2\,2p$	$^2P - {^2P^0}$	$3/2 - 3/2$	a
		$3/2 - 1/2$	b
		$1/2 - 3/2$	c
		$1/2 - 1/2$	d
	$^4P - {^2P^0}$	$5/2 - 3/2$	e
		$3/2 - 3/2$	f
		$3/2 - 1/2$	g
		$1/2 - 3/2$	h
		$1/2 - 1/2$	i
	$^2D - {^2P^0}$	$5/2 - 3/2$	j
		$3/2 - 1/2$	k
		$3/2 - 3/2$	l
	$^2S - {^2P^0}$	$1/2 - 3/2$	m
		$1/2 - 1/2$	n
$1s\,2s^2 - 1s^2\,2p$	$^2S - {^2P^0}$	$1/2 - 3/2$	o
		$1/2 - 1/2$	p
$1s\,2p\,2s - 1s^2\,2s$	$(^1P)\,{^2P^0} - {^2S}$	$3/2 - 1/2$	q
		$1/2 - 1/2$	r
	$(^3P)\,{^2P^0} - {^2S}$	$3/2 - 1/2$	s
		$1/2 - 1/2$	t
	$^4P^0 - {^2S}$	$3/2 - 1/2$	u
		$1/2 - 1/2$	v
$1s\,2p - 1s^2$	$^1P^0 - {^1S}$	$1 - 0$	w
	$^3P^0 - {^1S}$	$2 - 0$	x
		$1 - 0$	y
$1s\,2s - 1s^2$	$^3S - {^1S}$	$1 - 0$	z

Table 2.2. Wavelengths λ [Å] and radiative transition probabilities A_r [s^{-1}] for Fe XXV

Transition	$\lambda_{\text{experim.}}$	Source	$\lambda_{\text{theoret.}}$	$A_{r\ \text{theoret.}}$
$1s\,2p\,{}^1P_1-1s^2\,S_0$ (w)	1.8500 [2.12]	Solar flare	1.8500 [2.28]	4.7×10^{14} [2.37]
	1.8500 [2.21]	Spark	1.8505 [2.35]	4.59×10^{14} [2.36]
	1.8500 [2.27]	Tokamak	1.8500 [2.20]	4.75×10^{14} [2.35]
	1.8500 ± 0.0015 [2.38]	Beam-foil	1.8507 [2.32]	4.63×10^{14} [2.20]
	1.8504 ± 0.0004 [2.39]	Spark	1.85046 [2.40]	
$1s\,2p\,{}^1P_1-1s\,2s\,{}^3S_1$	–	–	194.0 [2.35]	3.14×10^{8} [2.35]
$1s\,2p\,{}^1P_1-1s\,2s\,{}^1S_0$	–	–	383.53 [2.35]	4.38×10^{8} [2.35]
$1s\,2p\,{}^3P_2-1s^2\,{}^1S_0$ (x)	1.8555 [2.12]	Solar flare	1.8551 [2.18]	6.5×10^{9} [2.37]
	1.8552 [2.27]	Tokamak	1.8555 [2.35]	6.55×10^{9} [2.36]
			1.8550 [2.20]	6.79×10^{9} [2.41]
			1.85540 [2.40]	
$1s\,2p\,{}^3P_2-1s\,2s\,{}^3S_1$	271.02 ± 0.09 [2.42]	Beam-foil	270.22 [2.35]	1.37×10^{9} [2.35]
$1s\,2p\,{}^3P_1-1s^2\,{}^1S_0$ (y)	1.8585 [2.12]	Solar flare	1.8591 [2.18]	3.7×10^{13} [2.37]
	1.8589 [2.21]	Spark	1.8595 [2.35]	4.26×10^{13} [2.36]
	1.8592 [2.27]	Tokamak	1.8591 [2.20]	4.34×10^{13} [2.35]
			1.85944 [2.40]	3.82×10^{13} [2.20]
$1s\,2p\,{}^3P_1-1s\,2s\,{}^3S_1$	–	–	393.5 [2.35]	4.05×10^{8} [2.35]
$1s\,2s\,{}^1S_0-1s\,2p\,{}^3P_1$	–	–	–	1.57×10^{13} [2.35]
$1s\,2s\,{}^3S_1-1s^2\,{}^1S_0$ (z)	1.868 [2.12]	Solar flare	1.8678 [2.20]	2.09×10^{8} [2.36]
	1.8681 [2.27]	Tokamak	1.8686 [2.35]	2.35×10^{8} [2.41]
			1.86819 [2.40]	2.08×10^{8} [2.20]

The problem is now reduced to the solution of the secular equation

$$\text{Det}\,\|H_{aa'}-E\,\delta_{aa'}\|=0. \tag{2.12}$$

Substituting its roots (eigenvalues) into the system of linear equations

$$\sum_{a'}(H_{aa'}-E_\gamma\,\delta_{aa'})\,V(a'|\gamma)=0, \tag{2.13}$$

one obtains the eigenvectors $V(a|\gamma)$ and the wave functions (2.9). The eigenvalues E_γ give the energy levels of an ion with the intermediate coupling and configuration interactions taken into account. They are denoted, however, by the quantum numbers SLJ, indicating that for $Z\to1$, they are close to the "pure" LS-levels. However, for $Z\gg1$ they are properly "mixed". In the Hartree-Fock approach [2.17–20], the matrix elements $H_{aa'}$ (2.11) are calculated with the wave functions (2.8), obtained by numerical integration of (2.7)

In the Coulomb-function expansion method [2.32–35], a perturbation-theory expansion in powers of Z_n^{-1} is used, the calculations being done with the hydrogenic wave functions. All the matrix elements $H_{aa'}$ can be obtained in the form

$$H_{aa'} = \langle a|\, H\, |a'\rangle = Z_{\rm n}^2 \left(H_{aa'}^{(0)} + \frac{H_{aa'}^{(1)}}{Z_{\rm n}} + \frac{H_{aa'}^{(2)}}{Z_{\rm n}^2} + \dots \right). \tag{2.14}$$

The corrections $H_{aa'}^{(1)}$ and $H_{aa'}^{(2)}$ in (2.14) include infinite sums over both discrete and continuous spectra and can be calculated in an explicit analytical form for any isoelectronic sequence.

Both approaches described above are based on the same physical background, and account for relativistic effects to the order $\alpha^2 = (e^2/\hbar c)^2$. The wave functions in the zero-order approximation are obviously different. The $Z_{\rm n}^{-1}$ expansion method provides an asymptotic solution of the problem and its validity requires fulfillment of the condition $Z_{\rm n} - N \gg 1$. The Hartree-Fock approach is valid for all values of $Z_{\rm n}$; however, elaborate numerical calculations are required for any specific ion, starting with integration of (2.7).

Tables 2.2–4 contain theoretical results and available experimental data for the resonance lines and related satellites of hydrogen-like and helium-like multicharged ions. In order to avoid complicated spectroscopic symbols, we use the line designations introduced by *Gabriel* [2.17], listed in Table 2.1.

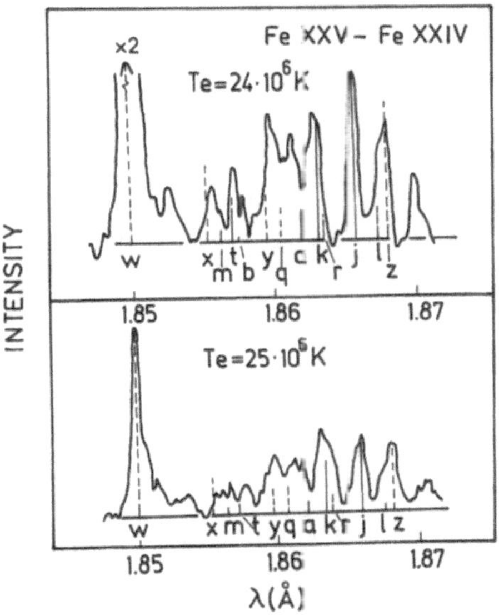

Fig. 2.2. Fe XXV spectra and associated satellites recorded at two times [2.13]. T_e is the electron temperatures of the x-ray solar flare

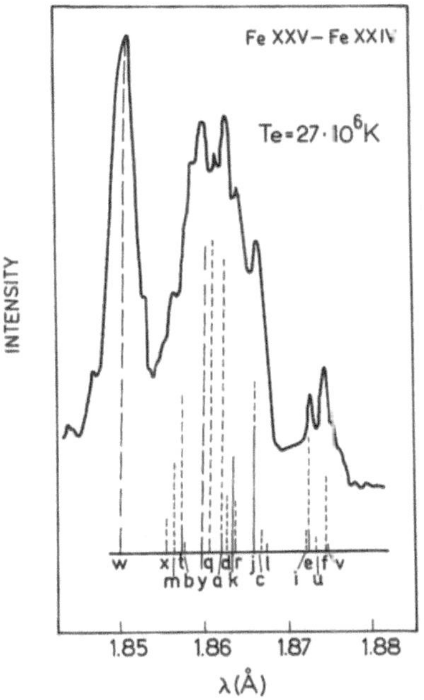

Fig. 2.3. Fe XXV spectrum and associated satellites of a low-inductance vacuum spark [2.21]

Table 2.3. Calculated wavelengths λ [Å], autoionization A_a and radiation A_r probabilities [$10^{13}\ s^{-1}$] for the resonance, intercombination and forbidden lines in He-like Mg, Ca, Fe ions and related satellites. Wavelengths are calculated with the help of the method [2.35, 41] including the Lamb shift effect[a]

Key letter	Mg ($Z_n = 12$)			Ca ($Z_n = 20$)			Fe ($Z_n = 26$)		
	λ	A_a	A_r	λ	A_a	A_r	λ	A_a	A_r
a	9.2983	0.158	2.40	3.2038	2.04	21.8	1.8622	4.14	62.2
b	9.2948	0.158	0.355	3.1997	2.04	1.75	1.8578	4.14	0.881
c	9.3016	1.16−3	0.882	3.2081	2.94−2	6.61	1.8672	9.04−2	16.4
d	9.2981	1.16−3	1.88	3.2040	2.94−2	17.8	1.8628	9.04−2	54.4
e	9.3841	2.32−2	1.13−3	3.2238	0.568	0.249	1.8727	2.63	3.50
f	9.3859	1.50−3	7.25−4	3.2258	3.16−2	0.103	1.8743	0.102	1.01
g	9.3823	1.50−3	7.32−6	3.2216	3.16−2	2.13−4	1.8699	0.102	2.30−3
h	9.3875	1.04−4	1.55−5	3.2277	4.53−3	2.53−3	1.8767	3.44−2	1.05−2
i	9.3839	1.04−4	4.97−4	3.2235	4.53−3	0.123	1.8706	3.44−2	2.11
j	9.3206	18.7	0.923	3.2102	18.1	8.03	1.8659	16.1	21.6
k	9.3163	18.5	0.878	3.2065	16.6	9.97	1.8631	14.4	32.7
l	9.3198	18.5	6.33−2	3.2107	16.6	0.161	1.8675	14.1	3.85
m	9.2226	2.61	0.591	3.1898	2.95	6.74	1.8567	3.26	24.4
n	9.2192	2.61	0.235	3.1857	2.95	1.06	1.8524	3.26	1.09
o	9.5797	15.5	5.48−2	3.2694	15.1	0.393	1.8968	14.7	0.870
p	9.5760	15.5	2.94−2	3.2650	15.1	0.275	1.8923	14.7	0.873
q	9.2844	0.357	1.71	3.2006	4.98−2	16.0	1.8610	3.36−2	48.7
r	9.2863	0.669	1.64	3.2028	1.62	13.1	1.8636	3.2	32.0
s	9.2355	11.8	0.142	3.1906	12.1	0.455	1.8563	12.1	7.27−2
t	9.2362	11.5	0.213	3.1915	10.6	3.52	1.8571	8.95	17.9
u	9.3920	5.20−4	6.81−4	3.2256	1.59−2	0.125	1.8738	8.12−2	1.58
v	9.3929	3.90−4	2.70−4	3.2268	6.92−3	4.35−2	1.8748	2.45−2	0.481
w	9.1697	0	1.97	3.1773	0	16.8	1.8505	0	47.5
x	9.2284	0	1.61−6	3.1893	0	9.60−5	1.8555	0	7.83−4
y	9.2314	0	3.29−3	3.1928	0	0.464	1.8595	0	4.34
z	9.3147	0	1.03−8	3.2112	0	1.70−6	1.8683	0	2.35−5

[a] 1.16−3 means 1.16×10^{-3}.

Table 2.4. The same quantities as in Table 2.3 for Mo, Xe and W ions.

Key letter	Mo ($Z_n = 42$)			Xe ($Z_n = 54$)			W ($Z_n = 74$)		
	λ	A_a	A_r	λ	A_a	A_r	λ	A_a	A_r
a	0.68898	6.10	413	0.40585	6.30	1180	0.20432	6.35	4800
b	0.68446	6.10	0.169	0.40142	6.30	1.87−2	0.20025	6.35	5.34−2
c	0.69417	6.47−2	109	0.41078	1.78−2	339	0.20879	1.45−3	1460
d	0.68958	6.47−2	320	0.40624	1.78−2	838	0.20454	1.45−3	3260
e	0.69013	8.98	92.1	0.40641	7.41	226	0.20453	6.63	846
f	0.69533	0.251	31.7	0.41123	0.152	144	0.20892	5.46−3	90.1
g	0.69072	0.251	0.617	0.40668	0.152	9.56	0.20467	5.46−3	218
h	0.69911	3.69−1	4.31−2	0.41554	0.592	0.173	0.21331	0.858	0.385
i	0.69445	0.369	115	0.41089	0.592	449	0.20888	0.858	2070
j	0.69482	9.7	97.5	0.41106	11.3	332	0.20891	12.1	1440
k	0.68923	12.3	255	0.40603	12.2	751	0.20443	12.3	2960
l	0.69381	12.3	64.9	0.41056	12.2	188	0.20868	12.3	563
m	0.68814	4.12	207	0.40553	4.56	596	0.20421	4.91	2420
n	0.68363	4.12	0.160	0.40110	4.56	7.64−2	0.20014	4.91	0.149
o	0.70487	13.5	1.82	0.41878	12.9	1.76	0.21555	12.3	1.24
p	0.70014	13.5	7.43	0.41406	12.9	21.8	0.21103	12.3	75.4
q	0.68913	0.74	293	0.40522	9.99	115	0.20408	9.31	718
r	0.69275	6.55	134	0.41006	7.37	321	0.20849	7.87	1130
s	0.68763	10.7	20.2	0.40604	1.03	725	0.20446	1.34	2570
t	0.68809	5.52	229	0.40552	4.65	724	0.20424	4.15	3100
u	0.69501	0.696	69.3	0.41115	1.15	288	0.20898	1.52	1390
v	0.69554	0.110	18.5	0.41140	0.151	82.9	0.20903	0.162	443
w	0.68635	0	296	0.40465	0	819	0.20389	0	3270
x	0.68739	0	2.50−2	0.40512	0	0.224	0.20406	0	3.98
y	0.69201	0	92.7	0.40980	0	326	0.20846	0	1450
z	0.69455	0	2.51−3	0.41142	0	5.01−2	0.20998	0	1.26

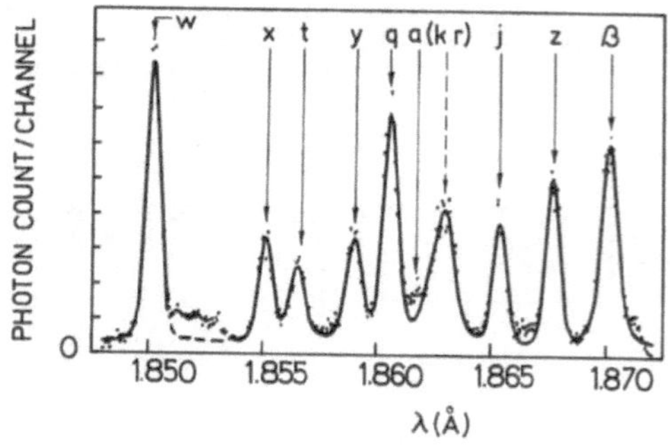

Fig. 2.4. X-ray spectrum of the resonance line (w) of Fe XXV ions and its satellites, obtained in the PLT tokamak [2.27]. The line β is a sum of some line intensities for Be-like ions

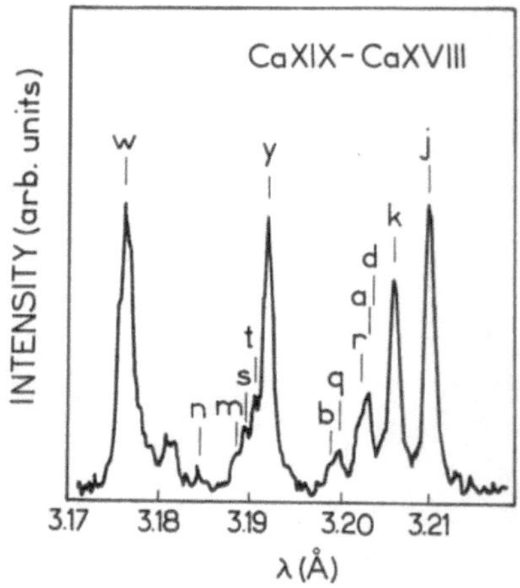

Fig. 2.5. Ca XIX – Ca XVIII ion spectra of laser-produced plasmas, recorded with a resolution of $\Delta\lambda/\lambda = 3\times10^{-4}$ [2.43]

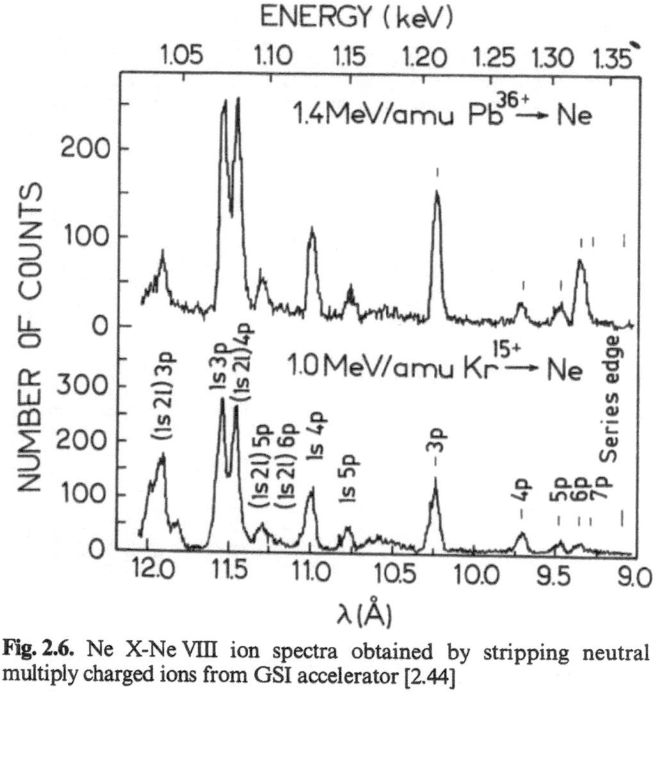

Fig. 2.6. Ne X-Ne VIII ion spectra obtained by stripping neutral neon atoms with fast multiply charged ions from GSI accelerator [2.44]

The comparison of experimental data with the results of the two theoretical methods shows that all three agree within $\Delta\lambda/\lambda \approx 10^{-4}$ for $\lambda \simeq 1-2$ Å. It is likely that this is the error limit for each of the sets. Further improvement to the accuracy of the theoretical results can be achieved by using relativistic wave functions in the zero-order approximation.

It is necessary to emphasize that the present experimental observations do not represent absolute wavelength measurements: usually the resonance line wavelength for a helium-like ion, w, is taken from theoretical calculations, and the distances between different lines are measured. Very recently the first experimental results on the absolute wavelengths were obtained. For basic information see, for example, [2.84–86]. These results are the most accurate at present, and they confirm the reliability of the earlier calculations and measurements. Rather good resolution has been obtained in recent experiments with astrophysical and laboratory sources. Some examples are given in Figs. 2.2–6.

Detailed information and extended references on ion spectra in the wavelength region $\lambda = 15-2500$ Å are contained in *Edlen's* review [2.45] and the compilation of *Reader* et al. [2.46].

2.4 Relativistic and Radiative Effects

It was mentioned above that the correct interpretation of x-ray spectra of multiply charged ions requires inclusion of the relativistic and radiative effects in the calculations. These effects increase as the ionic charge increases. Even in the case of a one-electron ion this problem presents many difficulties. For many-electron ions the problem becomes extremely complicated.

In the calculations of ion energy levels the relativistic effects are important practically for any value of the nuclear charge Z_n. When Z_n is small, one can use perturbation theory to account for these effects [2.47].

It is well known that the non-relativistic Schrödinger equations does not describe the level fine structure (splitting with respect to j), which takes place due to the relativistic dependence of the electron mass on velocity and the electron spin $s = 1/2$. The relativistic analogue of the Schrödinger equation for hydrogen-like ions is the four-dimensional Dirac equation [2.47],

$$HU = EU, \tag{2.15}$$

$$H = -e\varphi + \beta m c^2 + \boldsymbol{\alpha} \cdot (c\boldsymbol{p} + e\boldsymbol{A}), \tag{2.16}$$

where φ and A are the scalar and vector potentials of the external electromagnetic field, respectively, $\boldsymbol{p} = -\mathrm{i}\hbar\nabla_r$ is the momentum operator, $\boldsymbol{\alpha}$ is a vector operator associated with the Pauli spin matrix $\boldsymbol{\sigma}^{\mathrm{P}}$,

$$\boldsymbol{\alpha} = \begin{pmatrix} 0 & \boldsymbol{\sigma}^{\mathrm{P}} \\ \boldsymbol{\sigma}^{\mathrm{P}} & 0 \end{pmatrix}, \quad \text{and} \tag{2.17a}$$

$$\beta = \begin{pmatrix} I & 0 \\ 0 & -I \end{pmatrix}, \tag{2.17b}$$

I being a unit quadratic two-row matrix. For the Coulomb field,

$$\varphi = Ze/r\,, \quad A = 0\,,$$

and (2.15) has an analytic solution for each component of the wave function U. The corresponding eigenvalue E for the $n\,l$ bound state is given by

$$E = mc^2\left[1 + \left(\frac{\alpha Z}{n - k + \sqrt{k^2 - \alpha^2 Z^2}}\right)^2\right]^{-1/2}, \quad k = j + 1/2 \tag{2.18}$$

where $j = l \pm 1/2$ is the total momentum of the electron, including the spin. Expansion of E in powers of $(\alpha Z)^2$ gives

$$E = E_0 + E_1 + \dots\,, \tag{2.19}$$

where $E_0 = -Z^2\,\mathrm{Ry}/n^2$ is the usual non-relativistic part of the energy, E_1 is the fine structure splitting (in the Pauli approximation):

$$E_1 = -\frac{\alpha^2 Z^4\,\mathrm{Ry}}{n^3}\left(\frac{1}{j + 1/2} - \frac{3}{4n}\right) \tag{2.20}$$

and Ry is the Rydberg energy unit (1 Ry = 13.606 eV). It is seen that $E_1 \sim (Z\alpha)^2 E_0$. According to the Dirac theory, levels with different $l = j \pm 1/2$ have exactly the same energy, see (2.18). This degeneracy is, however, removed by including the radiative corrections (see below), i.e. corrections connected with quantum-electrodynamical effects, such as the Lamb shift, vacuum polarization, etc. The radiation corrections to the energy are of the order of $\alpha(\ln\alpha)\,E_1 \ll E_1$.

For non-hydrogenic ions, the difficulties involved in the problem increase significantly due to additional interactions between the electrons. The relativistic corrections for many-electron systems are accounted for usually in the framework of the Breit relativistic equation (or its modifications). In the absence of external fields ($A = 0$), the Breit equation takes the form [2.47]

$$H\Psi = E\Psi, \quad H = H_0 + H_1 + H_2 + H_3 + H_4 + H_5\,, \tag{2.21}$$

where H_0 is the non-relativistic Hamiltonian (2.6),

$$H_0=\sum_{i=1}^{N}\left(\frac{p_i^2}{2m}-\frac{Z_n e^2}{r_i}\right)+\sum_{i>j}\frac{e^2}{r_{ij}}, \quad r_{ij}=|r_i-r_j|, \tag{2.22}$$

H_1 is the relativistic correction, connected with the dependence of the electron mass on velocity (which does not depend on the nuclear spin)

$$H_1=-\frac{1}{8m^3c^2}\sum_{i=1}^{N}p_i^4, \tag{2.23}$$

H_2 is the orbit-orbit interaction or the classical relativistic correction to the electron interaction (connected with the retardation of the electromagnetic field caused by the electrons),

$$H_2=-\frac{e^2}{2m^2c^2}\sum_{i>j}\frac{1}{r_{ij}}\left(p_i\cdot p_j+\frac{r_{ij}\cdot(r_{ij}\cdot p_i)\,p_j}{r_{ij}^2}\right), \tag{2.24}$$

H_3 is the contact (or Darwin) interaction,

$$H_3=H_3'+H_3''=\frac{Z_n\pi e^2\hbar^2}{2m^2c^2}\sum_{i=1}^{N}\delta(r_i)-\frac{\pi e^2\hbar^2}{m^2c^2}\sum_{i>j}\delta(r_{ij}), \tag{2.25}$$

H_4 is the spin-orbit interaction,

$$H_4=\frac{e^2\hbar^2}{2m^2c^2}\left\{\left[\sum_{i=1}^{N}\frac{Z_n}{r_i^3}(r_i\times p_i)-\sum_{i>j}\frac{1}{r_{ij}^3}(r_{ij}\times p_i)+\sum_{i>j}\frac{2}{r_{ij}^3}(r_{ij}\times p_j)\right]\cdot s_i\right\}, \tag{2.26}$$

and H_5 is a sum of the spin-contact and spin-spin interactions between electrons,

$$H_5=H_5'+H_5''=-\frac{e^2\hbar^2}{m^2c^2}\frac{8\pi}{3}\sum_{i>j}(s_i s_j)\,\delta(r_{ij})$$
$$+\frac{e^2\hbar^2}{m^2c^2}\sum_{i>j}\frac{1}{r_{ij}^3}\left(s_i s_j-\frac{3(s_i r_{ij})(s_j r_{ij})}{r_{ij}^2}\right). \tag{2.27}$$

The one-electron operators H_1 and H_3' influence the major part of the ion energy and do not depend on the level fine structure. The two-electron operators H_2, H_3'' and H_5 are relativistic corrections to the electrostatic energy of the ion. They make it possible to improve the relative energies of levels. The two-electron operators H_4 and H_5 describe the fine structure energy levels.

As was mentioned in Sect. 2.3, the Breit-Pauli equation can be solved on the basis of perturbation theory, with the eigenfunctions of the non-relativistic Hamiltonian and using the Hartree-Fock or Coulomb wave functions as a zeroth approximation.

A more detailed analysis of the ion structure and spectra requires one also to take into account the radiative corrections arising from quantum electrodynamics. Two main problems must be considered in this context; the interaction of the electron with its radiation field, and the polarization of the vacuum [2.47]. The radiative corrections are mainly studied for atoms and ions with one or two electrons. It is clear that the basis of non-relativistic wave functions is not sufficient to deal with this problem. Therefore, it is necessary to use the Dirac equation (2.15) even in the zero-order approximation, and using the wave functions obtained in this way, to construct the perturbation theory expressions for the radiative corrections.

In the case of H-like ions the total radiative shift ΔE of the level $n\,l\,j$ can be written in the form [2.48]

$$\Delta E(n\,l\,j) = \Lambda(n\,l\,j) + \Pi(n\,l\,j) + \Delta E_{\mathrm{D}}(n\,l\,j) \tag{2.28}$$

where Λ and Π are the self-energy and the vacuum-polarization corrections, and ΔE_{D} is highest relativistic correction to the energy in the Pauli approximation[1]. The terms in (2.28) have the following orders of magnitude:

$$\Lambda \sim \alpha^3 Z^4, \tag{2.29a}$$

$$\Pi \sim \begin{cases} \alpha^3 Z^4, & l=0, \\ \alpha^5 Z^6, & l=1, \end{cases} \tag{2.29b}$$

$$\Pi(n\,s_{1/2}) = n^{-3}\,\Pi(1\,s_{1/2}), \tag{2.29c}$$

$$\Delta E_{\mathrm{D}} \sim \alpha^4 Z^6. \tag{2.29d}$$

For large values of Z, all three terms may equally contribute to the energy. However, they have different signs and therefore, at a certain Z, the correction ΔE may also change its sign. The Z-dependence of ΔE for the $1\,s_{1/2}$ and $2\,p_{1/2}$ levels is shown in Fig. 2.7. It can be seen that the value $\Delta E\ (1\,s_{1/2})$ changes sign at $Z = 36$. The analytical expressions for Λ, Π and ΔE_{D} for the $1\,s_{1/2}$ state have the following forms:

$$\Lambda(1\,s_{1/2}) = \alpha \frac{(\alpha Z)^4}{\pi} m\,c^2 F(\alpha Z), \tag{2.30}$$

---

1 The sum $\Lambda + \Pi$ is called the Lamb shift.

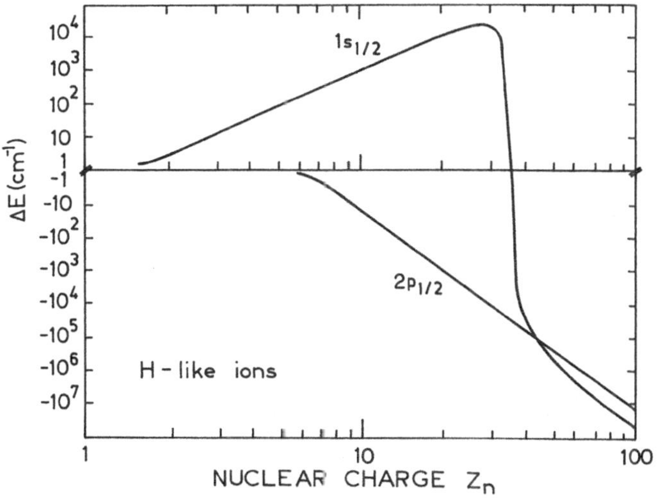

Fig. 2.7. The total radiative correction ΔE (2.28), to the $1s_{1/2}$ and $2p_{1/2}$ levels in H-like ions as a function of the nuclear charge [2.48].

$$\Pi(1s_{1/2}) = \frac{4\alpha}{3\pi}(\alpha Z)^4 mc^2 \left\{ -\frac{1}{5} - \frac{5\pi}{c^4}(\alpha Z) - \left[0.1 \ln(\alpha Z)^{-2} + \frac{1199}{2100}\right](\alpha Z)^2 + \left[\frac{5\pi}{128}\ln(\alpha Z)^{-2} + 0.5\right](\alpha Z)^3 \right\}, \quad (2.31)$$

$$\Delta E_D(1s_{1/2}) = -Z^2(\alpha Z)^4 \left[\frac{1}{16} + \frac{5}{128}(\alpha Z)^2 + \frac{7}{256}(\alpha Z)^4 + \frac{21}{1024}(\alpha Z)^6\right]. \quad (2.32)$$

Table 2.5. The function $F(\alpha Z)$ for the $1s_{1/2}$ state, (2.30), vs. nuclear charge

Z	$F(\alpha Z)$	Z	$F(\alpha Z)$
10	4.6540	70	1.5675
20	3.2461	80	1.5032
30	2.5519	90	1.4880
40	2.1351	100	1.5317
50	1.8644	110	1.6614
60	1.6838		

Table 2.6. The total radiative shift ΔE [GHz] in H-like ions for the level $n=2$ $(2s_{1/2}-2p_{1/2})$

Nuclear charge Z_n	Experiment	Theory [2.48]	Theory [2.49]	Theory [2.50]
1	1.057845 (9) [2.51]	1.059457	–	–
3	63.031 (327) [2.52]	62.7460	–	–
	63.765 (21) [2.52]	–	–	–
6	780.1 (80) [2.52]	780.113	–	–
8	2216 (8) [2.52]	2148.27	–	–
	2203 (11) [2.52]	–	–	–
9	3339 (35) [2.52]	3336.66	3342	3349
10	–	4751.67	4859	4889
15	$2.018(25)\times 10^4$ [2.53]	2.0085×10^4	–	2.0254×10^4
17	$3.119(22)\times 10^4$ [2.54]	3.0938×10^4	–	–
20	–	5.315×10^4	5.51×10^4	5.64×10^4
26	–	1.346×10^5	–	–
30	–	2.196×10^5	2.23×10^5	2.34×10^5
40	–	5.858×10^5	6.00×10^5	6.4×10^5
50	–	1.258×10^6	1.31×10^6	1.42×10^6
60	–	2.386×10^6	2.54×10^6	2.7×10^6
70	–	4.167×10^6		
80	–	6.925×10^6		
90	–	1.124×10^7		
100	–	1.793×10^7		

Table 2.7. Wavelengths [Å] of the resonance lines in H-like and He-like ions [a]

	H-like ions		He-like ions
Z_n	$2p_{1/2}-1s_{1/2}$	$2p_{3/2}-1s_{1/2}$	$2\,^1P_1-1\,^1S_0$
10	12.1371	12.1317	13.44783
13	7.17618	7.17077	7.75745
14	6.18569	6.18028	6.64805
16	4.73267	4.72726	5.03879
20	3.02385	3.01844	3.17720
22	2.49658	2.49116	2.61045
23	2.28300	2.27758	2.38201
26	1.78342	1.77799	1.85046
28	1.53575	1.53032	1.58848
30	1.33592	1.33049	1.37792
34	1.03683	1.03139	1.06419
40	0.74501	0.73956	0.75993
42	–	–	0.68658

[a] Calculated with relativistic and radiative corrections [2.40, 48]

The function $F(\alpha Z)$ is given in Table 2.5. It is worth noting that the values of $F(\alpha Z)$ given in Table 2.5 are calculated up to terms of the order of $(Z\alpha)^3$. The inclusion of terms of the order of $\ln(Z\alpha)^{-2}$ [2.47] overestimates the value of $F(\alpha Z)$ for $Z \gtrsim 20$.

In Table 2.6 the theoretical values of ΔE for the transition $2s_{1/2} - 2p_{1/2}$ in H-like ions are given and compared with the experimental results. The results of calculations of the wavelengths and level energies in He-like ions including the relativistic and radiative corrections are given in [2.40]. The wavelengths of resonance lines in H- and He-like ions as functions of the nuclear charge Z_n are given in Table 2.7.

In principle the accuracy of the wavelength and energy calculations can be improved by including the higher order corrections to ΔE with respect to the parameter $(Z\alpha)$, as well as the corrections due to the structure and motion of the ion nucleus. However, such calculations have not been performed as yet.

2.5 Decay of Excited Ionic States. Radiation and Autoionization

All the radiative transitions in multiply charged ions, arising from either singly- or doubly-excited states, increase with increasing ion charge $Z = Z_n - N + 1$. The general classification of radiative transitions in ions is the same as in neutral atoms. A detailed theory of multipole electric and magnetic radiative transitions is given, for example, in the monograph by *Sobel'man* [2.2]. It is well known that for light neutral atoms and ions with a small charge, the LS-coupling scheme is valid. With increasing charge, the relativistic effects become more and more significant, and their influence is described by introducing the intermediate coupling scheme. As an example, let us consider the electric dipole transition between the ion states a and a', accompanied by radiation of a photon with frequency $\omega_{aa'} = (E_a - E_{a'})/\hbar$. The transition probability is equal to

$$A_{aa'} = \frac{4\,\omega_{aa'}^3}{3\hbar c^3} |\boldsymbol{D}_{aa'}|^2 \quad [\mathrm{s}^{-1}], \tag{2.33}$$

where $\boldsymbol{D}_{aa'}$ is the matrix element of the electric dipole momentum of the ion, calculated in LS-coupling, i.e. with the wave functions (2.8). In the intermediate coupling scheme, the matrix element of $\boldsymbol{D}$ must be calculated with the wave functions (2.9), in accordance with (2.10). For the dipole matrix element of the transition $\gamma \to \gamma'$, one obtains

$$\boldsymbol{D}_{\gamma\gamma'} = \langle \gamma | D | \gamma' \rangle = \sum_{aa'} V(a|\gamma)\, \boldsymbol{D}_{aa'}\, V(a'|\gamma'), \tag{2.34}$$

where the eigenvectors $V(a \mid \gamma)$ can be found from the system of linear equations (2.13). The final expression for the transition probability has the form (2.34), replacing indices $a a'$ by $\gamma\gamma'$. The intermediate coupling leads to a change of the selection rules and the forbidden transitions in LS-coupling now become allowed. In the case of pure LS-coupling, all intercombination transitions ($\Delta S \neq 0$) are forbidden. For the $2\,^3P_1 \rightarrow 1\,^1S_0$ transition in helium-like ions, $A_{aa'} = 0$ ($\Delta S = 1$). Intermediate coupling mixes the $2\,^3P_1$ and $2\,^1P_1$ states, and $A_{\gamma\gamma'}$ increases with increasing charge: for the neutral He atom and the He-like ions Ne IX and Fe XXV the transition probabilities $A_{\gamma\gamma'}$ are equal to $1.80 \times 10^2\ \mathrm{s}^{-1}$, $5.43 \times 10^9\ \mathrm{s}^{-1}$ and $4.34 \times 10^{13}\ \mathrm{s}^{-1}$, respectively.

Calculations of the electric multipole (quadrupole and others) and magnetic multipole transitions are similarly influenced by the intermediate coupling.

Doubly-excited (autoionizing) states can also decay via the autoionization channel:

$$X_{Z-1}^{**}(\gamma_0, n\,l) \rightarrow X_Z(\gamma) + e\,(\varepsilon, l')\,. \tag{2.35}$$

One of the electrons of the initial state configuration $(\gamma_0, n\,l)$ transfers to a lower state γ and the second electron becomes free with a kinetic energy $\varepsilon = E(\gamma_0, n\,l) - E(\gamma) > 0$. The total energy of the system (including the nucleus + all the electrons) is conserved. The autoionization probability can be calculated in the following form:

$$A_\mathrm{a} = \frac{2\pi}{\hbar} \left| \left\langle \gamma_0, n\,l \left| \frac{1}{r_{12}} \right| \gamma, \varepsilon\,l' \right\rangle \right|^2 \quad [\mathrm{s}^{-1}]\,, \tag{2.36}$$

where r_{12} is the inter-electron distance and the wave functions of the initial and final states are calculated in the intermediate coupling approximation. Here the relativistic effects are taken into account through the wave functions only. This approximation is valid if

$$\hbar A_\mathrm{a} \ll \varepsilon\,. \tag{2.37}$$

For multiply charged ions this condition is always satisfied. The calculations of radiative and autoionization rates can be done using the Hartree-Fock or Coulomb-expansion methods, with the relativistic corrections included. The Hartree-Fock method with intermediate coupling has been used in [2.17–20] for systematic calculations of wavelengths, and radiative (A_r) and autoionization (A_a) probabilities. Similar calculations on the basis of the Coulomb expansion method have been done in [2.33, 35, 36]. Some results are given in Tables 2.2–4. We have already mentioned that some radiative transitions which are very weak (or forbidden) in the case of neutral atoms become allowed in the case of multiply charged ions due to the relativistic effects, and have rather large intensities. Detailed calculations for these transitions are given in [2.55–62]. A large amount of data on the wavelengths and radiative transition rates are compiled and systematized by *Reader* et al. [2.46].

In many applications, simple expressions for the radiation probabilities of intercombination and radiative transitions from ion states with principal quantum numbers $n=2$ and $n=3$ in He-like ions may be useful. Such semiempirical formulae are derived from numerical calculations and have the form [2.41, 55–58]

$$A_r(2\,^3P_2-1\,^1S_0)=0.039\,(Z+0.34)^8\quad[\mathrm{s}^{-1}]\,, \tag{2.38}$$

$$A_r(2\,^3S_1-1\,^1S_0)=\frac{2^5\,(Z-1)^{10}}{3^9\cdot 137^9}\left(\frac{\Delta\varepsilon}{(Z+1)^2\,\mathrm{Ry}}\right)^3\left(1-\frac{0.24}{Z+1}\right)^2\frac{1}{\tau_0}\,, \tag{2.39}$$

$$\tau_0=2.419\times10^{-17}\,\mathrm{s}\,,\quad \Delta\varepsilon=\text{transition energy}\,,$$

$$A_r(2\,^3P_1-1\,^1S_0)+A_r(2\,^1P_1-1\,^1S_0)=13.3\,Z^4\cdot10^8\quad[\mathrm{s}^{-1}]\,, \tag{2.40}$$

$$\frac{A_r(2\,^3P_1-1\,^1S_0)}{A_r(2\,^1P_1-1\,^1S_0)}=0.092\left(\frac{Z}{25}\right)^{4.9}, \tag{2.41}$$

$$\frac{A_r(3\,^3P_1-1\,^1S_0)}{A_r(3\,^1P_1-1\,^1S_0)}=0.11\left(\frac{Z}{25}\right)^{4.5}. \tag{2.42}$$

The radiation probability for two-photon decay is

$$A_r(2\,^1S_0-1\,^1S_0)=16.46\,(Z+0.2)^6\quad[\mathrm{s}^{-1}]\,. \tag{2.43}$$

2.6 Static and Dynamic Dipole Polarizabilities of Multiply Charged Ions

2.6.1 General Remarks

The dipole polarizabilities β characterize the magnitude of the dipole moment $\boldsymbol{d}$ induced in the ion (or atom) by an external electric field $\boldsymbol{\mathscr{E}}$,

$$\boldsymbol{d}=\beta\boldsymbol{\mathscr{E}}\,. \tag{2.44}$$

For the ith ionic state, the quantum-mechanical value of β is given by the second-order perturbation theory expression

$$\beta_i(\omega) = -\sum_{k \neq i} \int |\langle i | D_z | k \rangle|^2 \times \left[\frac{1}{E_i - E_k + \omega - i\eta} + \frac{1}{E_i - E_k - \omega - i\eta} \right]_{\eta \to 0}, \qquad (2.45)$$

where D_z is the z component of the dipole moment, E_i, E_k are the ionic state energies, and ω is the frequency of the external field. The sum and the integral in (2.45) mean summation over the discrete and integration over the continuum states.

The quantities $\beta(\omega)$ and $\beta(0)$ are called the dynamic and static polarizability of the ion (or atom), respectively. The static polarizability $\beta(0)$ describes the shift ΔE_i of the ionic energy level E_i in the presence of an external electric field (the Stark effect) and the polarized part of the energy of highly excited levels of multiply charged ions [2.63]. The real part of the dynamic polarizability describes "shaking" of the level E_i in an external electric field of frequency ω, i.e.

$$\Delta E_i = -\tfrac{1}{4} \operatorname{Re}\{\beta_i(\omega)\} \mathcal{E}^2 . \qquad (2.46)$$

For frequencies $\omega > |E_i|$, the imaginary part of $\beta_i(\omega)$ is related to the photoionization cross section $\sigma_i(\omega)$ of the ith ionic state by

$$\operatorname{Im}\{\beta_i(\omega)\} = \frac{1}{4\pi\alpha\omega} \sigma_i(\omega) , \qquad (2.47)$$

where $\alpha = e^2/\hbar c$ is the fine structure constant.

2.6.2 Static Polarizability

The properties of $\beta(0)$ are theoretically well investigated for multiply charged ions with $N \leq 4$ electrons. It is worth noting that the summation over the discrete spectrum in (2.45) gives the main contribution to $\beta(0)$. Therefore if the transitions with no change in the principal quantum numbers ($\Delta n = 0$) are dominant, then $\beta(0) \sim Z_n^{-3}$, while if the transitions with $\Delta n \neq 0$ dominate, then $\beta(0) \sim Z_n^{-4}$. For hydrogen-like ions in the ground state the value of $\beta(0)$ is [2.64]

$$\beta(0) = \tfrac{9}{2} Z_n^{-4} \quad [a_0^3] , \qquad (2.48)$$

a_0 being the Bohr radius of the electron orbit in the ground-state hydrogen atom. The inclusion of relativistic corrections up to the order of $(Z_n \alpha)^2$ gives [2.65]

$$\beta(0) = \frac{9}{2} Z_n^{-4} \left[1 - \frac{28}{27} (\alpha Z_n)^2 \right] \quad [a_0^3] . \tag{2.49}$$

From (2.49) it is seen that relativistic corrections lead to a decrease of $\beta(0)$ which is about 50% for $Z_n \approx 100$.

In the calculations of $\beta(0)$ for ions with $N \geq 2$ it is necessary to take into account the effect of the screening of the nuclear charge by the electrons. For He-like ions the result for $\beta(0)$ is [2.64]

$$\beta(0) = 9\,(Z_n - b)^{-4} \quad [a_0^3] , \quad b = 0.3594 . \tag{2.50}$$

The numerical values for $\beta(0)$ provided by this expression practically coincide with those obtained from the following one [2.66]:

$$\beta(0) = \frac{9}{Z_n^4} + \frac{12.94}{Z_n^5} + \frac{12.0}{Z_n^5} + \frac{25}{Z_n^7} \quad [a_0^3] . \tag{2.51}$$

The static polarizability of He-like ions in the first excited states is given by [2.66]

$$\beta(2\,^1S) = \frac{641.8}{Z_n^3} + \frac{2402}{Z_n^4} + \frac{4.3 \times 10^3}{Z_n^5} + \frac{1.7 \times 10^4}{Z_n^6} \quad [a_0^3] , \tag{2.52}$$

$$\beta(2\,^3S) = \frac{476.2}{Z_n^3} + \frac{1169.7}{Z_n^4} + \frac{1.63 \times 10^3}{Z_n^5} + \frac{5.3 \times 10^3}{Z_n^6} \quad [a_0^3] . \tag{2.53}$$

Relativistic calculations of $\beta(0)$ for He-like ions, including the electron correlations in the first order perturbation theory expressions using Dirac wave functions [2.67], have demonstrated that the relativistic corrections lead to a decrease of $\beta(0)$ with increasing Z_n. For $Z_n = 20$ and $Z_n = 100$, the relativistic corrections are about 6% and 20%, respectively.

Hartree-Fock calculations of $\beta(0)$ have been performed for the ground state B, C, N, O, F, Ne, Mg and Ar-like ions [2.68], and for the Be, Mg, Zn-like ions (configuration $n\,s^2$) [2.69]. An extensive compilation of data on $\beta(0)$ can be found in *Fraga* et al. [2.70].

Calculations of $\beta(0)$ can also be performed by using the following well-known formula [2.71]:

$$\beta_i(0) = 4 R^2 \sum_{k=i}^{\infty} f_{ik} \lambda_{ik}^2 , \tag{2.54}$$

Table 2.8. Experimental values of $\beta(0)$ for multiply charged ions [2.76, 77]

Ion	$\beta\,[a_0^3]$	Isoelectronic sequence	Ion	$\beta\,[a_0^3]$	Isoelectronic sequence
Mg VIII	0.29 ± 0.15	B	Kr XXVII	0.000931	Ne
Ca XI	0.0174	Ne	Sr XXIX	0.000728	
T XIII	0.0104		Zr XXXI	0.000576	
Cr XV	0.00665		Mo XXXIII	0.000462	
Fe XVII	0.00444		K IX	0.55 ± 0.07	Na
	0.0038 ± 0.0010				
Co XVIII	0.00369		Ca X	0.47 ± 0.06	
	0.0024 ± 0.0010				
Ni XIX	0.00308		Sc XI	0.40 ± 0.10	
	0.0040 ± 0.0015				
Cu XX	0.0026		Ti XII	0.31 ± 0.07	
	0.0021 ± 0.0010				
Zn XXI	0.0021		Fe IX	0.37 ± 0.07	Ar
Br XXVI	0.00106		Co X	0.26 ± 0.04	

Table 2.9. Parameters a, b and c in (2.57) for the dipole polarizabilities of multiply charged ions

Isoelectronic sequence	Outer shell configuration	Total number of electrons	a	b	c
H	$1s$	1	9/2	0	4
He	$1s^2$	2	9	0.359	4
Li	$2s$	3	165	2.84	3
Be	$2s^2$	4	255	2.16	3
B	$2p$	5	195	2.76	3
C	$2p^2$	6	146	3.49	3
N	$2p^3$	7	1817	2.60	4
O	$2p^4$	8	1613	3.20	4
F	$2p^5$	9	1201	4.26	4
Ne	$2p^6$	10	870	4.96	4
Na	$3s$	11	951	9.16	3
Mg	$3s^2$	12	1981	8.31	3
Ar	$3p^6$	18	1.63×10^4	11.7	4
Cr^+	$3d^5$	23	235	21.3	3
Cu^+	$3d^{10}$	28	6.6×10^4	18.8	4
Cu	$4s$	29	3200	23.7	3
Zn	$4s^2$	30	7713	22.8	3
Kr	$4p^6$	36	2.26×10^5	23.4	4
Xe	$5p^6$	54	1.7×10^6	35.6	4

where R is the Rydberg constant ($R = 109\,677.581\ \mathrm{cm}^{-1}$), f_{ik} is the oscillator strength, and λ_{ik} is the wavelength (in cm). The values of f_{ik} may be calculated using the Coulomb approximation [2.72]. Such calculations have been performed for the Li-, Na-, and Ca-isoelectronic sequences [2.73], and for the isoelectronic sequences with $Z_n - N = 5-25$, $N = 5-26$ by extrapolation of the low-Z_n values of $\beta(0)$ to high Z_n [2.74].

Another method for calculating $\beta(0)$ has been suggested in [2.75]. The method is based on the Thomas-Fermi-Dirac model for the atom and gives the result:

$$\beta(0) = r_0^3\left(1 + \frac{3\,U_p(r_0)}{r_0\,U_p'(r_0)}\right)^{-1}. \tag{2.55}$$

In (2.55), r_0 is the ionic radius in the Thomas-Fermi-Dirac model and $U_p(r)$ is the radial part of the polarization potential, induced by an external electric field. In the case of highly charged ions, $\beta(0)$ has a simple analytical form,

$$\beta(0) = \frac{63\,N^3}{16\,Z_n^4}\quad [a_0^3]\,. \tag{2.56}$$

This method provides satisfactory results for $\beta(0)$ in cases when the outer electronic shell is filled by more than one half of the maximum number of electrons for that shell.

Experimental data for $\beta(0)$ have been obtained for the isoelectronic sequences of Ne, Na and Ar, by measuring the energy levels of the highly excited ion states [2.76, 77] (Table 2.8).

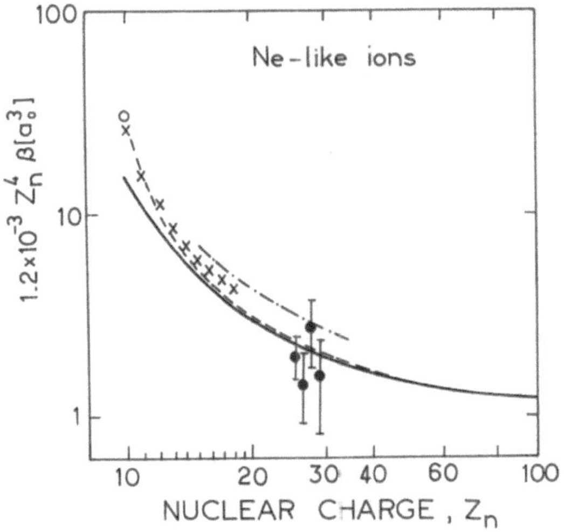

Fig. 2.8. Static dipole polarizability for Ne-like ions as a functions of nuclear charge
Experiment: (o) Ne I [2.79]; (ф) Fe XVII-Cu XX [2.77]; (---) Ne I-Mo XXXIII [2.76]
Theory: (×××) uncoupled Hartree-Fock method [2.68]; (—·—) screening approximation [2.74]; (—) Eq. (2.57) with $a = 870$ $b = 4.96$, $c = 4$

The analysis of the existing experimental and theoretical data on $\beta(0)$ for multiply charged ions suggests that all data may be represented approximately by the following semiempirical formula [2.78]:

$$\beta(0) = a\,(Z_n - b)^c \quad [a_0^3]\,, \tag{2.57}$$

where a and b are fitting parameters. If transitions with $\Delta n = 0$ give the main contributions to $\beta(0)$, then $c = 3$ in (2.57); if the transitions with $\Delta n \neq 0$ are dominant, then $c = 4$. The values of the constants a and b in (2.57) are determined by a fitting procedure using the experimental and theoretical data for the isoelectronic sequences from H through Xe. The values obtained are given in Table 2.9. A comparison of $\beta(0)$ calculated by (2.57) with the experimental data for Ne-like ions is given in Fig. 2.8.

2.6.3 Dynamic Polarizability

The properties of dynamic polarizabilities $\beta(\omega)$ are studied theoretically mostly for the H- and He-like ions. At present no experimental data on $\beta(\omega)$ exist for multiply charged ions.

Let us rewrite the expression (2.45) for $\beta(\omega)$ in a form which is more convenient for calculations. Neglecting the fine-structure intervals between the levels and summing over the momentum L_k and the spin S_k of final states, the following expression for $\beta_i(\omega)$ can be obtained for the state $i = n\,l^q LS$ (q is the number of equivalent electrons):

$$\beta_i(\omega) = \frac{4q}{3} \sum_{n'l' \neq nl} \frac{|\langle n\,l\,|\,\boldsymbol{r}\,|\,n'\,l'\rangle|^2}{(E_{n'l'} - E_i)\left[1 - \left(\dfrac{\hbar\,\omega}{E_{n'l'} - E_i}\right)^2\right]} \tag{2.58}$$

where E is expressed in Rydberg units. If the frequency of the external field lies in the interval $0 \le \omega < \omega_0$, where ω_0 is the frequency of the transition between the ground and the nearest dipole-allowed bound level $(\omega_0 = |\,E_{n_1'l_1'} - E_i\,|)$, $\beta_i(\omega)$ can be approximated by

$$\beta_i(\omega) \approx \frac{\beta_i(0)}{1 - (\omega/\omega_0)^2}\,, \quad 0 \le \omega < \omega_0\,, \tag{2.59}$$

where $\beta_i(0)$ is the static polarizability.

For the ground state H-like ions the scaling law for the dynamic polarizability can be written in the form

$$\beta(\omega) = Z^{-4}\,\beta_{\mathrm{H}}(\omega/Z^2)\,, \tag{2.60}$$

where $\beta_H(\omega)$ is the dynamic polarizability of the hydrogen atom. Explicit expressions for $\beta_H(\omega)$ are available [2.80] for any arbitrary nl-state. Thus, for the ground state, one has

$$\beta_H^{1s}(\omega) = f(\omega) + f(-\omega)\,,$$
$$f(\pm\omega) = \frac{2^{11}x^3}{(x+1)^{10}}\left\{(3-1/x)^{-1}(2-1/x)^{-1}\right.$$
$$\cdot\,{}_2F_1\left(5, 2-1/x; 4-1/x; \left[\frac{x-1}{x+1}\right]^2\right) + \frac{5x^2}{(x+1)^2(3-1/x)}$$
$$\left.\cdot\,{}_2F_1\left(6, 3-1/x; 4-1/x; \left[\frac{x-1}{x+1}\right]^2\right)\right\}, \quad x = (1\pm 2\omega)^{-1/2}. \quad (2.51)$$

In the static limit $\omega = 0$, $x = 1$, ${}_2F_1 = 1$ and $\beta_H^{1s}(0) = \frac{9}{2}a_0^3$. In the vicinity of the ionization threshold $\beta_H^{1s}(\omega)$ has the following value [2.81]:

$$\beta_H^{1s}(\omega) = -4.9107\cot\left(\frac{\pi}{\sqrt{1-2\omega}}\right) - 4.3105\,, \quad (2.62)$$
$$\omega > \omega_0 = \tfrac{3}{8} \quad \text{a. u.}$$

This formula also describes qualitatively the behaviour of $\beta_H^{1s}(\omega)$ in the frequency region above the first resonance, i.e. at $\omega > |E(n=1) - E(n=2)|$. According to (2.59), the function $\beta(\omega)/\beta(0)$ in the region $\omega < \omega_0$ is a universal function, i.e. it does not depend on the nuclear charge Z_n.

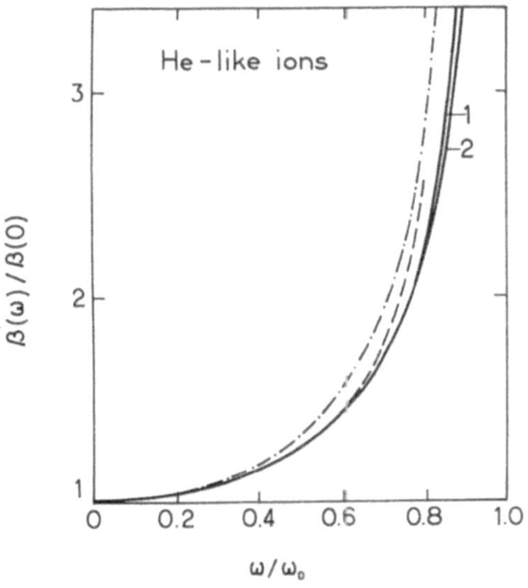

Fig. 2.9. Ratio of the dynamic to the static dipole polarizability for He-like ions below the first resonance ω_0 Theory: (---) perturbation-variation method [2.82]; (—) relativistic calculations [2.83] for $Z_n = 20$ (*curve 1*) and $Z_n = 80$ (*curve 2*); (-·-) Eq. (2.59)

Calculations of $\beta(\omega)$ for the He-like ions have also been performed (see, for example, [2.82]). For the ground state He-like ions with $Z_n = 20$ and $Z_n = 80$ calculations of $\beta(\omega)$ have been carried out [2.83] including the relativistic corrections. It has been demonstrated that these corrections lead to an increase of $\beta(\omega)$ when Z_n and ω increase. The ratio $\beta(\omega)/\beta(0)$ for the He-like ions, calculated in different approximations in the region $\omega < \omega_0$, is presented in Fig. 2.9. It follows from (2.58) that the value of $|\beta(\omega)|$ at the resonance frequencies tends to infinity, which has no physical meaning. In these narrow frequency regions it is necessary to take into account the effects of the finite width of the resonance levels.

3. Radiative Processes in the Continuous Spectrum

In this chapter we shall consider the radiative processes in which either the initial or the final state of an ion, or both, belong to the continuous spectrum. Since all the ions (except the bare nuclei and H-like ions) have autoionizing states lying in the continuum, autoionization decay should also be included in our consideration.

The most important processes which will be considered here, are the following ones:

I) transitions between different states of the continuous spectrum (bremsstrahlung),
II) electron transitions from continuous spectrum states into discrete states (radiative recombination),
III) dielectronic recombination.

Dielectronic recombination may be considered as a specific case of radiative recombination in which an incident electron has an energy equal to the energy of an autoionizing level of the ion. Radiative processes inverse to the above are: inverse bremsstrahlung for the process (I), photoionization for the radiative recombination (II), and radiative excitation of the autoionizing states for the dielectronic recombination (III).

3.1 Bremsstrahlung

In this process an electron with a positive kinetic energy E_0 and momentum $\boldsymbol{p}_0$ collides with an ion (or bare nucleus), radiates a photon with momentum $\boldsymbol{k}$ (energy $\hbar\omega = \hbar k c$) and polarization $\boldsymbol{e}_{k\varrho}$ ($\varrho = 1, 2$), perpendicular to the direction $\boldsymbol{k}$, and then has an energy E_1 ($< E_0$) and momentum $\boldsymbol{p}_1$ after the collision. The photon and electron energies are related to each other by the equation

$$\hbar\omega = E_0 - E_1 = \frac{1}{2m}(p_0^2 - p_1^2) = \hbar k c\,. \tag{3.1}$$

In the case of a multiply charged ion, one can assume that in this collision the bound ionic electrons are not perturbed and, consequently, the

Sommerfeld theory can be used to describe the process [3.1]. For the non-relativistic electron-ion collisions, the general expression for the probability of a dipole transition with radiation of a photon into a unit solid angle dO_k has the form

$$dA = \frac{\omega^3}{2\pi\hbar c^3} |\boldsymbol{e}_{k\varrho} \boldsymbol{D}_{01}|^2 \, dO_k \tag{3.2}$$

where $\boldsymbol{D} = e\boldsymbol{r}$ and e is the electron charge. Note that after integration over all directions of the photon momentum $\boldsymbol{k}$ and summation over the polarization directions $\varrho = 1, 2$, one obtains from (3.2) the expression (2.33) of Sect. 2.5 for the transition probability between two discrete states. The differential cross section for bremsstrahlung with a photon frequency in an interval ω, $\omega + d\omega$ and into the solid angle dO_k is equal to

$$d\sigma_{\boldsymbol{p}_0,\boldsymbol{p}_1\boldsymbol{k}} = \frac{m^2 e^2 k^3}{(2\pi)^4 \hbar^3} \cdot \frac{p_1}{p_0} \left| \boldsymbol{e}_{k\varrho} \int (\psi^-_{\boldsymbol{p}_1})^* \, \boldsymbol{r} \, \psi^+_{\boldsymbol{p}_0} \, d\boldsymbol{r} \right|^2 d\omega \, dO_k \, d\boldsymbol{p}_1 \,. \tag{3.3}$$

The wave functions of the initial state ($\psi^+_{\boldsymbol{p}_0}$) and the final state ($\psi^-_{\boldsymbol{p}_1}$) are exact Coulomb wave functions for the attractive Coulomb field with the asymptotic behaviour of a plane wave + outgoing (incoming) wave in the initial (final) channel. They have the form:

$$\psi^+_{\boldsymbol{p}_0} = N^+(p_0) \, e^{i\boldsymbol{p}_0\boldsymbol{r}/\hbar} F\left(i\eta_0, 1; i[p_0 r - \boldsymbol{p}_0 \boldsymbol{r}]/\hbar\right) , \tag{3.4}$$

$$\eta_0 = \frac{(Z-1)e^2}{\hbar v_0} , \quad v_0 = p_0/m ,$$

$$\psi^-_{\boldsymbol{p}_1} = N^-(p_1) \, e^{i\boldsymbol{p}_1\cdot\boldsymbol{r}/\hbar} F\left(-i\eta_1, 1; -i[p_1 r + \boldsymbol{p}_1 \cdot \boldsymbol{r}]/\hbar\right) , \tag{3.5}$$

$$\eta_1 = \frac{(Z-1)e^2}{\hbar v_1} , \quad v_1 = p_1/m ,$$

where $F(a, b; x)$ is the confluent hypergeometric function and

$$N^\pm(p) = e^{\pi\eta/2} \, \Gamma(1 \mp i\eta) \,. \tag{3.6}$$

For calculations of the Sommerfeld integral (3.3) one can use the well-known equality (see, for example, [3.2])

$$\boldsymbol{D}_{01} = e\boldsymbol{r}_{01} = \frac{1}{i\omega_{01}} \frac{e}{m} \boldsymbol{p}_{01} = \frac{e\hbar}{\omega_{10} m} (\nabla_r)_{01} \,. \tag{3.7}$$

The most simple and general way of calculating integrals containing Coulomb wave functions has been developed by *Nordsieck* [3.3] using a contour representation for the confluent hypergeometric function. The result for the integral in (3.3) is

$$\int (\psi_{p_1}^-)^* \, \boldsymbol{r} \, \psi_{p_0}^+ \, d\boldsymbol{r} = \frac{\mathrm{i}\,\hbar}{\omega\, m} N^+(p_0)\, N^-(p_1) \frac{8\,\pi\, \mathrm{e}^{-\pi\eta_0}}{(p_0-p_0)^3 (p_0+p_1)} \left(\frac{p_0+p_1}{p_0-p_1}\right)^{-\mathrm{i}(\eta_0+\eta_1)}$$

$$\times (1-x)^{\mathrm{i}(\eta_0+\eta_1)-1} [\mathrm{i}\,\eta_0\, p_0\, \boldsymbol{q}\, F(x) + (1-x)\, F'(x)\,(p_1\, \boldsymbol{p}_0 - p_0\, \boldsymbol{p}_1)] \tag{3.8}$$

with the following designations:

$$\boldsymbol{q} = \boldsymbol{p}_1 - \boldsymbol{p}_0, \quad x = -2\frac{p_0\, p_1 - \boldsymbol{p}_0 \cdot \boldsymbol{p}_1}{(p_0-p_1)^2}, \quad q^2 = (p_0-p_1)^2(1-x)\,, \tag{3.9a}$$

$$F(x) = F(\mathrm{i}\,\eta_1, \mathrm{i}\,\eta_0; 1; x), \quad F' = \frac{d}{dx} F(x)\,, \tag{3.9b}$$

where $F(\mathrm{i}\,\eta_1, \mathrm{i}\,\eta_0; 1; x) = F(\mathrm{i}\,\eta_0, \mathrm{i}\,\eta_1; 1; x)$ is the Gauss hypergeometric function (see, for example, [3.2, 4]). Substituting (3.8) into (3.3), one can obtain the differential cross section for bremsstrahlung. However, the general formula is too cumbersome. It is better to proceed to an analysis of the spectral distribution of radiation which can be obtained after integration of the differential cross section over all directions of the photon and final-state electron.

Integration over dO_k and summation over the photon polarization directions $\boldsymbol{e}_{k\varrho}$, leads to the substitution

$$\int dO_k \sum_{\varrho=1,2} \boldsymbol{e}_{k\varrho} = \frac{8\,\pi}{3}\,, \tag{3.10}$$

and for the integration over scattered electron directions dO_{p_1}, one can use the exact expressions for the Gauss confluent hypergeometric functions [3.1–5]. The final result for the bremsstrahlung cross section in a frequency interval ω, $\omega + d\omega$, after integration over all the directions of the electron and photon, has the form

$$d\sigma_{E_0;\,E_1\omega} = g(\eta_0, \eta_1) \frac{16\,\pi}{3\sqrt{3}} \alpha^2 a_0^2 \left(\frac{e^2 (Z-1)^2}{\hbar\, v_0}\right) \frac{d\omega}{\omega}\,. \tag{3.11}$$

Here α is the fine structure constant, a_0 is the Bohr radius, $v_0 = p_0/m$ is the initial electron velocity and, in the non-relativistic case of the Sommerfeld theory, the g-factor is

$$g(\eta_0, \eta_1) = \frac{\pi\sqrt{3}}{(\mathrm{e}^{2\pi\eta_0}-1)(\mathrm{e}^{2\pi\eta_1}-1)} x_0 \frac{d}{dx_0} |F(\mathrm{i}\,\eta_0, \mathrm{i}\,\eta_1; 1; x_0)|^2, \tag{3.12}$$

$$x_0 = -\frac{4\,\eta_0\,\eta_1}{(\eta_1-\eta_0)^2}\,, \tag{3.13}$$

$$\eta_0 = \frac{(Z-1)\, e^2}{\hbar\, v_0}\,, \quad \eta_1 = \frac{(Z-1)\, e^2}{\hbar\, v_1}\,, \quad \omega = \frac{(Z-1)^2\, e^4\, m}{2\,\hbar^3} \left(\frac{1}{\eta_0^2} - \frac{1}{\eta_1^2}\right). \tag{3.14}$$

Equations (3.11–14) have been used for extensive calculations of differential bremsstrahlung cross sections [3.6]. It is of interest, however, to consider the simpler limiting cases investigated in [3.1, 2, 5, 7, 8], in particular, the Born approximation and the classical limit. For fast electrons, $\eta_0 \ll 1$, the bremsstrahlung spectrum is described by the Born-Elvert approximation in which the g-factor is given by

$$g(\eta_0, \eta_1) = \frac{\sqrt{3}}{\pi} \frac{\eta_1}{\eta_0} \frac{1 - \exp(-2\pi\eta_0)}{1 - \exp(-2\pi\eta_1)} \ln\left(\frac{\eta_1 + \eta_0}{\eta_1 - \eta_0}\right), \quad \eta_0 \ll 1 . \tag{3.15}$$

If the final electron velocity is also large, $2\pi\eta_1 \ll 1$, one obtains the Born approximation:

$$g(\eta_0, \eta_1) = \frac{\sqrt{3}}{\pi} \ln\left(\frac{\eta_1 + \eta_0}{\eta_1 - \eta_0}\right), \quad 2\pi\eta_0 < 2\pi\eta_1 \ll 1 . \tag{3.16}$$

This limiting case corresponds to large velocities of the incident and scattered electron, for which the Coulomb wave functions (3.4, 5) can be replaced by plane waves: $\psi_p^{\mp}(\boldsymbol{r}) = \exp(\mathrm{i}\,\boldsymbol{p}\cdot\boldsymbol{r}/\hbar)$. For multiply charged ions, (3.15, 16) have to be used, however, at very large velocities, for which one must take into account the relativistic and retardation effects.

For multiply charged ions ($Z \gg 1$), the more realistic case in many applications is $\eta_0 \gg 1$. If $\eta_1 > \eta_0 \gg 1$, (3.12) becomes

$$g = \frac{\pi\sqrt{3}}{4} \mathrm{i}\,\nu H_{\mathrm{i}\nu}^{(1)}(\mathrm{i}\,\nu)\, H_{\mathrm{i}\nu}^{(1)\prime}(\mathrm{i}\,\nu), \quad \nu = \frac{(Z-1)e^2}{m v^3}\omega, \tag{3.17}$$

where $\mathrm{H}^{(1)}$ is the Hankel function and $\mathrm{H}^{(1)\prime}$ is its derivative with respect to the argument. This is the classical limit, and the result (3.17) can be obtained with the use of classical electrodynamics methods [3.7].

The function (3.17) is monotonic; its limiting expressions have the form

$$g = 1, \quad \nu \gg 1 \tag{3.18a}$$

$$= \frac{\sqrt{3}}{\pi} \ln(2/\gamma\nu), \quad \nu \ll 1 , \tag{3.18b}$$

where $\gamma = 1.781\ldots$ is the Euler constant. The case $\nu \ll 1$ describes the bremsstrahlung near the low frequency edge, $\omega = 0$, when the scattered electron has a kinetic energy value almost equal to the initial value, $\eta_1 \approx \eta_0$.

For the multiply charged ions, it is of interest to consider the formula obtained by *Gervids* and *Kogan* [3.9] using the semiclassical method,

$$g = \frac{\pi\sqrt{3}}{4} \mathrm{i}\,\nu\left(1 + \frac{1}{\gamma\eta_0}\right) \mathrm{H}_{\mathrm{i}\nu}^{(1)}\left[\mathrm{i}\,\nu\left(1 + \frac{1}{\gamma\eta_0}\right)\right] \mathrm{H}_{\mathrm{i}\nu}^{(1)\prime}\left[\mathrm{i}\,\nu\left(1 + \frac{1}{\gamma\eta_0}\right)\right]. \tag{3.19}$$

This expression includes the main features of both limiting cases (the Born approximation and the classical limit). For $\eta_0 \gg 1$, or, more precisely, for $\eta_0/\nu \gg 1$, the result (3.19) concides with (3.17). Its connection with the Born approximation can easily be seen from the limiting case $\omega \to 0$. Taking into account that, for small ν, the Hankel function is $H_{i\nu}^{(1)}(x) \approx H_0^{(1)}(x)$ and that for small x it has the form

$$\mathrm{i}\, H_0^{(1)}(\mathrm{i}\, x) \approx \left(\frac{2}{\pi}\right) \ln\left(\frac{2}{\gamma x}\right),$$

one obtains

$$g \approx \frac{\sqrt{3}}{\pi} \ln\left(\frac{2}{\nu(\gamma + 1/\eta_0)}\right), \tag{3.20}$$

where η_0 has an arbitrary value and $\nu \ll 1$. The classical limit (3.18) is obvious from (3.20). In the Born approximation, one can obtain from (3.15, 16) for $\omega \to 0$

$$g \approx \frac{\sqrt{3}}{\pi} \ln(2\,\eta_0/\nu), \quad \eta_1 \approx \eta_0. \tag{3.21}$$

For multiply charged ions the semiclassical result (3.19) is a good approximation to the exact Sommerfeld formula (3.12) for all values of the parameters, except in the so-called short wavelength limit where $\nu \to 0$, $\eta_0 \to 0$, $\nu/\eta_0 \to 1$. This limiting case, however, corresponds to large electron velocities for which relativistic effects are important and the Sommerfeld formula (3.12) is not valid.

The relativistic theory of bremsstrahlung is discussed in many books on quantum electrodynamics (see, for example, [3.2, 5]), and in the original [3.10–13] and review [3.14, 15] papers; [3.14] contains results of numerical calculations.

Now we shall consider the process of inverse bremsstrahlung (or continuum photoabsorption). While the frequencies of emitted photons in the bremsstrahlung process fall into the interval $0 < \omega < E_0/\hbar$, the frequencies of absorbed photons in the inverse bremsstrahlung process belong to the interval $0 < \omega < \infty$. The bremsstrahlung differential cross section, $d\sigma_{p_0;p_1k}$, and that for continuum photoabsorption, $d\sigma_{p_1k;p_0}$, are mutually connected through the detailed balance principle,

$$\frac{d\sigma_{p_1k;p_0}}{p_0^2\, dO_{p_0}} = (2\,\pi)^3 \frac{d\sigma_{p_0;p_1k}}{k^2\, d\omega\, dO_k\, p_1^2\, dO_{p_1}}. \tag{3.22}$$

For the cross sections, integrated over all electron and photon scattering directions, one obtains

$$\frac{k^2}{\pi^2} p_1^2 \, \sigma_{E_1\omega;\,E_0} = p_0^2 \frac{d\sigma_{E_0;\,E_1\omega}}{d\omega}\,, \quad \hbar\, k\, c = \hbar\, \omega = E_0 - E_1\,. \tag{3.23}$$

The continuum photoabsorption cross section is thus equal to

$$\begin{aligned}\sigma_{E_1\omega;\,E_0} &= \frac{p_0^2}{p_1^2} \frac{\pi^2 c^2}{\omega^2} \frac{d\sigma_{E_0;\,E_1\omega}}{d\omega} \\ &= g(\eta_0, \eta_1) \frac{16\,\pi^3 c^2}{3\sqrt{3}\,\omega^3} \eta_0^2 \,\alpha^3 a_0^2\,,\end{aligned} \tag{3.24}$$

where the g-factor is given by (3.12). As can be seen from (3.24), the continuum photoabsorption cross section is measured in units of $\mathrm{cm}^4\,\mathrm{s}$. Note that the continuum photoabsorption processes are not important for most applications since they correspond to triple collisions: electron + photon + ion. Their rates (per unit time) are proportional to the product of photon and electron density fluxes, and, therefore, are much smaller than the rates of the binary processes.

Strictly speaking, the Sommerfeld formula (3.11, 12) has been obtained for bremsstrahlung on bare nuclei. Simple estimates (both classical and quantal) show that one can also use these results for multiply charged ions, if the ion charge $Z-1 = Z_n - N$ is large and the electron energy range is $E_0 \lesssim I$, where I is the ion ionization potential. Corrections due to the screening effects at larger energies are discussed in [3.2, 8, 9]. In some applications, the total intensity of the bremsstrahlung, integrated over the whole radiation spectrum, is required. One can obtain an estimate for this quantity under the following assumptions: (I) the electron velocity distribution $F(v)$ is Maxwellian and (II) the main contribution is due to the electrons with $g(\eta_0, \eta_1) \approx 1$. The intensity in the spectral interval ω, $\omega + d\omega$ is then given by

$$Q(\omega)\, d\omega = N_e\, N_i\, \hbar\, \omega \int_{\sqrt{2\hbar\omega/m}}^{\infty} \frac{d\sigma}{d\omega} v\, F(v)\, dv\, d\omega \tag{3.25}$$

$$= \frac{32\,\pi}{3\sqrt{3}} \alpha^3 a_0^2 (Z-1)^2 \frac{e^4}{\hbar} \left(\frac{m}{2\,\pi\,\varkappa\, T}\right)^{1/2} N_e\, N_i\, \mathrm{e}^{-\hbar\omega/\varkappa T} d\omega\,, \tag{3.26}$$

where N_e and N_i are the concentrations of electrons and ions with charge $Z-1$, T is the electron temperature, and $\varkappa$ is the Boltzmann constant. The total bremsstrahlung intensity is equal to

$$Q_t = \int Q(\omega)\, d\omega = \frac{32\,\pi}{3\sqrt{3}} \alpha^2 a_0^2 (Z-1)^2\, \mathrm{Ry}\, N_e\, N_i \left(\frac{2\,\varkappa\, T}{\pi\, m}\right)^{1/2}. \tag{3.27}$$

If T is measured in electronvolts, one gets

$$Q_t = 1.54 \times 10^{-25} (Z-1)^2 \, T^{1/2} \, N_e \, N_i \left[\frac{\text{erg}}{\text{cm}^3 \, \text{s}}\right]. \tag{3.28}$$

It is known [3.8] that the use of the Born approximation for the bremsstrahlung cross section gives a factor of $2\sqrt{3}/\pi = 1.1$ on the right-hand side of (3.28). It does not mean, however, that Q_t has a monotonic dependence on T. In the region $\eta_0 \sim 1$, Q_t has a wide maximum, which differs from (3.28) by a factor of 1.43. Thus, the result (3.28) can be used for estimates of Q_t to within a factor of two accuracy.

3.2 Photoionization and Photorecombination

In electron-ion collisions, photon radiation may also take place when the final electronic state is bound. This process is called radiative (or photo-) recombination. The probability of such a process is described by (3.2) as well, where the matrix element is calculated with a final-state wave function that belongs to the discrete spectrum, and the energy conservation law has the form

$$\hbar k c = \hbar \omega = \frac{p^2}{2m} - E_b = \frac{p^2}{2m} + |E_b|. \tag{3.29}$$

In (3.29), $E_b (< 0)$ is the energy of a bound electron in the final state. The differential photorecombination cross section for photon radiation in a solid angle dO_k, in the dipole approximation is given by

$$d\sigma^{rv}_{p,bk} = \frac{1}{2\pi} \frac{e^2 m k^3}{\hbar p} \left| \boldsymbol{e}_{\kappa\varrho} \int \psi_b^* \, \boldsymbol{r} \, \psi_p^+ \, d\boldsymbol{r} \right|^2 dO_k \tag{3.30}$$

where ψ_b is the electron wave function of the final bound state and other symbols are the same as in Sect. 3.1. The inverse process to photorecombination is the photoionization process, consisting in photon absorption and emission of a bound electron into the continuum. Since we deal here with binary collisions

$$\hbar \omega + X_{Z+1} \rightleftarrows X_Z + e, \tag{3.31}$$

the cross sections of both processes have the same dimensions and are mutually related by the detailed balance principle,

$$\frac{1}{k^2} \frac{d\sigma^{rv}_{p,bk}}{dO_k} = \frac{\hbar^2}{p^2} \frac{d\sigma^{v}_{bk;p}}{dO_p} \tag{3.32}$$

where $d\sigma^{\nu}_{bk,p}$ is the differential photoionization cross section. For the total cross sections, integrated over all the possible directions of the emitted particles and summed over the polarization directions, one can use the relation

$$(p^2/\hbar^2)\, g_{Z+1}\, \sigma^{r\nu} = k^2\, g_Z\, \sigma^{\nu}\,, \tag{3.33}$$

where g_{Z+1} and g_Z are statistical weights of the ions X_{Z+1} and X_Z in (3.31).

The theory of photoionization and photorecombination processes can be constructed for ions with an arbitrary number of electrons. For ions with $Z \gg 1$, the non-relativistic Coulomb functions (3.4, 5) can be used as wave functions of the electron in the continuum. For the bound states, it is convenient to use intermediate coupling. General formulae for the separation of the angular and radial parts in matrix elements of the type (3.30) are given by *Sobel'man* [3.8].

Let us start with the photoionization of H-like ions (the inverse process corresponds to photorecombination with bare nuclei). In this case, formulae for σ^{ν} and $\sigma^{r\nu}$ can be written in a closed analytical form. For example, for the 1s state, one has [3.8]

$$\sigma^{r\nu} = \frac{2^8\pi}{3}\alpha^3 \frac{E_0^3}{E_\omega^3(E_\omega - E_0)} \frac{\exp(-4\,\varkappa \operatorname{arccot} \varkappa)}{1-\exp(-2\,\pi\,\varkappa)} [\pi\, a_0^2]\,, \tag{3.34}$$

$$\sigma^{\nu} = \frac{2^9\pi}{3}\frac{\alpha}{Z^2}\left(\frac{E_0}{E_\omega}\right)^4 \frac{\exp(-4\,\varkappa \operatorname{arccot} \varkappa)}{1-\exp(-2\,\pi\,\varkappa)} [\pi\, a_0^2]\,, \tag{3.35}$$

$$\varkappa^2 = E_0/(E_\omega - E_0)\,, \quad E_0 = Z^2\,\mathrm{Ry}\,, \quad E_\omega = \hbar\,\omega\,, \tag{3.36}$$

where α is the fine structure constant. The cross section σ^{ν} has a maximum at $E_\omega \sim E_0$, with

$$\sigma^{\nu}_{\max} = \frac{2\,\pi}{3}\left(\frac{4}{e}\right)^4 \frac{\alpha}{Z^2} [\pi\, a_0^2]\,, \tag{3.37}$$

whereas $\sigma^{r\nu} \to \infty$, when $E_\omega \to E_0$.

In the case $E_\omega - E_0 \gg E_0$, one has

$$\sigma^{\nu} \approx \frac{2^8}{3}\frac{\alpha}{Z^2}\left(\frac{E_0}{E_\omega}\right)^{7/2} [\pi\, a_0^2], \tag{3.38}$$

$$\sigma^{r\nu} \approx \frac{2^7}{3}\alpha^3\left(\frac{E_0}{E_\omega}\right)^{5/2} [\pi\, a_0^2]\,. \tag{3.39}$$

For an arbitrary hydrogen-like ion state $n\,l$, the cross-sections σ^{v} and σ^{rv} can be written in the form [3.16]

$$\sigma^{v}(n\,l) = \frac{64}{3\cdot 137}\,\frac{1}{Z^2}\,\frac{1}{(2\,l+1)}\,\frac{1}{n^3 k^7}\sum_{\lambda=l\pm 1}\mathscr{D}_\lambda(k)\quad[\pi\,a_0^2]\,, \tag{3.40}$$

$$\sigma^{rv}(n\,l) = \frac{32}{3\,(137)^3\,n^3\,k^5}\sum_{\lambda=l\pm 1}\mathscr{D}_\lambda(k)\quad[\pi\,a_0^2]\,, \tag{3.41}$$

$$\mathscr{D}_\lambda(k) = \frac{1}{l_{\max}}\left[k\,(l_{\max}^2 + k^{-2})^{1/2}\,F_l\left(\frac{1}{k},\frac{2}{k}\right) - F_\lambda\left(\frac{1}{k},\frac{2}{k}\right)\right]^2, \tag{3.42}$$

$$\lambda = l\pm 1\,,\quad l_{\max} = (l+\lambda+1)/2\,,\quad k = \sqrt{\frac{\varepsilon}{Z^2\,\mathrm{Ry}}} \gg \frac{1}{n}\,, \tag{3.43}$$

$$\varepsilon = \hbar\,\omega - E_0\,,\quad E_0 = Z^2\,\mathrm{Ry}/n^2\,, \tag{3.44}$$

where $F_l(a,x)$ is the Coulomb wave function (see, for example [3.4]), which satisfies the equation

$$\frac{d^2F_l(a,x)}{dx^2} + \left(1 + \frac{2a}{x} - \frac{l\,(l+1)}{x^2}\right)F_l(a,x) = 0\,. \tag{3.45}$$

The cross sections σ^{v} and σ^{rv}, given by (3.40, 41), when summed over all the orbital quantum numbers l, give the classical Kramers cross sections:

$$\sigma^{v}_{\mathrm{Kr}}(n) = \frac{64}{3\sqrt{3}\cdot 137\,n^2}\,\frac{|E_0|^3}{(|E_0|+\varepsilon)^3}\quad[\pi\,a_0^2]\,, \tag{3.46}$$

$$\sigma^{rv}_{\mathrm{Kr}}(n) = \frac{32}{3\sqrt{3}\,(137)^3}\,\frac{Z\,|E_0|^{3/2}}{(E_0+\varepsilon)\,\varepsilon}\quad[\pi\,a_0^2]\,. \tag{3.47}$$

The Kramers classical cross sections are connected by the relation

$$\sigma^{v}_{\mathrm{Kr}}(n) = \frac{2\,m\,c^2\,\varepsilon}{\hbar^2\,\omega^2}\,\frac{1}{2\,n^2}\,\sigma^{rv}_{\mathrm{Kr}}\,. \tag{3.48}$$

The cross sections σ^{v} and σ^{rv} can also be written in the form

$$\sigma^{v} = Q_v\,G\,\sigma^{v}_{\mathrm{Kr}}(n)\,,\quad \sigma^{rv} = Q_{rv}\,G\,\sigma^{rv}_{\mathrm{Kr}}(n)\,, \tag{3.49}$$

where $\sigma^{v}_{\mathrm{Kr}}(n)$, $\sigma^{rv}_{\mathrm{Kr}}(n)$ are defined by (3.46, 47). The quantity G is called the Gaunt factor. Since $\sigma^{v,rv}_{\mathrm{Kr}}$ contains the main dependence on Z and n, the Gaunt factor has a weaker dependence on the parameters Z, E_0 and ε, than the cross sections themselves.

In general, for non-hydrogenic systems the following formulae are used (within the dipole approximation):

$$\sigma^{v} = \frac{4}{3\cdot 137}\,\frac{Q_v}{2l+1}\,E_\omega\,[l\,R_{l-1}^2 + (l+1)\,R_{l+1}^2] \quad [\pi a_0^2]\,, \tag{3.50a}$$

$$\sigma^{rv} = \frac{2}{3\cdot(137)^3}\,Q_{rv}\,\frac{E_\omega}{\varepsilon}\,[l\,R_{l-1}^2 + (l+1)\,R_{l+1}^2] \quad [\pi a_0^2]\,, \tag{3.50b}$$

$$R_{l\pm 1} = \int_0^\infty P_{nl}(r)\,P_{\varepsilon,l\pm 1}(r)\,r\,dr\,, \tag{3.51}$$

$$E_\omega = |E_0| + \varepsilon = \hbar\,\omega/\mathrm{Ry}\,, \tag{3.52}$$

where $P_{nl}(r)$ and $P_{\varepsilon l}(r)$ are the wave functions of the optical electron in the discrete and continuous spectra, respectively, with the normalizations:

$$\int_0^\infty P_{nl}^2(r)\,dr = 1\,, \tag{3.53}$$

$$\int_0^\infty P_{\varepsilon l}(r)\,P_{\varepsilon' l}(r)\,dr = \pi\,\delta(\varepsilon - \varepsilon')\,. \tag{3.54}$$

In the above equations, $E_0\,(<0)$ is the binding energy of the electron in the ion X_Z (in the state $n\,l$), and $\varepsilon\,(>0)$, $l \pm 1$ are the energy and the momentum of the electron in the continuum. The angular coefficients Q_v and Q_{rv} depend on the type of transition. For example, for the transition $l^q\,S_0\,L_0 \to l^{q-l}\,S_1\,L_1$,

$$Q_v = q\,|G_{S_1L_1}^{S_0L_0}|^2\,, \tag{3.55}$$

$$Q_{rv} = q\,\frac{(2\,S_0+1)\,(2\,L_0+1)}{2\,(2\,l+1)\,(2\,S_1+1)\,(2\,L_1+1)}\,|G_{S_1L_1}^{S_0L_0}|^2\,, \tag{3.56}$$

where q is the number of equivalent electrons and $G_{S_1L_1}^{S_0L_0}$ are the Racah fractional parentage coefficients. For photoionization from the shell $n\,l^q$, one has

$$Q_v = q\,,\quad Q_{rv} = 1 - \frac{(q-1)}{2\,(2\,l+1)}\,. \tag{3.57}$$

The factors Q and G are discussed in [3.17].

The photoionization cross section is maximum at the threshold, and for $E_\omega - E_0 \gg E_0$ it has the following asymptotic behaviour:

$$\sigma_l^v \sim \varepsilon^{-(3+l+1/2)}\,,\quad \sum_l \sigma_l^v \sim \varepsilon^{-7/2}\,. \tag{3.58}$$

For very high energies of the ejected electron ($\varepsilon > 50\, Z^2$ Ry), one has

$$\sum_l \sigma_l^v \sim \varepsilon^{-3}, \quad \varepsilon > 50\, Z^2\, \mathrm{Ry}. \tag{3.59}$$

The photoionization cross section σ^v can be expressed in terms of the continuum oscillator strength

$$\frac{df}{d\varepsilon} = \frac{137}{4\pi}\, \sigma^v \quad [\pi\, a_0^2], \tag{3.60}$$

where ε is expressed in Rydberg units. In the discrete spectrum, the quantity $df/d\varepsilon$ is connected with the usual oscillator strength $f_{0\to nl}$ (for the transition $0 \to nl$) by the relation:

$$\left.\frac{df}{d\varepsilon}\right|_{\varepsilon\to 0} = \left.\frac{n^{*3} f_{0\to nl}}{2Z^2}\right|_{n\to\infty} = \frac{137}{4\pi}\, \sigma^v \quad [\pi\, a_0^2], \tag{3.61}$$

$$n^* = Z\, (\varepsilon_{nl}/\mathrm{Ry})^{-1/2}, \tag{3.62}$$

where n^* is the effective quantum number of the level nl and ε_{nl} is the corresponding energy, measured from the ionization limit.

To calculate the spectral line intensities and the ionization equilibrium in hot plasmas, one often requires the photorecombination rate $\langle v\, \sigma^{rv}\rangle$, i.e. the photorecombination cross sections, averaged over the Maxwellian velocity distribution of incident electrons. This quantity can be written as

$$\langle v\, \sigma^{rv}\rangle = \frac{2\sqrt{\pi}\, \hbar\, a_0}{m} \left(\frac{E_0}{\mathrm{Ry}}\right)^{1/2} \beta^{3/2} \int_0^\infty u\, \frac{\sigma^{rv}(u)}{\pi\, a_0^2}\, \exp(-\beta u)\, du, \tag{3.63}$$

$$u = \varepsilon/E_0, \quad \beta = E_0/kT. \tag{3.64}$$

Approximate formulae can be derived for $\langle v\, \sigma^{rv}\rangle$. Expressions for some transitions in highly charged ions are given in [3.17].

Calculations of photoionization cross sections of multiply charged ions have been performed elsewhere e.g., [3.18–22]. Values of σ^v for all ionization stages of O^{Z+} and Fe^{Z+} ions are calculated in [3.18, 19], allowing for the inner-shell ionization. In [3.21] the cross sections σ^v for Ne-like ions are calculated up to $Z_n = 50$.

As has been shown in [3.19, 22], the photoionization cross section for an innershell electron $n_0 l_0 j_0$ from the ion X_Z has a weak Z dependence at a given frequency ω, i.e. it does not depend much on the removal of electrons with $n > n_0$. Only the photoionization threshold increases with increasing ion charge Z. For example, according to [3.22], the photoionization cross sections of the inner $2p_{1/2}$ electron in Th I and Th^{80+} at $\omega = 81.635$ keV are

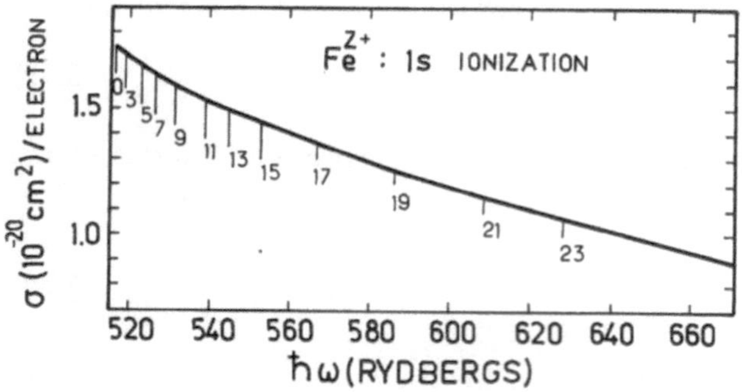

Fig. 3.1. Calculated photoionization cross section for the $1s$ electron of Fe^{Z+} ions ($Z=0-23$) [3.19]. Vertical lines are theoretical thresholds for a given stage of ionization.

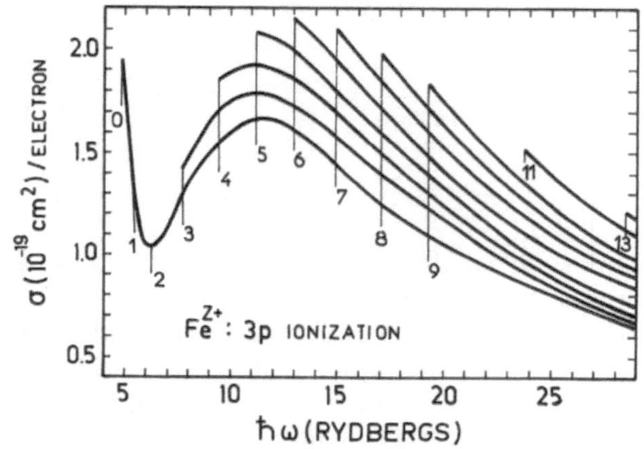

Fig. 3.2. Calculated photoionization cross section for the $3p$ electron of Fe^{Z+} ($Z=0-13$) [3.19]. Vertical lines are theoretical thresholds for a given stage of ionization

approximately equal to 227.8 and 235.1 barns, respectively. This effect is connected with the fact that the binding energy of innershell electrons has a weak dependence on the outer electrons (with $n > n_0$), but is very sensitive to the charge of the ionic core (i.e. to shells with $n \leq n_0$). The situation is illustrated in Figs. 3.1, 2, where the photoionization cross sections of $1s$- and $3p$ electrons from the iron ions are shown.

3.3 Dielectronic Recombination

3.3.1 Resonance Character of the Process

The origin of the process of dielectronic recombination (as well as autoionization) lies in the capture of a free electron by the ion, with simultaneous excitation of an ion electron (see also Sect. 2.5):

$$X_{Z+1}(\alpha_0) + e \rightarrow X_Z^{**}(\gamma), \quad \gamma = \alpha n l L S J. \tag{3.65}$$

The doubly excited ion X_Z^{**} thus created may decay in two different ways (see Fig. 2.1), either by autoionization, with emission of an electron into the continuum,

$$X_Z^{**}(\gamma) \rightarrow X_{Z+1}^{*}(\alpha') + e\,, \tag{3.66}$$

or by radiative decay, with a photon emission, and ion transition to a "stable" state γ', which lies below the ionization limit:

$$X_Z^{**}(\gamma) \rightarrow X_Z^{*}(\gamma') + \hbar\,\omega\,. \tag{3.67}$$

The process (3.65, 67) is called dielectronic recombination. The autoionization (3.65, 66) leads to excitation $\alpha_0 \rightarrow \alpha'$ ($\neq \alpha_0$) of the ion X_{Z+1} via the intermediate level γ. Such an additional excitation channel with respect to the usual one is called resonance excitation, and will be considered in more detail in Chap. 4. The process of electron capture (3.65) by the ion X_{Z+1} has a resonance nature. This process is possible only within the width δ_γ of the level γ, around the electron energy ε defined by

$$\varepsilon = E_\gamma - E_{\alpha_0} \approx E_\alpha - E_{\alpha_0} - \frac{Z^2\,\mathrm{Ry}}{n^2}\,. \tag{3.68}$$

The width δ_γ of the level γ is equal to

$$\delta_\gamma = \hbar\,[A_a(\gamma) + A_r(\gamma)] \quad \text{or} \tag{3.69}$$

$$\delta_\gamma = 4.83 \times 10^{-17}\,(A_a[\mathrm{s}^{-1}] + A_r[\mathrm{s}^{-1}])\;\;[\mathrm{Ry}]\,, \tag{3.70}$$

where A_a and A_r are the autoionization and the radiative probability for decay of the level γ, respectively.

The effective cross section σ_d for the electron capture is described by a dispersion formula with a width δ_γ. However, for most applications the cross section σ_d itself is not of interest but rather the rate $\tilde{\varkappa}_d$ of dielectronic capture. If the electron energy distribution is $F(\varepsilon)$ (not necessarily Maxwellian), the capture rate is given by

$$\tilde{\varkappa}_d = F(\varepsilon)\,(2\,\varepsilon)^{1/2} \int_0^\infty \sigma_d(\varepsilon')\,d\varepsilon' \quad [\mathrm{cm}^3\,\mathrm{s}^{-1}]\,, \tag{3.71}$$

where ε is defined by (3.68). Here it is assumed that $\delta_\gamma \ll T$, where T is the electron temperature. By using the detailed balance principle and assuming a Maxwellian electron velocity distribution, one can derive the following relation:

$$g_{\alpha_0}\tilde{\varkappa}_d(\gamma) = 4\,\pi^{3/2}\,a_0^3 \left(\frac{\mathrm{Ry}}{T}\right)^{3/2} \mathrm{e}^{-\varepsilon/T}\, g_\gamma A_a(\gamma, \alpha_0)\,, \tag{3.72}$$

where g_{α_0} and g_γ are, respectively, the statistical weights of the levels α_0 and γ, and $A_a(\gamma, \alpha_0)$ is the autoionization decay probability for the ionic state α_0. The total autoionization probability of the level γ is

$$\sum_{\alpha_0} A_a(\gamma, \alpha_0) = A_a(\gamma) . \tag{3.73}$$

The quantity $A_a(\gamma, \alpha_0)$ will be discussed in more detail in Sect. 3.4.

3.3.2 Theory and Experiment

The general problems connected with dielectronic recombination are considered elsewhere (see [3.23–27]). As mentioned above, the process takes place in two steps:

$$X_{Z+1}(\alpha_0) + e \rightarrow X_Z^{**}(\gamma) \rightarrow X_Z^{*}(\gamma') + \hbar\omega , \tag{3.74}$$

where $\gamma = \alpha\, n\, l\, (LSJ)$ and the level γ lies below the ionization limit of the ion X_Z. The spectral lines corresponding to the photon emission $\hbar\omega$ are called dielectronic satellites; they play a very important role in the creation of the spectra of multiply charged ions. Dielectronic recombination itself is a process which strongly influences the ionization equilibrium in hot plasmas.

As a rule, the radiative transition in dielectronic recombination is due to the inner electron α, i.e. $\gamma' = \alpha'\, n\, l\, L'\, SJ$. However, in some cases transition of an outer electron nl ($\gamma' = \alpha\, n'\, l'\, L'\, SJ'$) is also important, although the probability of such a process is proportional to n^{-3}.

The dielectronic recombination rate $\varkappa_d(\gamma)$, via the level γ, is defined by the following expression:

$$\varkappa_d(\gamma) = \tilde{\varkappa}_d(\gamma) \frac{A_r(\gamma)}{A_a(\gamma) + A_r(\gamma)} , \tag{3.75}$$

where $\tilde{\varkappa}_d$ is the rate of electron capture (3.71), and A_r and A_a are, respectively, the probabilities of radiative and autoionization decay from the level γ. The total dielectronic recombination rate is equal to

$$\varkappa_d^t = \sum_\gamma \varkappa_d(\gamma) . \tag{3.76}$$

Using the detailed balance principle (3.72), it is possible to connect $\varkappa_d^t$ with the autoionization probability. Actually (see [3.27]):

$$\varkappa_d^t = \sum_\alpha \varkappa_d^\alpha\,, \quad \varkappa_d^\alpha = 10^{-3} B_d(\alpha)\, \beta^{3/2}\, e^{-\beta x}\, [\mathrm{cm^3\, s^{-1}}]\,, \tag{3.77}$$

$$B_d(\alpha) = C \sum_\gamma q(\gamma, \alpha_0)\, e^{\delta\beta}\,, \quad \gamma = \alpha\, n\, l\, LS\,, \tag{3.78}$$

$$q(\gamma, \alpha_0) = \frac{g_\gamma A_r(\gamma) A_a(\gamma, \alpha_0)}{A_r(\gamma) + A_a(\gamma)}\,, \tag{3.79}$$

$$C = 10^{13} \frac{4\pi^{3/2} a_0^3}{g_{\alpha_0}(Z+1)^3}\,, \quad \beta = \frac{(Z+1)^2\,\mathrm{Ry}}{T}\,, \quad \beta x = E_{\alpha\alpha_0}/T\,, \tag{3.80}$$

where $A_a(\gamma, \alpha_0)$ is the autoionization probability from the state γ to α_0, $E_{\alpha\alpha_0} = E_\alpha - E_{\alpha_0}$ and $B_d \sim 1$. In accordance with (3.69),

$$\delta\beta = (E_{\alpha\alpha_0} - E_{\gamma\alpha_0})/T \approx \frac{I_Z(n\,l)}{T} \simeq \frac{Z^2\,\mathrm{Ry}}{n^2 T}\,, \tag{3.81}$$

where $I_Z(n\,l)$ is the electron binding energy in the nth state of the ion X_Z.

In the majority of cases, dielectronic recombination takes place to highly excited levels, $n \gg 1$, for which $\delta\beta$ is small and the coefficient B_d has a weak dependence on the temperature T. However, this rule is not valid for $1s-2p$ transitions in H- and He-like ions with $Z \gtrsim 20$.

The calculation of $\varkappa_c^t$ using the general formulae (3.77–81) is a rather complicated problem, because it requires calculation of $A_a(\gamma, \alpha_0)$ for each state $\gamma = \alpha\, n\, l\, LS$. In practical calculations, simple models are usually employed. For example, a simple semiempirical formula for $\varkappa_d^t$ has been suggested by *Burgess* [3.23]:

$$\varkappa_d^t = \sum_\alpha \varkappa_d^\alpha\,, \quad \varkappa_d^\alpha = 10^{-13} B_d\, \beta^{3/2} \exp(-\beta\, \chi_d)\,, \tag{3.82}$$

$$B_d = 480 f_{\alpha\alpha_0} \left[\frac{Z\chi}{Z^2 + 13.4}\right]^{1/2} \times [1 + 0.105\,\chi(Z+1) + 0.015\,(Z+1)^2 \chi^2]^{-1}\,, \tag{3.83}$$

$$\beta = \frac{(Z+1)^2\,\mathrm{Ry}}{T}\,, \quad \chi_d = \chi\left[1 + 0.015 \frac{Z^3}{(Z+1)^2}\right]^{-1}\,, \quad \chi = \frac{E_{\alpha\alpha_0}}{(Z+1)^2\,\mathrm{Ry}}\,, \tag{3.84}$$

where $f_{\alpha\alpha_0}$ is the oscillator strength for the transition $\alpha - \alpha_0$ in the ion X_{Z+1}.

A more exact model for the calculation of $\varkappa_d$ for multiply charged ions has been suggested by *Vainshtein* [3.26], where the partial excitation cross section $\sigma(\alpha_0, \alpha\, l)$ for the transition $\alpha_0 - \alpha$ is used instead of the autoionization probability $A_a(\alpha\, n\, l, \alpha_0)$. A comparison of the *Burgess* [3.23] and *Vainshtein* [3.26] results shows that, in general, they agree well with each other. However, for some transitions the Burgess formula has a significant

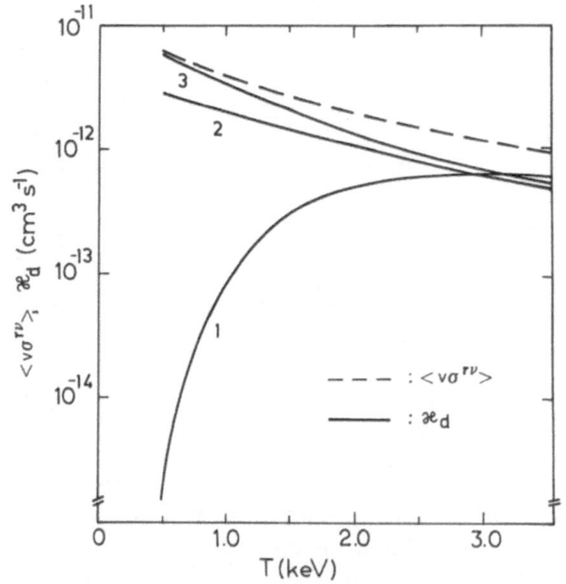

Fig. 3.3. Calculated radiative recombination rates, $\langle v\,\sigma^{rv}\rangle$, and dielectronic recombination rates, $\varkappa_d$, for multiply charged iron ions [3.26].
Curve 1 is $\varkappa_d$ for the [He] → [Li] process,
Curve 2 is $\varkappa_d$ for the [Li] → [Be] process,
Curve 3 is $\varkappa_d$ for the [Be] → [B] process,
Dashed curve is $\langle v\,\sigma^{rv}\rangle$ for each of processes [He] → [Li], [Li] → [Be] and [Be] → [B]

error, because it does not take into account the individual peculiarities of a transition, such as additional channels of autoionization decay (which are observed, for example, for the $2s-2p$ transitions in Li- and Be-like ions), etc.

Let us now consider the dependence of the dielectronic recombination rate on the electron temperature T and density N. According to (3.82 – 84), one has

$$\begin{aligned}\varkappa_d &\sim e^{-1/T}, \quad T\to 0\\ &\sim T^{-3/2}, \quad T\to\infty\\ &\sim \varkappa_{d,\max}, \quad \text{at} \quad T_{\max}\simeq 0.67\,\chi_d\,(Z+1)^2\,\mathrm{Ry}\,.\end{aligned} \tag{3.85}$$

A comparison of calculated values of $\varkappa_d$ and the photorecombination rate $\langle v\,\sigma_v\rangle$ for iron ions is shown in Fig. 3.3. It is seen that for the recombination of He- to Li-like ions at $T\simeq 1$ keV, the value $\langle v\,\sigma_v\rangle$ is one order of magnitude larger than $\varkappa_d$, and for the reactions [Li] → [Be][1], [Be] → [B] both rates are equal. The probability of electron recombination $N_e^2\langle\!\langle v\,\sigma_r\rangle\!\rangle$ is much less than the sum $N_e\,(\varkappa_d+\langle v\,\sigma_v\rangle)$, at $T \gtrsim 0.5$ keV, $N_e > 10^{21}$ cm^{-3}.

All the formulae for $\varkappa_d$ considered in this section, are obtained in the zero-density approximation. In real plasmas, $N_e \neq 0$ and there always exists an excited state n_1 of the ion X_Z, starting from which the probabilities of

1 [A] denotes an A-like ion.

collisional transitions are larger than those of radiative transitions from the level n_1. In this case ionization from excited states is possible, which may lead to a decrease in the dielectronic recombination rate. In other words, in dense plasmas the following reaction begins to play an important role:

$$X_Z(\alpha\, n\, l) + e \rightarrow X_{Z+1}(\gamma') + 2e\,. \qquad (3.86)$$

This process has to be taken into account when calculating the ionization equilibrium in dense high-temperature plasmas [3.16].

Let us now discuss the experimental results on dielectronic recombination. As was noted above, the main difficulty in the calculations of $\varkappa_d$ is connected with inclusion of the large number of transitions via intermediate states.

In connection with this, the absolute or relative measurements of dielectronic recombination rates are of considerable interest. In [3.28, 29], the values of $\varkappa_d$ were measured for the ions Fe IX–X in θ-pinch plasmas with parameters $T = 50-200$ eV, $N_e = 3\times10^{16}$ cm^{-3}. In [3.30, 31] values of $\varkappa_d$ were estimated for the ions Mo XXXI–XXXII and Fe XXV in high-temperature tokamak plasmas. In these papers the absolute values of $\varkappa_d$ were obtained by solving the system of equations for ionization balance using certain models for the radiating plasmas. It was necessary to introduce some simplifying considerations regarding the dependence of $\varkappa_d$ on T. Another method was used in [3.32] to obtain $\varkappa_d$ in laser-produced plasmas of Ca XIX and Ti XXI ions with parameters $T = 700-900$ eV, $N_e \simeq 10^{21}$ cm^{-3}.

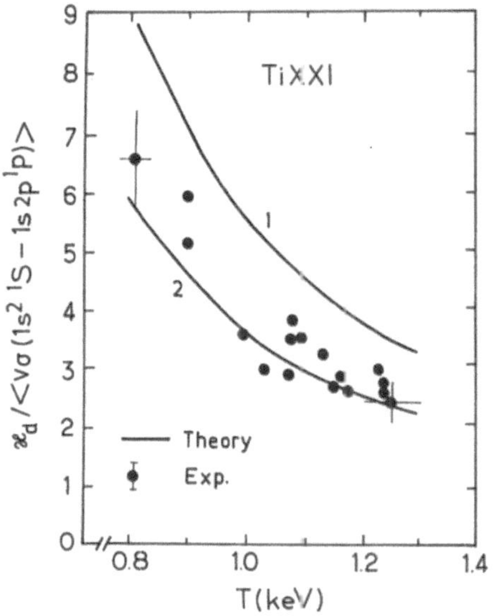

Fig. 3.4. Ratio of the rates for dielectronic recombination, $\varkappa_d$, and resonance line excitation, $\langle v\,\sigma\rangle$, of the He-like Ti XXI ion. Dots are experimental data [3.32], solid curves are calculations: (*1*) [3.23], (*2*) [3.26] (see text)

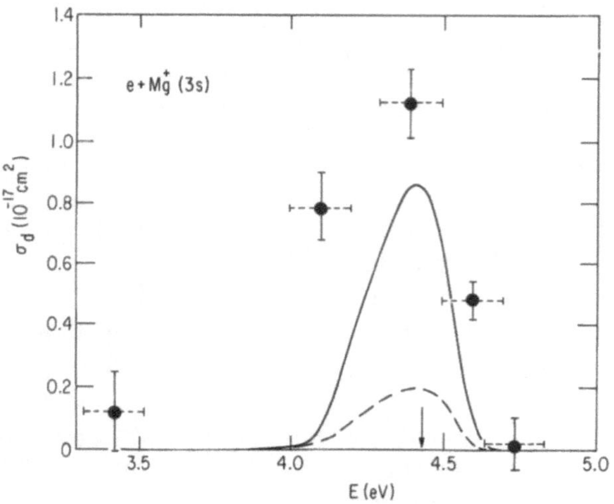

Fig. 3.5. Dielectronic recombination cross section for $e + Mg^+ (3s)$. Experiment: (⧫) [3.34]. Theory [3.35] (---) convoluted cross section for $n \leq 64$ with experimental electron energy distribution, (—) same as dashed curve, but including all n. Arrow indicates threshold energy

In this paper the relative values $\varkappa_d / \langle v\,\sigma \rangle_w$ were measured, where $\langle v\,\sigma \rangle_w$ is the electron excitation rate of the resonance $1s^2 - 1s\,2p\,{}^1P_1$ line in the He-like ion. One of the advantages of this method is a rather simple interpretation of experimental data and a weak dependence of the results on different theoretical models. In Fig. 3.4, the measured dependence $\varkappa_d(T)$ for Ti XXI ions is compared with the Burgess formula and the results of the Vainshtein model [3.26].

Quite recently, direct measurements of dielectronic recombination cross sections have been reported for the $e + C^+ (2p)$ [3.33] and $e + Mg^+ (3s)$ [3.34] systems, by using the merged beam and crossed beam method, respectively. The data for $e + Mg^+ (3s) \rightarrow Mg^{**} (3pnl) \rightarrow Mg^* (3snl) + \hbar\omega$ recombination are shown in Fig. 3.5. They are compared with the theoretical cross section averaged over the resonances [3.35], and convoluted with the experimental electron energy distribution. The dashed curve is the theoretical result which includes only the electron capture to Rydberg states with $n \leq 64$, actually populated in the experiment. When the absolute experimental uncertainties of the above-mentioned experimental data are taken into account (about $\pm 58\%$ in the $e + Mg^+$ case), the data are higher by a factor of about 3 or more than the theoretical predictions.

3.4 Autoionization

The decay of an excited ionic state via electron emission is called autoionization:

$$X_Z^{**}(\gamma) \rightarrow X_{Z+1}(\alpha') + e\,. \tag{3.87}$$

The initial state γ may be of different types, including doubly excited states, states with an inner-shell electron hole, etc. Here we shall consider only doubly excited states, i.e. the process

$$X_Z^{**}(\gamma) \rightarrow X_{Z+1}(\alpha') + e, \quad \gamma = \alpha\, n\, l\, LSJ, \tag{3.88}$$

with a given genealogy scheme

$$\alpha = \alpha_p [S_p L_p]\, n_\alpha\, l_\alpha\, L_\alpha\, S_\alpha . \tag{3.89}$$

The index p refers to the parent ion. The emitted electron has an energy $\varepsilon > 0$, defined by (3.68), and a momentum λ'. The case $\alpha = \alpha'$ corresponds to resonance electron scattering (see Sect. 4.2).

The probability of the autoionization process (3.87) in first-order perturbation theory is defined by matrix elements of the type (2.36)

$$A_a = \frac{2\pi}{\hbar} \left| \left\langle \alpha' \left| \frac{1}{r_{12}} \right| \alpha \right\rangle \right|^2, \tag{3.90}$$

where $|\alpha'\rangle$, $|\alpha\rangle$ are the ion wave functions in the continuum, and r_{12}^{-1} is the electrostatic interaction between electrons.

The general problems of autoionization processes have been discussed elsewhere [3.35–40]. For highly excited states, $n \gg 1$, the autoionization probability $A_a(\alpha\, n\, l, \alpha')$ is connected with the electron excitation cross section $\sigma(\alpha', \alpha\, l)$ at the threshold by [3.16]

$$A_a(\alpha\, n\, l, \alpha') = \frac{1}{\tau_0} \left(\frac{Z}{Z+1}\right)^2 \frac{\varepsilon}{\pi n^3} \frac{g_{\alpha'}}{g_{\alpha n l}} (Z+1)^4\, \sigma(\alpha', \alpha\, l), \tag{3.91}$$

$$\sigma(\alpha', \alpha\, l) = \sum_{\lambda'} \sigma(\alpha' \lambda', \alpha\, l), \tag{3.92}$$

$$\tau_0 = a_0 \hbar / e^2 = 2.419 \times 10^{-17} \quad [\mathrm{s}], \tag{3.93}$$

where λ' is the angular momentum of the emitted electron, and $g_{\alpha'}$, $g_{\alpha n l}$ are the statistical weights of the ions X_{Z+1} and X_Z, respectively; σ is expressed in units of πa_0^2 and ε in Rydbergs.

Among all the processes (3.88), the dipole transitions $\alpha - \alpha'$ are of special interest. For such transitions the value $A_a(\alpha\, n\, l, \alpha')$ is also connected with the photoionization cross section $\sigma_v(n\, l, \varepsilon)$ (see Sect. 3.2):

$$A_a(\alpha\, n\, l, \alpha') = \frac{3 \cdot 137}{16\pi} Z^2 \varepsilon^2 f_{\alpha\alpha'}\, \sigma_v(n\, l, \varepsilon) \tag{3.94}$$

where $f_{\alpha\alpha'}$ is the oscillator strength for the transition $\alpha-\alpha'$. For $n \gg 1$, the expression for σ_ν and, consequently, that for A_a can be written in a closed analytical form (see Sect. 3.2):

$$A_a(\alpha\, n\, l, \alpha') = \frac{4 f_{\alpha\alpha'}}{\tau_0\, n^3\, (2l+1)\, k^3} \sum_{\lambda'=l\pm 1} \mathcal{D}_{\lambda'}(k)\,, \tag{3.95}$$

$$\mathcal{D}_{\lambda'}(k) = \frac{1}{l_{\max}} \left[k\,(l_{\max} + k^{-2})^{1/2}\, F_l\left(\frac{1}{k}, \frac{2}{k}\right) - F_{\lambda'}\left(\frac{1}{k}, \frac{2}{k}\right)\right]^2, \tag{3.96}$$

$$l_{\max} = (l+\lambda'+1)/2\,, \quad k = \sqrt{\varepsilon/(Z^2\,\mathrm{Ry})} \gg n^{-1}, \quad n \gg 1\,, \tag{3.97}$$

where $F_l(x,y)$ is the Coulomb wave function. The numerical calculations have shown that, for $n_\alpha = 2$, the approximation $A_a(\alpha\, n\, l, \alpha') \sim n^{-3}$ is valid for $n > 4$. The dipole approximation (3.95) usually holds for $l > 3$. It is worth noting that this approximation is rather useful since numerical calculations of $A_a(\alpha\, n\, l, \alpha')$ [for example, using (3.91)] are rather complicated and present considerable difficulties in the numerical integration.

For transitions within the same principal quantum number ($n_{\alpha'} = n_\alpha$) it is necessary to pay attention to the problem of proper normalization of the autoionization probability (e.g. for the $2s-2p$ transitions in ions with $Z \leq 10$). The normalized probability $W^{\mathrm{N}}(\alpha\, n\, l, \alpha')$ can be obtained by using in (3.91) normalized cross sections $\sigma^{\mathrm{N}}(\alpha', \alpha\, l)$. In its turn, $\sigma^{\mathrm{N}}(\alpha', \alpha\, l)$ can be calculated using some of the standard normalization procedures (Seaton's K-matrix method [3.40], for instance).

The values A_a for the transitions in He-like ions are given in Tables 2.3 and 2.4.

4. Electron Collisions with Highly Charged Ions: General Theory and Excitation Processes

The inelastic collision processes between electrons and ions are considerably more complex than radiative processes, considered in the preceding chapter. In the present chapter, we consider the excitation processes in electron-ion collision at non-relativistic energies of the incident electron. After some introductory remarks, we shall give a detailed account on the existing theoretical descriptions of these processes and compare their results with experimental data.

4.1 Introductory Remarks

The long-range attractive Coulomb interaction in an electron-ion collision leads to characteristic features in the excitation cross sections at the collision energies corresponding to the thresholds of new reaction channels. These threshold phenomena are associated with the strong coupling of the reaction channels. For instance, the direct (potential) electron-impact excitation of an ion

$$X_Z(\gamma_i) + e \rightarrow X_Z^*(\gamma_f) + e\,, \tag{4.1}$$

is accompanied by resonance excitation,

$$X_Z(\gamma_i) + e \rightarrow X_{Z-1}^{**}(\gamma n l) \rightarrow X_Z^*(\gamma_f) + e\,, \tag{4.2}$$

where γ_i and γ_f denote the initial and final states of the ion X_Z. The process (4.2) means electron capture into a doubly excited non-(quasi-)stationary state with principal quantum number n and orbital quantum number l. The autoionization decay of the doubly excited state gives an additional contribution to the excitation cross section. Further, the radiative decay of the doubly excited state,

$$X_Z(\gamma_i) + e \rightarrow X_{Z-1}^{**}(\gamma n l) \rightarrow X_{Z-1} + \hbar\omega\,, \tag{4.3}$$

(i.e., the dielectronic recombination process, see Sect. 3.3) leads to radiation of satellite lines (Sect. 2.2). It is clear that all the processes (4.1–3) must be considered simultaneously. From a certain point of view the physical picture of resonance scattering in a Coulomb field is similar to the well-known *Breit-Wigner* [4.1–3] scattering in a short-range field with isolated resonances corresponding to formation and decay of compound states of the system of colliding particles. In the electron-ion scattering problem, the long-range Coulomb field plays a dominating role. Some results of its influence are well known. For example, the total cross section for elastic scattering of a charged particle in a Coulomb field diverges at small scattering angles (the Rutherford formula); the impact excitation cross sections for electron-positive ion collisions have finite values at the reaction threshold [4.4], etc.

To solve the electron-ion scattering problem one has to remember that each ionic level is the limit of a denumerable set of resonances for the $e + X_Z$ system, and this information is contained in the analytic properties of Coulomb wave functions. Thus, the investigation of general features of electron-ion scattering can be based on the analysis of the properties of the Coulomb wave function in the complex energy plane, especially with respect to energy singularities. At present, there exist three methods for analysis of the electron-ion scattering which lead essentially to equivalent results. The first method is based on the effective-range theory [4.5–9], the second one is a generalization of quantum defect theory [4.10–13], and the third one is the correct asymptotic expansion of the electron-ion scattering wave function [4.14, 15] in powers of $1/Z$ (where $Z-1$ is the ionic charge). All three methods allow one to investigate both potential (4.1) and resonance (4.2) scattering. The third method also takes into account the mutual influence of dielectronic recombination (4.3) and resonance scattering (4.2).

4.2 Basic Equations

Consider an N-electron ion + incident-electron system. The Hamiltonian of such a system has been discussed in Sect. 2.3, 4. Using atomic units with $\hbar = m = e^2 = 1$ (where e and m are the electron charge and mass), we can write the Schrödinger equation in the form

$$[-\tfrac{1}{2}\nabla_r^2 + H_\mathrm{i}(\xi) + V(\boldsymbol{r}, \xi) - E]\,\Psi(\boldsymbol{r}, \xi) = 0\,. \tag{4.4}$$

Here ξ is a set of variables of the non-perturbed ion and $\boldsymbol{r}$ is the coordinate of the incident electron. The ionic Hamiltonian, $H_\mathrm{i}(\xi)$, and the electron-ion interaction potential, $V(\boldsymbol{r}, \xi)$, should contain the relativistic terms discussed in Sect. 2.4.

We now replace the Schrödinger equation (4.4) by an equivalent system of coupled radial equations using the expansion of the total wave function in terms of the eigenfunctions of the unperturbed ion in the total momentum representation (see, for example [4.16]). Let $\Gamma = \gamma \frac{1}{2} l S_T L_T$ be the complete set of quantum numbers of the colliding system, which includes the quantum numbers of the ion, $\gamma = (n_i l_i) n_\gamma l_\gamma LS$, and of the incident electron, $(\frac{1}{2} l)$, S_T and L_T being the total spin and angular momenta of the system. The solution of (4.4) can be written as

$$\Psi = \hat{A} \sum_{\Gamma} \frac{1}{r} \mathscr{F}_\Gamma(r) \Phi_\Gamma , \tag{4.5}$$

$$\Phi_\Gamma = \sum_{M_L M_S m_l m_s} (\Gamma \mid \gamma M_S M_L l m_s m_l) \psi_{\gamma M_S M_L}(\xi_1, \dots \xi_N) Y_{l m_l}(\theta, \varphi) \chi_{m_s}(\lambda) , \tag{4.6}$$

where ψ are the eigenfunctions of the N-electron ion, $\xi = \{\boldsymbol{r}_i, \lambda\}$ is the set of coordinate ($\boldsymbol{r}_i$) and spin (λ) variables of the ion's electrons and $\hat{A}$ is the electronic antisymmetrization operator. The functions $\mathscr{F}_\Gamma$ satisfy the usual boundary conditions

$$\begin{aligned} &\mathscr{F}_\Gamma(r)\,|_{r=0} = 0 , \\ &\mathscr{F}_\Gamma(r) \underset{r \to \infty}{\sim} k^{-1/2} \{\delta_{\Gamma_0 \Gamma} \sin(k_{\Gamma_0} r + d_{\Gamma_0}) + T_{\Gamma_0 \Gamma} \exp[\mathrm{i}(k_\Gamma r + d_\Gamma)]\} , \end{aligned} \tag{4.7}$$

$$d_\Gamma = -\frac{\pi l}{2} + \frac{(Z-1)}{k_\Gamma} \ln(2 k_\Gamma r) + \arg \Gamma (l + 1 - \mathrm{i}[Z-1]/k_\Gamma) , \tag{4.8}$$

where $Z-1$ is the ion charge ($Z = 1$ for a neutral atom), $Z = Z_\mathrm{n} - N + 1$, Z_n being the nuclear charge, and

$$k_\Gamma = \sqrt{2(E - \Delta E_{\Gamma \Gamma_0})} , \quad \mathrm{Im}\{k_\Gamma\} \geq 0 . \tag{4.9}$$

The angular and spin coefficients in (4.6), $(\cdots | \cdots)$, are given, for example in [4.12, 16] for different types of momentum coupling. The general form of the radial equations does not depend on the type of momentum coupling and can be written as follows:

$$\left(\frac{d^2}{dr^2} - \frac{l(l+1)}{r^2} + \frac{2(Z-1)}{r} + U_{\Gamma \Gamma} + k_\Gamma^2 \right) \mathscr{F}_\Gamma = \sum_{\Gamma' \neq \Gamma} U_{\Gamma \Gamma'} \mathscr{F}_{\Gamma'} . \tag{4.10}$$

All the matrix elements $U_{\Gamma \Gamma'}$ are, generally speaking, integral operators and can be expressed in terms of radial (direct + exchange) integrals. It is essential that in (4.10), the long-range Coulomb interaction $2(Z-1)/r$ is expressed in explicit form, and that other potentials, $U_{\Gamma \Gamma'}$, decrease faster than r^{-1}, with increasing r. Expressions for $U_{\Gamma \Gamma'}$ are given in [4.12] within the LS-coupling scheme, and in [4.16] for more general cases (intermediate coupling, and others).

Let the complete set of quantum numbers of a channel be denoted by $\Gamma = \gamma l \Gamma_T$, where γ specifies the ion state and l is the orbital momentum of the incident electron. Γ_T includes the total momentum and other quantum numbers depending on coupling (LS, jj, or intermediate). The partial cross section for the transition $\Gamma_0 \to \Gamma_1$ is

$$\sigma(\Gamma_0 \to \Gamma_1) = \sigma(\gamma_0 l_0 \to \gamma_1 l_1) = \frac{g_{\Gamma_1}}{g_0 E} \, |T_{\Gamma_0 \Gamma_1}|^2 \, \pi a_0^2 , \tag{4.11}$$

where a_0 is the Bohr radius, and g_0 and g_{Γ_1} are statistical weights of the initial state and the channel Γ_1, respectively. The total cross section is given by

$$\sigma(\gamma_0 \to \gamma_1) = \sum_{l_0 l_1 \Gamma_T} \sigma(\Gamma_0 \to \Gamma_1) . \tag{4.12}$$

Let us recall the fact that for elastic scattering of an electron in the Coulomb field only the partial cross sections have finite values and the total cross section diverges.

In general scattering theory some other matrices can be useful along with the $\underline{T}$ matrix defined by (4.7). The $\underline{S}$ matrix (scattering matrix) is connected with the $\underline{T}$ matrix (transition matrix) by the relation

$$\underline{S} = \underline{I} - \mathrm{i}\, 2 \underline{T} , \tag{4.13}$$

where $\underline{I}$ is the unit matrix, i.e.

$$S_{\Gamma_0 \Gamma} = \delta_{\Gamma_0 \Gamma} - \mathrm{i}\, 2 T_{\Gamma_0 \Gamma} . \tag{4.14}$$

The solutions of the radial equations (4.10) should now satisfy the following boundary conditions:

$$\mathscr{F}_\Gamma(r) \underset{r \to \infty}{\sim} k_\Gamma^{-1/2} \{\delta_{\Gamma_0 \Gamma} \exp[-\mathrm{i}\,(k_{\Gamma_0} r + d_{\Gamma_0})] - S_{\Gamma_0 \Gamma} \exp[\mathrm{i}\,(\mathrm{k}_\Gamma r + d_\Gamma)]\} . \tag{4.15}$$

The $\underline{S}$ matrix is unitary, i.e.

$$\sum_\Gamma |S_{\Gamma_0 \Gamma}|^2 = 1 . \tag{4.16}$$

In many respects it is also useful to introduce the reactance matrix ($\underline{K}$ matrix), defined by

$$\underline{S} = \frac{\underline{I} + \mathrm{i}\underline{K}}{\underline{I} - \mathrm{i}\underline{K}} . \tag{4.17}$$

The $\underline{K}$ matrix is symmetric and real, but not unitary. In terms of this matrix, the solutions of (4.10) have the following asymptotic form:

$$\mathscr{F}_{\Gamma}(r) \underset{r\to\infty}{\sim} k_{\Gamma}^{-1/2}\left[\delta_{\Gamma_0\Gamma}\sin(k_{\Gamma_0}r+d_{\Gamma_0})+K_{\Gamma_0\Gamma}\cos(k_{\Gamma}r+d_{\Gamma})\right]. \quad (4.18)$$

Here we have to mention one important point. *Seaton* and co-workers (see, for example, [4.10–13]) denote the reactance matrix as the $\underline{R}$ matrix. Elsewhere, especially in general scattering theory (see, for example [4.1–9]), $\underline{R}$ matrix means the matrix connected with the logarithmic deriatives of the total wave function at the edge of the interaction region (see Sect. 4.3).

For a description of the process, one can also use the dimensionless collision strength, which is related to the cross section by

$$\Omega(\Gamma_0\to\Gamma_1)=\frac{g_0 k_0^2}{\pi}\sigma(\Gamma_0\to\Gamma_1). \quad (4.19)$$

The collision strength is symmetric with respect to direct and inverse processes,

$$\Omega(\Gamma_0\to\Gamma_1)=\Omega(\Gamma_1\to\Gamma_0). \quad (4.20)$$

Thus, the electron-ion scattering (elastic and inelastic) problem requires the solution of the infinite set of coupled radial equations (4.10) with the boundary conditions (4.7, 15 or 18). As mentioned in Sect. 4.1, this problem can be solved within the framework of one of the three presently existing methods. Now we discuss these methods in more detail.

4.3 Effective-Range Theory for Attractive Coulomb Field

The effective-range theory was developed originally for scattering (both elastic and inelastic) of neutral atomic and nuclear particles [4.1–4]. Its generalization to the scattering problem in a Coulomb field has been made by *Baz'* [4.5], *Fonda* and *Newton* [4.6–8], and *Gailitis* [4.9]. *Baz'* [4.5] has shown that the partial cross section for elastic scattering has a pronounced resonance structure and changes sharply at the threshold of a new (energetically closed) channel. For collision energies below the threshold, the partial elastic cross section has an oscillatory behaviour; its oscillation frequency increases as it approaches the threshold. At the threshold of a new channel the elastic scattering cross section (averaged over the resonances) decreases with a jump, which has a value equal to the inelastic threshold cross section in the new open channel. Thus, the sum of the averaged cross sections, both elastic and inelastic, has a constant value below and above the new channel threshold. *Fonda* and *Newton* [4.6–8] investigated the cross section behaviour in the vicinity of higher level thresholds. They expressed the scattering matrix below the threshold in terms of its value above the threshold and also established the continuity of the sum of partial cross sections (elastic + inelastic). *Gailitis* [4.9] obtained a generalization of these results to the case of inelastic processes, and gave expressions for the cross sections, widths and shifts of resonances. A qualitative picture of the resonance structures for a given partial cross section is shown in Fig. 4.1. We should note that neither the effective-range theory nor the quantum defect method take into consideration the radiative decay of resonances. This effect will be discussed in Sect. 4.5.

Let us now analyze the system (4.10). According to the effective-range theory, we assume that there exists an electron-ion distance $r = r_0$ such that for $r \geqq r_0$ all non-Coulomb potentials $U_{\Gamma\Gamma'}$ are negligibly small in comparison with the Coulomb field $2(Z-1)/r$ [i.e. $U_{\Gamma\Gamma'}(r) = 0$ for $r \geqq r_0$]. For the region $r \geqq r_0$, a general solution of the system (4.10) can be written in form of a quadratic matrix

$$\underline{\mathscr{F}}(r) = k^{-1/2}\,[\underline{F}(k\,r) + \underline{G}(k\,r)\cdot\underline{K}], \tag{4.21}$$

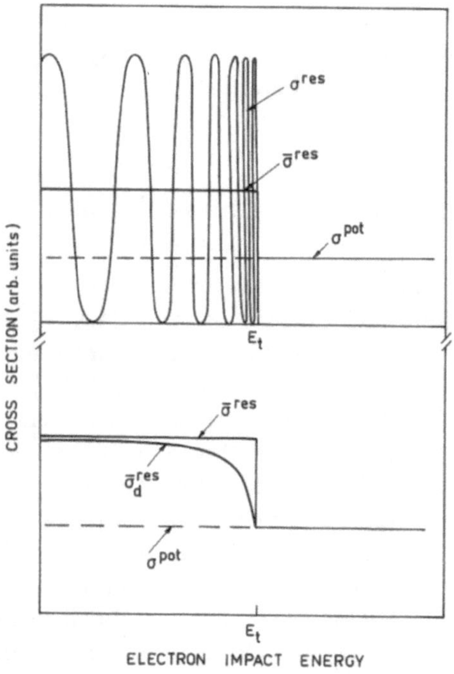

Fig. 4.1. (a) Qualitative picture of resonance structures near the threshold of a new channel. E_t is the threshold energy. (b) Influence of radiative decay of resonances (dielectronic recombination) on the averaged cross section; $\bar{\sigma}_d^{res}$ and $\bar{\sigma}^{res}$ represent the resonance contributions with and without inclusion of resonance radiative decay

where $F(kr)$ and $G(kr)$ are the regular and irregular Coulomb wave functions [4.17, 18] and $\underline{K}$ is the reactance matrix (4.17). Different columns in (4.21) represent different linearly independent solutions of (4.10), i.e. solutions with different initial channels. Different rows in (4.21) are components of the total wave function for different initial channels. The expression (4.21) contains only open channels for which k_Γ (4.9) is real and $E - \Delta E > 0$, and those closed channels for which the collision energy lies in the immediate vicinity of the threshold, $E - \Delta E < 0$, but $E \approx \Delta E$. The wave function components for the other closed channels with $E \ll \Delta E$ decrease rapidly when r increases, and for sufficiently large values of r_0 their contributions are negligible.

The reactance $\underline{K}$ matrix is connected with the $\underline{T}$ matrix by the relation

$$\mathrm{i}\,\underline{T} = (\mathrm{i}\,\underline{K}^{-1} - \underline{I})^{-1}. \tag{4.22}$$

The $\underline{K}$ matrix as a function of the energy E has a branching point at the threshold of a new channel E_t. It is necessary to separate the irregular terms of $\underline{K}$ from the terms which are analytic in the vicinity of the threshold. For this purpose one can employ the usual form of $\underline{R}$ matrix theory [4.1] which enables one to separate out the analytic singularities. The general definition for the $\underline{R}$ matrix is

$$\underline{\mathscr{F}}(r_0) = \underline{R}\, r_0 \frac{d}{dr} \underline{\mathscr{F}}(r)\,|_{r=r_0}\,. \tag{4.23}$$

Substituting (4.21) into (4.23), we obtain an expression for the $\underline{K}$ matrix in terms of the $\underline{R}$ matrix [4.9]:

$$\begin{aligned} \underline{K}^{-1} &= \frac{\underline{G}}{\underline{F}} + \frac{1}{\underline{\dot{F}}\,\underline{F}} - \frac{1}{\underline{\varrho}^{1/2}\,\underline{\dot{F}}}\left[\underline{\varrho}^{-1}\frac{\underline{F}}{\underline{\dot{F}}} - \underline{R}\right]^{-1} \cdot \frac{1}{\underline{\dot{F}}\,\underline{\varrho}^{1/2}} \\ &= -[\underline{\varrho}^{-1/2}\,\underline{F} - \underline{R}\,\underline{\varrho}^{1/2}\,\underline{\dot{F}}]^{-1}\,[\underline{\varrho}^{-1/2}\,\underline{G} - \underline{R}\,\underline{\varrho}^{1/2}\,\underline{\dot{G}}]\,. \end{aligned} \tag{4.24}$$

Here $\underline{F}$, $\underline{G}$, $\underline{\varrho}$ are diagonal matrices with elements $F_j(\varrho_j)$, $G_j(\varrho_j)$, $\varrho_j = k_j r_0$ respectively, $\underline{\dot{F}} = d\underline{F}/d\varrho$ and $\underline{\dot{G}} = d\underline{G}/d\varrho$.

From the general properties of the Schrödinger equation, it follows that for $r \geqq r_0$ the $\underline{R}$ matrix is an analytic function of the energy and has only simple poles on the real E axis [4.1, 3]. The branching points in (4.24) are due to the Coulomb wave functions.

In Sect. 4.5 we shall discuss in more detail the analytic properties of the Coulomb wave functions with respect to the energy. Here we recall that the threshold of a new channel, at which $k_{\Gamma_c} = [2\,(E - \Delta E_{\Gamma_c\Gamma})]^{1/2} = 0$, is a branching point. Below the threshold, i.e. for $E < \Delta E_{\Gamma_c\Gamma}$, the Coulomb wave functions $F(\varrho)$ and $G(\varrho)$ contain a denumerable set of poles, converging to the point $k_{\Gamma_c}=0$. Separating out these singularities in (4.24) and substituting the result into (4.23), *Gailitis* [4.9] obtained the following result which connects the elements of the $\underline{T}$ matrix below the threshold with their values above the threshold:

$$T^{<}_{\Gamma\Gamma_0}=T^{>}_{\Gamma\Gamma_0}-\frac{T^{>}_{\Gamma\Gamma_c}T^{>}_{\Gamma_c\Gamma}}{T^{>}_{\Gamma_c\Gamma_c}}+\frac{T^{>}_{\Gamma\Gamma_c}T^{>}_{\Gamma_c\Gamma_0}}{(T_{\Gamma_c\Gamma_c})^2}\,\frac{1}{y+\mathrm{i}+(T_{\Gamma_c\Gamma_c})^{-1}}\,. \tag{4.25}$$

Here $T^{<}_{\Gamma\Gamma'}$ and $T^{>}_{\Gamma\Gamma'}$ denote the $\underline{T}$ matrix elements calculated below and above the threshold of a new channel Γ_c, respectively. In the right-hand side of (4.25) there is just one function, y, which varies rapidly in the vicinity of the threshold. It is equal to

$$y=\cot\pi\nu,\quad \nu=(Z-1)(\Delta E_{\Gamma_c\Gamma_0}-E)^{-1/2}\,. \tag{4.26}$$

Since y is a periodic function, all the matrix elements of open channels, $T_{\Gamma\Gamma_0}$, have an oscillatory behaviour for energies $E<\Delta E_{\Gamma_c\Gamma_0}$, for which the channel Γ_c is closed. The sum of the second and third terms on the right-hand side of (4.25) gives resonance contributions on the background of the first term which describes potential (or direct) scattering. When the widths and the shifts of resonances are small in comparison with the distance between the resonances (this is the case for multiply charged ions), maximum values of resonance contributions are reached at the energies for which the ν-values are close to any integer ($\nu\approx n=$ integer). In the immediate vicinity of the threshold of a new channel, the resonances converge and it is reasonable to introduce a cross section averaged over resonances:

$$\bar{\sigma}=\frac{1}{D}\int\limits_{E-D/2}^{E+D/2}\sigma(E)\,dE\,, \tag{4.27}$$

where D is the energy distance between the resonances. Substituting (4.25) and (4.11) into (4.27) one obtains the Gailitis formula [4.9]:

$$\bar{\sigma}(\Gamma_0\to\Gamma)=\sigma^{>}(\Gamma_0\to\Gamma)+\frac{\sigma^{>}(\Gamma_0\to\Gamma_c)\,\sigma^{>}(\Gamma_c\to\Gamma)}{\sum\limits_{\Gamma_\alpha}\sigma^{>}(\Gamma_c\to\Gamma_\alpha)}\,, \tag{4.28}$$

where all cross sections on the right-hand side are taken at the energy values at which the corresponding channels become open. We have to remember, however, that the channel Γ_c is closed, and when it becomes open, the second term on the right-hand side of (4.28) vanishes. The sum in the denominator corresponds to transitions into all open channels from the given closed one. Equation (4.28) is written for a partial cross section. Summing (4.28) over all open final channels Γ and over all initial channels Γ_0, including the elastic one, it is easy to demonstrate that at the threshold of a new channel all partial cross sections decrease in jumps; the sum of all jump values is equal to the partial cross section of a new channel, i.e. the cross section summed over all channels is continuous.

In reality, at the same threshold energy several new channels become open with different values of the orbital momentum of colliding particles, and the total momentum of the whole system is conserved. This means that the closed channels are degenerate, and one has to perform an orthogonalization of the wave functions (or $\underline{T}$ matrix elements) that have the same energy values. For narrow non-overlapping resonances, (4.28) has the form [4.9]

$$\bar{\sigma}(\Gamma_0 - \Gamma) = \sigma^{>}(\Gamma_0 \rightarrow \Gamma) + \sum_{\Gamma_c} \frac{\sigma^{>}(\Gamma_0 \rightarrow \Gamma_c)\,\sigma^{>}(\Gamma_c \rightarrow \Gamma)}{\sum_{\Gamma_\alpha} \sigma^{>}(\Gamma_c \rightarrow \Gamma_\alpha)}, \tag{4.29}$$

where the sum over Γ_c includes all closed channels with equal values of the threshold energy.

In practical calculations based on $\underline{R}$ matrix theory, one has also to know the wave function of the system (electron + ion) in the region $r < r_0$. This problem can only be solved using numerical methods. Another possibility is to insert into (4.28, 29) the values of the partial cross sections obtained by some other approximate calculation (say, by the Coulomb-Born-exchange approximation). It is necessary, however, to prove the applicability of such a procedure. This problem has been solved by *Seaton* [4.11] on the basis of the quantum defect method [4.10].

4.4 Quantum Defect Method. The Seaton Theory

Electron-ion scattering theory was developed by *Seaton* [4.10, 11] using the generalized quantum defect method (GQDM), which is now well known and widely used. For practical calculations *Seaton* and co-workers developed two modifications of GQDM, based on either a numerical solution [4.12] of the system of integro-differential equations (4.10) or using the more simple distorted-wave (DW) approximation [4.13]. In both cases all singularities connected with the resonance scattering are separated out in analytic form before performing the numerical calculations.

As we have already mentioned, Seaton and his co-workers use the designation $\underline{R}$ for the reactance matrix $\underline{K}$, which is related to the $\underline{S}$ matrix by the relation (4.17)

$$\underline{S} = (\mathrm{i}\,\underline{I} - \underline{K})\,(\mathrm{i}\,\underline{I} + \underline{K})^{-1} \tag{4.30}$$

where $\underline{I}$ is the diagonal unit matrix ($I_{\Gamma\Gamma'} = \delta_{\Gamma\Gamma'}$). On the basis of an investigation of the general analytic properties of Coulomb wave functions, Seaton has proposed an analytic continuation of the $\underline{S}$ matrix in the energy region where some channels are closed:

$$\underline{S} = (\mathrm{i}\,\underline{I} - \underline{K})\,(\underline{t}\,\underline{I} + \underline{K})^{-1} \tag{4.31}$$

where $\underline{t}$ is also a diagonal matrix, having elements $t_{\Gamma\Gamma} = \mathrm{i}$ for the open channels, and $t_{\Gamma_c\Gamma_c} = \tan\pi\,\nu_c$ for the closed ones; ν_c is given by (4.26). When all channels are open, (4.31) coincides with (4.30). When some of the channels are closed, resonance structures appear due to the terms $t_{\Gamma_c\Gamma_c}$ in (4.31). Introducing a $\underline{\chi}$ matrix, defined as

$$\underline{\chi} = (\mathrm{i}\,\underline{I} - \underline{K})\,(\mathrm{i}\,\underline{I} + \underline{K})^{-1} \tag{4.32}$$

for all (both open and closed) channels, and eliminating $\underline{K}$ from (4.31, 32), one obtains

$$\underline{S}\,[(\underline{t} + \mathrm{i}\,\underline{I}) + (\underline{t} - \mathrm{i}\,\underline{I})\,\underline{\chi}] = \mathrm{i}\,2\,\underline{\chi}\,. \tag{4.33}$$

Let us separate the sub-matrices $\underline{S}_{oo}$ for the open-open channel coupling, $\underline{S}_{cc}$ for the closed-closed channel coupling, and $\underline{S}_{oc}$ for the open-closed channel coupling, and let us separate the similar sub-matrices in the $\underline{\chi}$ matrix. Then we have

$$\underline{S} = \begin{pmatrix} \underline{S}_{oo} & \underline{S}_{oc} \\ \underline{S}_{co} & \underline{S}_{cc} \end{pmatrix}, \quad \underline{\chi} = \begin{pmatrix} \underline{\chi}_{oo} & \underline{\chi}_{oc} \\ \underline{\chi}_{co} & \underline{\chi}_{oc} \end{pmatrix}. \tag{4.34}$$

Substituting (4.34) into (4.33), one obtains [4.10, 11]

$$\underline{S}_{oo} = \underline{\chi}_{oo} - \underline{\chi}_{oc}\,(\underline{\chi}_{cc} - \mathrm{e}^{\mathrm{i}2\pi\nu_c})\,\underline{\chi}_{co}\,. \tag{4.35}$$

Taking into account the relation (4.13) connecting the $\underline{S}$ and $\underline{T}$ matrices, it is possible to see that (4.35) gives (in a matrix form) a result analogous to (4.25), i.e. the expressions for excitation amplitudes including resonance contributions.

The matrix equation (4.35) can be simplified significantly if one uses the expansion

$$\underline{\chi} = \underline{I} + \mathrm{i}\,2\underline{K} - 2\underline{K}^2 + \cdots \tag{4.36}$$

and restricts oneself to terms of second order. *Seaton* [4.11] showed that the expansion (4.36) corresponds to the DW approximation (DWA). In the DWA the result of QDM (4.35) coincides with (4.25) and the Gailitis formula (4.29) can be obtained for the averaged cross sections. Substituting (4.36) into (4.35), and using the collision strengths (4.19), one obtains after averaging over resonances,

$$\bar{\Omega}(\Gamma_0 \to \Gamma) = \bar{\Omega}(\Gamma \to \Gamma_0) = \Omega^>(\Gamma_0 \to \Gamma) + \sum_{\Gamma_c} \frac{\Omega^>(\Gamma_0 \to \Gamma_c)\,\Omega^>(\Gamma_c \to \Gamma)}{\sum_{\Gamma_\alpha} \Omega^>(\Gamma_c \to \Gamma_\alpha)} . \quad (4.37)$$

In accordance with (4.29), the sum over Γ_c in (4.37) means summation over the degenerate closed channels, and the sum over Γ_α means summation over all open channels.

Substituting (4.36) into (4.35), one can also calculate the cross sections and collision strengths without averaging over the resonances. In Sect. 4.6 we give some results of the calculations done by Seaton and co-workers, and other authors.

4.5 Correct Asymptotic Expansion. Influence of Dielectronic Recombination

In the problem of excitation of multiply charged ions by electron impact, there exists a natural small parameter, Z^{-1}, where $(Z-1)$ is the charge of the ion. The influence of resonance effects, however, makes it impossible to use a straightforward expansion similar to the Born series. A proper expansion has been obtained by *Presnyakov* and *Urnov* [4.14, 15] on the basis of the approach developed by *Feshbach* [4.2], i.e. on the basis of integral equations which are equivalent to (4.10). In order to solve them one needs to know the analytic properties of the Green's function for the Coulomb field. The representation used here for the Coulomb Green's function allows separation in explicit form of the effects of resonance scattering, which provide a contribution to the cross section of the same order in the parameter Z^{-1} as the direct (potential) scattering; in a number of cases it is even dominant.

Another problem which has to be taken into account is the radiative decay of resonances, i.e. the influence of dielectronic recombination on the excitation cross section. This effect is important for resonances with values of $\nu \gg 1$ [see (4.26)] in all cases, and it is of special interest for the case of multiply charged ions ($Z \gg 1$).

We start with the system of coupled-channel equations (4.10), written in Coulomb units[1]:

$$\left(\frac{d^2}{dr^2} - \frac{l(l+1)}{r^2} + \frac{2a}{r} + k_\Gamma^2\right)\mathscr{F}_\Gamma = \frac{1}{Z}\sum_{\Gamma'} U_{\Gamma\Gamma'}\mathscr{F}_{\Gamma'}, \quad (a = 1 - 1/Z) . \quad (4.38)$$

---

1 The mass, length and time units are m, $\hbar^2/m e^2 Z$, $\hbar^3/m e^4 Z^2$, respectively. The unit of energy is $m e^4 Z^2/\hbar^2$.

It is worth noting that, for $Z \gg 1$, all the $U_{\Gamma\Gamma'}$ do not depend on Z. We now rewrite system (4.38) in an integral form, designating the open channels by $\Gamma = \alpha$ and the closed ones by $\Gamma = \mathrm{c}$:

$$\mathscr{F}_\alpha = \delta_{\alpha\Gamma_0} P_{\Gamma_0} + \frac{1}{Z} \sum_\Gamma G_\alpha^{(+)} U_{\alpha\Gamma} \mathscr{F}_\Gamma , \tag{4.39}$$

$$\mathscr{F}_\mathrm{c} = \frac{1}{Z} \sum_\Gamma G_\mathrm{c}^{(+)} U_{\mathrm{c}\Gamma} \mathscr{F}_\Gamma . \tag{4.40}$$

The function P_{Γ_0} is the regular solution of the homogeneous equation (4.38) for the initial channel. The integral operators $G_\Gamma^{(+)}$ are defined by

$$G_\Gamma^{(+)} X = \int_0^\infty dr' \, G_{k_\Gamma}^{(+)}(r, r') \, X(r') , \tag{4.41}$$

$$k_\Gamma = [2(E - \Delta E_{\Gamma\Gamma_0})]^{1/2} , \quad \mathrm{Im}\,\{k_\Gamma\} \geq 0 , \tag{4.41 a}$$

where E and $\Delta E_{\Gamma\Gamma_0}$ are the energy of the incident electron and the excitation energy for the channel Γ, respectively. $G_{k_\Gamma}^{(+)}$ is the radial Green's function for the scattering problem in the Coulomb field.

It is obvious that the analytic properties of the Coulomb Green's function we need have their origin in the analytic properties of the Coulomb wave functions, which are considered in detail elsewhere [4.16–20].

4.5.1 Radial Green's Function

In the case of central-field scattering, the function $G_k^{(+)}$ is solution of the equation

$$\left[\frac{d^2}{dr^2} - \frac{(\mu^2 - 1/4)}{r^2} + 2V(r) + k^2\right] G_k^{(+)}(r, r') = \delta(r - r') , \quad (\mu \to l + 1/2) \tag{4.42}$$

and can be written as

$$G_k^{(+)}(r, r') = -\frac{\phi(\mu, k, r_<) \, f_+(\mu, k, r_>)}{w_+(\mu, k)} , \tag{4.43}$$

$$r_< = \min\{r, r'\} , \quad r_> = \max\{r, r'\} . \tag{4.44}$$

Both ϕ and f are solution of (4.42) without the right-hand side. The function ϕ is regular at the origin:

$$\lim_{r \to 0} \phi(\mu, k, r) \, r^{-\mu - 1/2} = 1 , \tag{4.44 a}$$

and $f_\pm$ behaves asymptotically as

$$f_\pm(\mu, k, r) \underset{r\to\infty}{\simeq} \exp\{\pm \mathrm{i}[k\,r + (a/k)\ln(2k\,r)]\}\,. \tag{4.45}$$

The second term in square brackets in (4.45) exists if the scattering potential $V(r)$ is a Coulomb potential, a/r. The denominator in (4.43) does not depend on r and has the form

$$w_\pm(\mu, k) = \phi(\mu, k, r)\frac{d}{dr}f_\pm(\mu, k, r) - f_\pm(\mu, k, r)\frac{d}{dr}\phi(\mu, k, r)\,. \tag{4.46}$$

The function $G_k^{(+)}$ is an analytical function of k for all the complex values given by (4.41 a). On the imaginary k axis the Green's function has poles corresponding to the bound states of the potential $V(r)$. We can separate the singular part of $G_k^{(+)}$ as follows. We rewrite $\phi(\mu, k, r)$ in terms of the linearly independent functions $f_\pm(\mu, k, r)$:

$$\phi(\mu, k, r) = \frac{1}{2\,\mathrm{i}\,k}[w_-(\mu, k)\,f_+(\mu, k, r) - w_+(\mu, k)\,f_-(\mu, k, r)]\,. \tag{4.47}$$

The function $f_-(\mu, k, r)$ is irregular at the origin and may be expressed in terms of the linearly independent functions $\phi(\pm\mu, k, r)$:

$$f_-(\mu, k, r) = -\frac{1}{2\mu}[w_-(-\mu, k)\,\phi(\mu, k, r) - w_+(\mu, k)\,\phi(-\mu, k, r)]\,. \tag{4.48}$$

Eliminating $f_-(\mu, k, r)$, we obtain

$$\begin{aligned} f_+(\mu, k, r) = &-\frac{2\,\mathrm{i}\,k}{w_-(\mu, k)}\varphi(\mu, k, r) \\ &-\frac{w_+(\mu, k)}{2\mu}\left[\frac{w_-(-\mu, k)}{w_-(\mu, k)}\phi(\mu, k, r) - \phi(-\mu, k, r)\right] \end{aligned} \tag{4.49}$$

and

$$\begin{aligned} G_k^{(+)}(r, r') = &-\frac{2\,\mathrm{i}\,k}{w_+(\mu, k)\,w_-(\mu, k)}\phi(\mu, k, r)\,\phi(\mu, k, r') \\ &+\frac{1}{2\mu}\phi(\mu, k, r_<)\left[\frac{w_-(-\mu, k)}{w_-(\mu, k)}\phi(\mu, k, r_>) - \phi(-\mu, k, r_>)\right]. \end{aligned} \tag{4.50}$$

In the case of the Coulomb field $V = a/r$, the radial Green's function has the form

$$G_k^{(+)}(r, r') = \frac{\pi\,\nu^3}{2\,a^2}\cot(\pi\,\nu)\,P_{\nu,l}(r)\,P_{\nu,l}(r') + G_k^{(\mathrm{P})}(r, r')\,, \tag{4.51}$$

$$G_k^{(p)}(r, r') = \frac{\pi \nu^3}{2a^2} \frac{P_{\nu l}(r_<)}{\sin(\pi \mu)} [\cos(2\pi\mu) P_{\nu,l}(r_>) - P_{\nu,-l-1}(r_>)] , \tag{4.52}$$

$$\mu \to l + 1/2 , \quad \nu = \mathrm{i}\, a/k , \quad \mathrm{Im}\{k\} \geq 0 . \tag{4.53}$$

The functions $P_{\nu,l}(r)$ coincide with the hydrogen-like eigenfunctions when ν is an integer, i.e.

$$\lim_{\nu \to n = 1,2,3,\dots} P_{\nu,l}(r) = P_{n,l}(r) , \quad \int_0^\infty dr\, P_{n,l}(r)\, P_{n',l}(r) = \delta_{n n'} . \tag{4.54}$$

For the complex values of ν given by (4.53), the functions $P_{\nu,l}(r)$ are defined as

$$P_{\nu,l}(r) = C_\nu P_{k,l}(r) , \quad C_\nu = \left[\frac{2a^2}{\pi \nu^3 [\mathrm{i} \cot(\pi \nu) + 1]}\right]^{1/2} , \tag{4.55}$$

whereas for real k, $P_{k,l}(r)$ is normalized as

$$\int_0^\infty dr\, P_{k,l}(r)\, P_{k',l}(r) = \frac{\pi}{2k} \delta(k - k') . \tag{4.56}$$

The Green's function (4.51) has two different parts. The first part contains a singular factor with respect to energy. The singularities correspond to resonances converging to the threshold for the opening of a new channel ($E \to \Delta E_{\Gamma\Gamma_0}$, $\nu = n \to \infty$). The second part of the Green's function, $G_k^{(p)}$ in (4.52), has no singularities. When $k \to 0$, the function $G_k^{(p)}$ does not depend on energy, being a regular function of both r and r'.

It is essential that the principal (singular) part of the Green's function (4.51) is factorized in the coordinate variables.

4.5.2 Solution of Integral Equations

It is easy to see that in the case of an open channel, $G_k^{(+)}$ (4.51) has no singularities. If all channels are open, a formal solution of the integral equations (4.39) may be expressed using perturbation theory as a series in powers of $1/Z$. In the limiting case of large Z we obtain small corrections to the first-order approximation. When some channels are closed any term of the perturbation series (beginning from the second one) contains singularities, and one must not use the standard perturbation theory.

We can separate out the singular part of the integral equation using the Green's function (4.51) already obtained. Substituting (4.51) into (4.40), we have

$$\mathscr{F}_\alpha = \delta_{\alpha\Gamma_0} P_{\Gamma_0} + \frac{1}{Z} \sum_{\Gamma} G_\alpha^{(+)} U_{\alpha\Gamma} \mathscr{F}_\Gamma , \tag{4.57}$$

$$\mathscr{F}_c = P_c \Lambda_c + \frac{1}{Z} \sum_{\Gamma} G_c^{(p)} U_{c\Gamma} \mathscr{F}_\Gamma , \tag{4.58}$$

where the quantities Λ_c do not depend on coordinates,

$$\Lambda_c = \frac{\pi \nu_c^3}{2a^2} \cot(\pi \nu_c) \sum_{\Gamma} \langle P_c U_{c\Gamma} \mathscr{F}_\Gamma \rangle . \tag{4.59}$$

Here we introduce the notation

$$P_\alpha = P_{k_\alpha, l}(r) , \quad P_c = C_{\nu_c} P_{k_c, l}(r) . \tag{4.60}$$

The integral operators $G_\alpha^{(-)}$ (for open channels) and $G_c^{(p)}$ (for closed channels) do not contain the resonance singularities, which are given by the quantities Λ_c. For $\nu \neq n$, the integral equations (4.57, 58) have regular kernels and may be solved by iteration. The final results are valid for integer values of ν, as well. The solution is given by

$$\mathscr{F}_\alpha = \delta_{\alpha\Gamma_0} P_{\Gamma_0} + \frac{1}{Z} G_\alpha^{(+)} W_{\alpha\Gamma_0} P_{\Gamma_0} + \frac{1}{Z} \sum_{c} G_\alpha^{(+)} W_{\alpha c} P_c \Lambda_c , \tag{4.61}$$

$$\mathscr{F}_c = P_c \Lambda_c + \frac{1}{Z} G_c^{(p)} W_{c\Gamma_0} P_{\Gamma_0} + \frac{1}{Z} \sum_{c'} G_c^{(p)} W_{cc'} P_{c'} \Lambda_{c'} , \tag{4.62}$$

where the quantities Λ_c (4.59) can be obtained from the following algebraic equations:

$$\sum_{c'} \left[Z \frac{2a^2}{\pi \nu_c^2} \tan(\pi \nu_c) \delta_{cc'} - \langle P_c W_{cc'} P_{c'} \rangle \right] \Lambda_{c'} = \langle P_c W_{c\Gamma_0} P_{\Gamma_0} \rangle . \tag{4.63}$$

For the W-matrix elements we have an asymptotic expansion,

$$W_{\Gamma\Gamma'} = U_{\Gamma\Gamma'} + \sum_{n=1}^{\infty} \sum_{j=1}^{n} \frac{1}{Z^n} U_{\Gamma\Gamma_1} G_{\Gamma_1} U_{\Gamma_1\Gamma_2} \dots G_{\Gamma_j} U_{\Gamma_j\Gamma'} , \tag{4.64}$$

where

$$G_\Gamma = G^{(+)} \quad \text{if} \quad \Gamma = \alpha , \tag{4.65}$$

$$G_\Gamma = G^{(p)} \quad \text{if} \quad \Gamma = c . \tag{4.66}$$

Taking into account (4.63, 64), one can see that each term in (4.61, 62) is a regular function of energy for all ν including $\nu = n$. It is worth noting that for complex values of k, the integrals $\langle P_\Gamma W_{\Gamma\Gamma'} P_{\Gamma'}\rangle$ are usually calculated assuming a finite radius of interaction [4.2]. It is possible to avoid this assumption if we use the general procedure for calculating integrals with non-stationary-state wave functions [4.21]. In our case we can use a simpler regularization procedure. Taking into account that all the $U_{\Gamma\Gamma'}$ are also integrals with the ion eigenfunctions we can change the integration sequence in the $\langle P_\Gamma W_{\Gamma\Gamma'} P_{\Gamma'}\rangle$ and obtain well-defined quantities.

4.5.3 *T* Matrix and Cross Sections

The wave function normalization (4.56) we used corresponds to the following relation between the $\underline{S}$ and $\underline{T}$ matrices:

$$\underline{S} = \underline{I} + \frac{2\mathrm{i}}{Z}\underline{T}, \quad S_{\Gamma\Gamma_0} = \delta_{\Gamma\Gamma_0} + \frac{2\mathrm{i}}{Z} T_{\Gamma\Gamma_0}. \tag{4.67}$$

From (4.61) it follows that for any open channel,

$$T_{\alpha\Gamma_0} = T^{\mathrm{pot}}_{\alpha\Gamma_0} + T^{\mathrm{res}}_{\alpha\Gamma_0}, \quad \text{where} \tag{4.68}$$

$$T^{\mathrm{pot}}_{\alpha\Gamma_0} = -\langle P_\alpha W_{\alpha\Gamma_0} P_{\Gamma_0}\rangle \quad \text{and} \tag{4.69}$$

$$T^{\mathrm{res}}_{\alpha\Gamma_0} = -\sum_{\mathrm{c}} \langle P_{\mathrm{c}} W_{\alpha\mathrm{c}} P_{\mathrm{c}}\rangle \Lambda_{\mathrm{c}}. \tag{4.70}$$

The quantity $T^{\mathrm{pot}}_{\alpha\Gamma_0}$ corresponding to potential (or direct) excitation is defined by the generalized potential (4.64). The term $T^{\mathrm{res}}_{\alpha\Gamma_0}$ exists when some channels are closed and contains infinite series of resonances converging to the closed-channel levels. The quantities Λ_{c} are given by the system of algebraic equations (4.63). In the case of non-overlapping resonances we may neglect non-diagonal matrix elements on the left-hand side of (4.63) and obtain

$$\Lambda_{\mathrm{c}} = \frac{\langle P_{\mathrm{c}} W_{\mathrm{c}\Gamma_0} P_{\Gamma_0}\rangle}{Z(2a^2/\pi\nu_{\mathrm{c}}^3)\tan(\pi\nu_{\mathrm{c}}) - \langle P_{\mathrm{c}} W_{\mathrm{cc}} P_{\mathrm{c}}\rangle}. \tag{4.71}$$

In more general cases one can use the diagonalization procedure discussed by *Feshbach* [4.2], *Gailitis* [4.9] and *Seaton* [4.11].

In the limiting case $Z \gg 1$ we retain the first non-vanishing terms for the real and imaginary parts of the $W_{\Gamma\Gamma'}$ (4.64). The potential part of the T matrix coincides with the perturbation theory result

$$T^{\mathrm{pot}}_{\alpha\Gamma_0} = -\langle P_\alpha U_{\alpha\Gamma_0} P_{\Gamma_0}\rangle = T^{(1)}_{\alpha\Gamma_0}. \tag{4.72}$$

For the resonance part, we obtain

$$T^{\mathrm{res}}_{\alpha\Gamma_0} = -\frac{1}{Z}\sum_{\mathrm{c}} \frac{\tau_{\alpha\mathrm{c}}(\nu_{\mathrm{c}})\,\tau_{\mathrm{c}\Gamma_0}(\nu_{\mathrm{c}})}{\tan(\pi\nu_{\mathrm{c}}) - Z^{-1}\tau_{\mathrm{cc}}(\nu_{\mathrm{c}}) + (\mathrm{i}/Z^2)\sum_{\alpha'}\tau^2_{\mathrm{c}\alpha'}(\nu)}, \tag{4.73}$$

where the real quantities $\tau_{\Gamma\Gamma'}$ are given by

$$\tau_{cc'} = \delta_{cc'}\left(\tfrac{1}{2}\right) \pi \nu_c^3 \langle P_c U_{cc} P_c \rangle , \quad (4.74)$$

$$\tau_{\alpha c} = \tau_{c\alpha} = (\pi \nu_c^3/2)^{1/2} \langle P_\alpha U_{\alpha c} P_c \rangle , \quad (4.75)$$

and the functions P_Γ are defined by (4.60). To evaluate these integrals for non-integer ν, one can use either the regularization procedure mentioned in Sect. 4.5.2 or the analytic continuation discussed by *Eissner* et al. [4.12] and *Hershkovitz* and *Seaton* [4.13]. Both methods give identical results. When $\nu_c \gg 1$, the quantities $\tau_{\Gamma\Gamma'}(\nu_c)$ practically do not depend on ν_c.

The general results of this theory, (4.69, 70), have an analytic form similar to the quantum-defect theory results [4.11]. For $Z \gg 1$, the first term of the asymptotic expansion (4.73) corresponds to the distorted-wave method [4.13]. For the partial cross section averaged over the resonances one obtains

$$\begin{aligned} \bar{\sigma}(\Gamma_0 \to \Gamma_1) &\equiv \bar{\sigma}(\gamma_0 l_0 \to \gamma_1 l_1) = \frac{1}{\Delta E} \int_{\Delta E} \sigma(\gamma_0 l_0 \to \gamma_1 l_1)\, dE \\ &= \sigma^{\mathrm{pot}}(\gamma_0 l_0 \to \gamma_1 l_1) + \bar{\sigma}^{\mathrm{res}}(\gamma_0 l_0 \to \gamma_1 l_1) , \end{aligned} \quad (4.76)$$

where

$$\sigma^{\mathrm{pot}} = \frac{g_{\Gamma_1}}{g_0 Z^4 E} |T^{(1)}_{\Gamma_1 \Gamma_0}|^2 \quad \text{and} \quad (4.77)$$

$$\begin{aligned} \bar{\sigma}^{\mathrm{res}}(\gamma_0 l_0 \to \gamma_1 l_1) &\equiv \bar{\sigma}^{\mathrm{res}}(\Gamma_0 \to \Gamma_1) \\ &= \frac{g_{\Gamma_1}}{2 g_0 Z^4 E} \sum_c \frac{\tau^2_{c\Gamma_0}(\nu_c)\, \tau^2_{c\Gamma_1}(\nu_c)}{\sum_{\alpha'} \tau^2_{c\alpha'}(\nu_c)} [1 - \theta(E - \Delta E_{c\Gamma_0})] , \end{aligned} \quad (4.78)$$

$$\theta(x) = \begin{cases} 1, & x > 0 , \\ -1, & x < 0 . \end{cases} \quad (4.79)$$

4.5.4 Radiative Decay of Resonances

The formula (4.78) has an obvious physical interpretation. Resonance excitation means capture of an electron into doubly excited non-stationary states followed by autoionization decay:

$$X_Z(\gamma_0) + e \to X^{**}_{Z-1} \to X^*_Z(\gamma_1) + e . \quad (4.80)$$

The quantities $\tau^2_{c\alpha}$ characterizing the efficiency of electron capture are proportional to the autoionization decay probabilities $A^a_{c\alpha}$,

$$A^a_{c\alpha} = \frac{2}{\pi \nu_c^3 Z^2} \tau^2_{c\alpha}(\nu_c) , \quad A^a_c = \sum_\alpha A^a_{c\alpha} . \quad (4.81)$$

Simultaneously, the radiative decay of doubly excited non-stationary states, i.e. dielectronic recombination, must be taken into account:

$$X_Z(\gamma_0) + e \to X_{Z-1}^{**} \to X_{Z-1} + \hbar\omega . \tag{4.82}$$

The result (4.78) was obtained assuming that the autoionization probabilities A_c^a are larger compared to the radiative decay probabilities A_c^r. In fact, the ratio A_c^r/A_c^a increases as the ion charge Z increases (as Z^4 for optically allowed transitions). A_c^r is almost independent of ν_c because the main contribution to A_c^r is due to the transition $\gamma \to \gamma'$ of an initially bound electron. Therefore, an increase in ion charge Z leads to a decrease in the value of $\bar{\nu}_c$, for which

$$A_c^a(\bar{\nu}_c) \simeq A_c^r . \tag{4.83}$$

Inclusion of dielectronic recombination (4.82) implies that we should replace the autoionization probability by the total decay probability

$$A_c^a \to A_c^a + A_c^r \tag{4.84}$$

in the denominator of (4.78). Using (4.78, 81, 84), we have

$$\bar{\sigma}^{\text{res}}(\Gamma_0 \to \Gamma_1) = \frac{g_{\Gamma_1}}{2 g_0 Z^4 E} \sum_c \frac{\tau_{c\Gamma_0}^2(\nu_c)\, \tau_{c\Gamma_1}^2(\nu_c)}{\sum_{\alpha'} \tau_{c\alpha'}^2(\nu_c) + (1/2)\,\pi Z^2 \nu_c^3 A_c^r} \times [1 - \theta(E - \Delta E_{c\Gamma_0})] . \tag{4.85}$$

This expression can be obtained on the basis of the detailed balance principle.

A more rigorous quantum-mechanical evaluation is as follows. We introduce in the right-hand side of (4.39 and 40) the electric dipole interactions of the radiation field with the ion. Retaining the first non-vanishing terms for the resonance width we obtain (4.85). It is worth noting that similar arguments for including the radiation field were used in the case of nuclear reactions [4.1], as well as in the case of dielectronic recombination [4.22].

Concluding this section we compare the results obtained here with those obtained in Sect. 4.3, 4. It is easy to see that the expressions for partial cross sections, without including radiative decay of resonances, (4.29), are the same as (4.37) obtained by *Gailitis* [4.9], and by *Seaton* [4.11], i.e.

$$\bar{\sigma}(\Gamma_0 \to \Gamma_1) = \sigma^{\text{pot}}(\Gamma_0 \to \Gamma_1) + \sum_c \frac{\sigma^{\text{pot}}(\Gamma_0 \to c)\, \sigma^{\text{pot}}(c \to \Gamma_1)}{\sum_\alpha \sigma^{\text{pot}}(c \to \alpha)} . \tag{4.86}$$

The cross section above the threshold, $\sigma^>$, is the potential cross section with a monotonic energy dependence. Radiative decay of resonances gives an additional contribution to the denominator in (4.86). The final results (4.76 and 85) of the asymptotic theory may be expressed in a different form which has a more simple physical meaning:

$$\bar{\sigma}^{\text{res}}(\Gamma_0 \to \Gamma_1) = \sum_{c} \sigma^{\text{pot}}(\Gamma_0 \to c) \frac{A^{\text{a}}_{c\Gamma_1}}{A^{\text{a}}_{c} + A^{\text{r}}_{c}} \left(\frac{1 - \theta(x)}{2} \right), \tag{4.87}$$

with $x = E - \Delta E_{c\Gamma_0}$.

When A^{r}_{c} is omitted, (4.87) coincides with (4.86). In accordance with (4.12) the total cross section is equal to

$$\bar{\sigma}(\gamma_0 \to \gamma_1) = \sum_{l_0 l_1 \Gamma_T} \left[\sigma^{\text{pot}}(\Gamma_0 \to \Gamma_1) + \sum_{c} \sigma^{\text{pot}}(\Gamma_0 \to c) \frac{A^{\text{a}}_{c\Gamma_1}}{A^{\text{a}}_{c} + A^{\text{r}}_{c}} \left(\frac{1 - \theta(x)}{2} \right) \right]. \tag{4.88}$$

Since the transition probabilities for radiative and autoionization decay for states with small principal quantum numbers are well investigated and tabulated (see Chap. 2), and those for large values of the principal quantum numbers can be obtained as an analytic continuation of the former, (4.88) may also be useful for practical applications.

4.6 Results of Calculations and Experimental Data

At present, experimental information on electron-ion excitation cross sections is rather scarce. Most of the results for this process are obtained by theoretical calculations. All three methods discussed above, namely, the *R* matrix (Sect. 4.3), the GQDM (Sect. 4.4), and the asymptotic expansion (AE) (Sect. 4.5), lead to equivalent results when radiative decay of resonances (the dielectronic recombination channel) is ignored. It follows from (4.88) that the dielectronic recombination channel can easily be taken into account for the cross sections averaged over the resonances. However, no adequate theoretical method has been developed, as yet, for a detailed treatment of the individual resonance structures taking into account radiative decay, which would also be suitable for numerical calculations.

The presence of the long-range Coulomb field means that all the electron-multiply charged ion calculations can be significantly simplified with respect to the electron-neutral atom ones. As was shown above for $Z \gg 1$, the first-order approximation is the Coulomb-Born-exchange (CBE) interaction + resonance excitation. The contribution of all other open channels is much smaller and in most cases can be neglected. *Hershkowitz* and *Seaton* [4.13] have shown that results of acceptable accuracy for resonance struc-

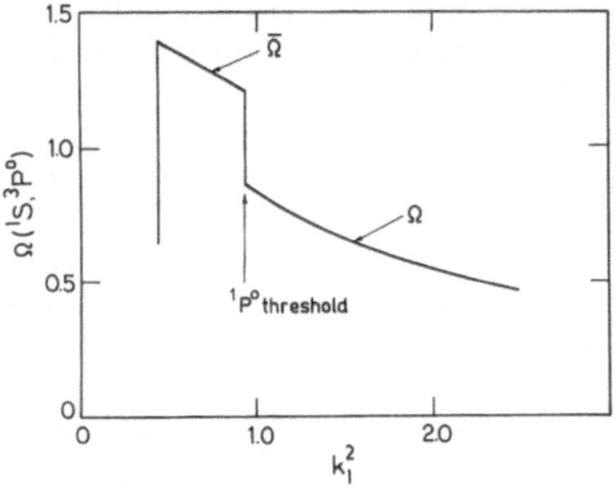

Fig. 4.2. Total collision strength $\Omega\,(^2S, ^3P^0)$ for C III, calculated in the DW approximation [4.13] including the target states ^1S, $^3P^0$ and $^1P^0$. Below the $^1P^0$ threshold, $\bar{\Omega}$ is the collision strength averaged over the resonances

Fig. 4.3. Partial collision strength $\Omega_{^2P^0}\,(^1S_p, ^3P^0_s)$ for C III. (—) calculations in the DW approximation; (---) calculations in the CC-ID approximation [4.13]

▼

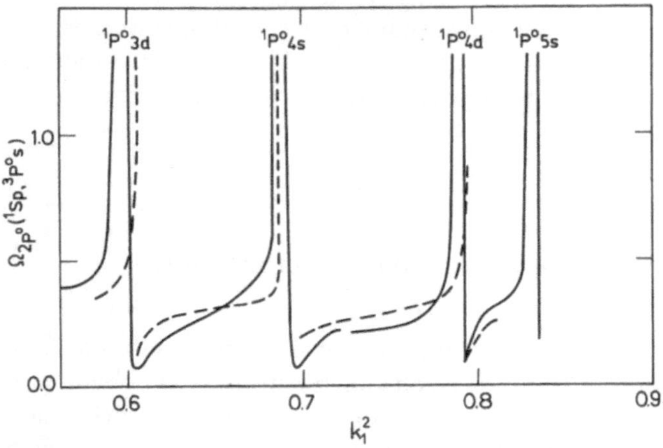

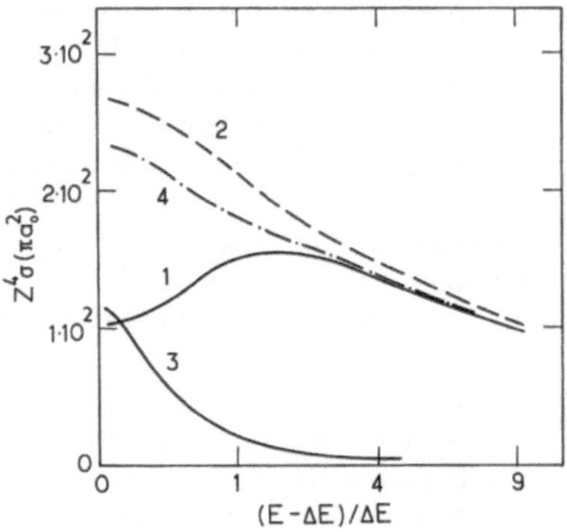

Fig. 4.4. Influence of the exchange effects on the excitation cross sections for O VII ion [4.25]. Transitions $1s^2 \to 1s\,2p$: (*1*) for $2\,^1P$ level with exchange, (*2*) the same without exchange, (*3*) for $2\,^3P$ level with exchange, (*4*) the sum of the cross sections 1 and 3

tures and total cross sections can be obtained using the first-order distorted-wave approximation with exchange (DWE) even in the case of C III. They also used the more accurate method for solving the close-coupled integro-differential equations (CC-ID), developed by *Eissner* and *Seaton* [4.12]. Some results of the GQDM are given in Figs. 4.2, 3.

For $Z \geq 5$, the difference between the DWE and CBE methods practically vanishes. Being the simplest among the existing methods, the CBE method requires an adequate treatment of the exchange effects. It is well known that the Born-Oppenheimer approximation for neutral atoms and the Coulomb-Born approximation for ions are not valid in the threshold region [4.23, 24]. *Vainshtein* [4.25] has shown that the exchange effects have a noticeable influence on the excitation cross sections even for the optically allowed transitions in the He-like ions (Fig. 4.4).

We would like to emphasize that there exist cases where the resonance excitation contribution is significantly larger than the potential (direct) excitation. Some examples are given in Figs. 4.5, 6, calculated using the AE method [4.14, 15]. These examples show that resonance excitation is important when the excitation threshold is a few times smaller than the one-electron ionization threshold. In this case the effect of electron exchange has a small influence on the potential part of a cross section (for example, excitation of Li-, Be- and Ne-like ions from the ground state). If the excitation threshold is comparable with the ionization threshold, the exchange effect is important, and the resonance excitation is comparatively small.

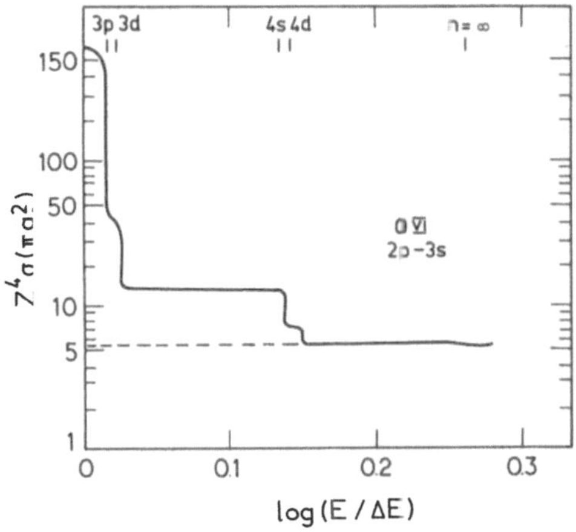

Fig. 4.5. Total excitation cross section $\bar{\sigma}(1s^2 2p \rightarrow 1s^2 3s)$ for the O VI ion averaged over the resonances. The broken curve is the potential part of the cross section [4.14, 15]

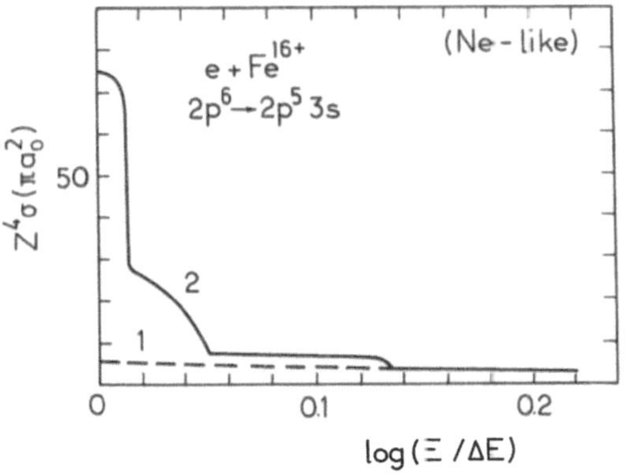

Fig. 4.6. Total excitation cross section $\bar{\sigma}(2p^6 \rightarrow 2p^5 3s)$ for Fe XVII (——) averaged over the resonances. The broken curve is the potential part of the cross section [4.26]

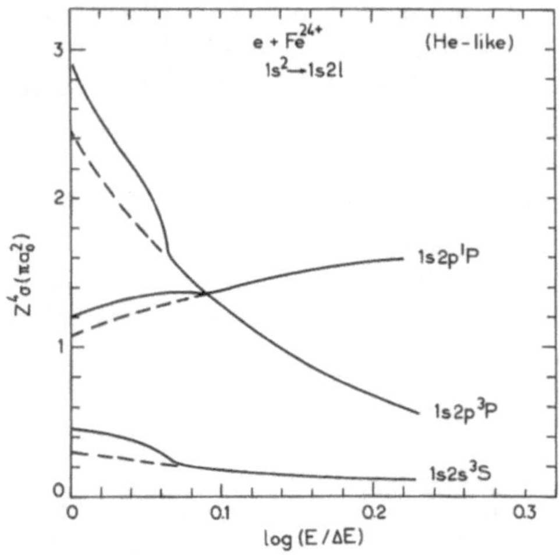

Fig. 4.7. Total cross sections for excitation of singlet and triplet states in helium-like iron ion calculated taking into consideration the exchange effects and the resonance contribution. The broken curves represent the potential part of the cross sections [4.26]

Nevertheless, the resonance contribution exists even for He-like ions, and an example is given in Fig. 4.7. The calculations [4.26] have been made using the AE method with intermediate coupling, which is important for ions with $Z > 10$. Intermediate coupling conserves the general scaling law, $\sigma \sim Z^{-4}$; however, some details may be different. Comparison of the results for 2^3P-level excitation of O VII (Fig. 4.4) and Fe XXV (Fig. 4.7) shows that relativistic effects (intermediate coupling) increase the scaled cross section.

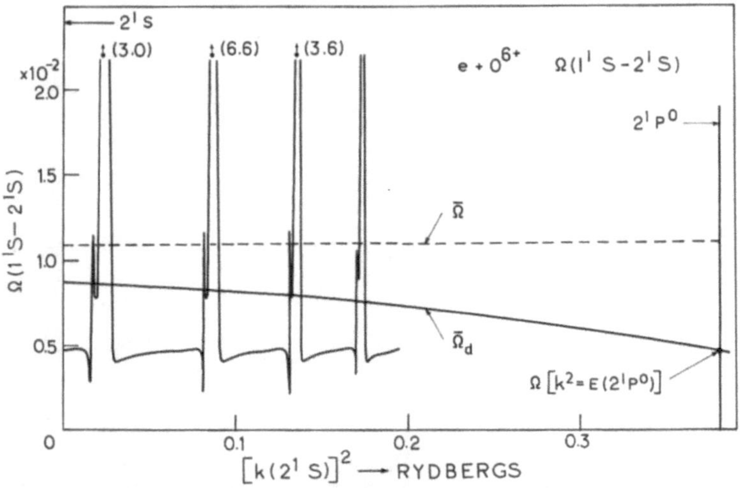

Fig. 4.8. Resonance structures and total collision strength Ω $(1\,{}^1S - 2\,{}^1S)$ for O VII. (---) Calculations without radiative decay of resonances; (——) calculations with radiative decay of resonances [4.28]

5. Electron-Impact Ionization of Highly Charged Ions

In this chapter, we shall consider the electron-ion ionization processes. We will devote our attention mainly to the direct and exchange channels of single-electron ionization. From the ionization processes involving electron transitions of several electrons, only the excitation-autoionization process will be discussed. Other multi-electron ionization processes, such as direct multiple ionization, inner-shell ionization followed by Auger decay of the autoionizing state created, will be omitted from our consideration, since the understanding of strongly correlated multi-electron processes at the present time is rather limited.

5.1 One-Electron Ionization. General Remarks

The electron-impact ionization process

$$e + X^{Z+} \rightarrow X^{(Z+1)+} + 2\,e \tag{5.1}$$

and its inverse three-body recombination,

$$2\,e + X^{(Z+1)+} \rightarrow e + X^{Z+}, \tag{5.2}$$

are the simplest representatives of a broader class of many-electron reactions

$$e + X^{Z+} \rightleftarrows X^{(Z+m)+} + (m+1)\,e, \tag{5.3}$$

which take place in hot laboratory and astrophysical plasmas. As a rule, the processes (5.1 and 2) are also the most effective ones in determining the degree of ionization, and together with other recombination processes (such as radiative and dielectronic recombination, charge exchange "recombination", etc.), substantially influence the ionization equilibrium of hot plasmas. The multi-electron processes (5.3) ($m \geqq 2$) are of lesser practical importance due to their small cross sections, but they are of significant fundamental interest for understanding the structure of ion electronic shells and their interactions with an incident electron.

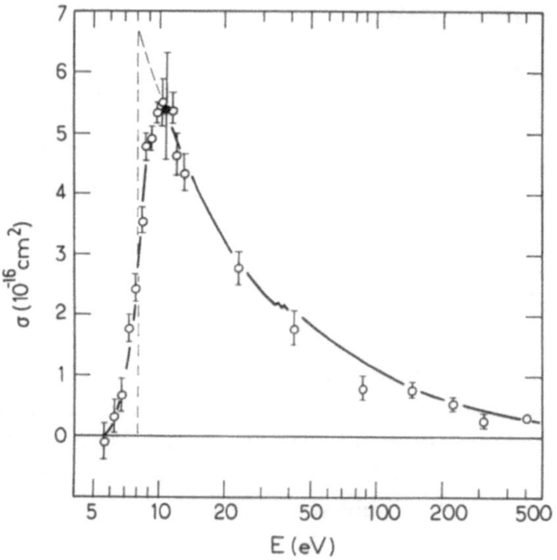

Fig. 4.9. Absolute electron excitation cross section of $2s-2p$ transition in C^{3+} ions. (---) Two-state close-coupling calculations with exchange [4.33]. (—) a convolution of the experimental results [4.31] with the calculations [4.33]

Further practical applications of GQDM have been reported by *Pradhan* et al. [4.27, 28]. One of the results [4.28] is given in Fig. 4.8, where the detailed resonance structure is calculated without including the effects of dielectronic recombination, and for the averaged collision strength the substitution (4.84) [4.14] is used. It is necessary to point out that the $\underline{R}$-matrix theory is also used for practical calculations of total cross sections and resonance structures (see, for example, *Berrington* et al. [4.29]). From a general point of view, the $\underline{R}$-matrix calculations are as accurate as the solution of the close-coupled equations. However, there is presently no information on how the dielectronic recombination channels can be taken into account within the $\underline{R}$-matrix theory, except in the expressions for the averaged cross sections.

As we have already pointed out, there are no crossed-beam experimental results for ions with $Z = Z_n - N + 1 > 5$. However, precise measurements have been made by the Oak Ridge group [4.30–32] for the transitions $1s^2 2s \rightarrow 1s^2 2p$ in Be II, C IV, N V (Be^+, C^{3+}, N^{4+}) in the Li-isoelectronic sequence. Some results are shown in Fig. 4.9. There is no evidence of resonance excitation, which is understandable. The analysis of (4.88) shows that the resonance contribution is proportional to the excitation cross sections of levels lying above the $2p$ level in Li-like ions. These cross sections are, however, much smaller than the $2p$ level cross sections.

The energy conservation law for the process (5.1) has the form (in atomic units $\hbar = m = e^2 = 1$)

$$\tfrac{1}{2}k^2 + \tfrac{1}{2}v^2 = \tfrac{1}{2}k_0^2 - I_Z = E_t, \tag{5.4}$$

where k_0 is the momentum of the incident electron, I_Z is the ionization potential of the X^{Z+} ion, and k and v are, respectively, the momenta of the scattered and the ejected electron in the final reaction channel.

Strictly speaking, the indistinguishability of electrons in the final channel gives rise to specific quantum-mechanical exchange effects which have to be taken into account in the theoretical description of the ionization process. On the other hand, within a wide range of variation of the parameters of the problem (collision energy, different regions of configuration space, etc.), the electron motion in the continuous spectrum of the ion can be described in terms of classical mechanics. Therefore, both quantum and classical treatments have been successfully applied in the past to the electron-impact ionization of ions.

It is worth underlining that the first result for the ionization cross section, the famous *Thomson* formula [5.1], was obtained in 1912 within the framework of classical mechanics, without introducing the concept of electrons. Since that time, an immense amount of both theoretical and experimental information about the electron-impact ionization process has been accumulated, mainly for atomic targets. In the last two decades, however, significant progress has also been made in the problem of electron-impact ionization of positive ions. Many of the achievements have already been reviewed [5.2–5]. Of special interest are the papers by *Crandall* [5.4] and *Bazylev* and *Chibisov* [5.5], devoted to the ionization of multiply charged ions. The most important aspects of the electron-impact ionization theory are given in the monograph by *Peterkop* [5.6], especially regarding the exchange effects and threshold behaviour.

5.2 Exchange Effects in Ionization. The Peterkop Theory

The importance of exchange effects in ionization processes was first pointed out by *Peterkop* [5.7]. Here, we give the main results of the theory [5.6, 7], restricting ourselves to the ionization of a hydrogen-like ion in the non-relativistic approximation. For this case the Schrödinger equation has the form

$$\left(-\frac{1}{2}\nabla_1^2 - \frac{1}{2}\nabla_2^2 - \frac{Z_n}{r} - \frac{Z_n}{r_2} + \frac{1}{r_{12}} - E_t\right)\Psi(\boldsymbol{r}_1, \boldsymbol{r}_2) = 0, \tag{5.5}$$

where $\boldsymbol{r}_1$ and $\boldsymbol{r}_2$ are the electron positions with respect to the nucleus, Z_n is the nuclear charge, and E_t is the total energy defined by (5.4). The ionization problem can be considered as a continuation of the excitation problem into the continuous spectrum. Therefore, the boundary conditions for the excitation problem can also be applied to the continuous spectrum and have the form

$$\Psi(\boldsymbol{r}_1, \boldsymbol{r}_2) \underset{r_1 \to \infty}{\sim} \psi^0(\boldsymbol{r}_1, \boldsymbol{r}_2) + \frac{1}{r_1} \sum_{nlm} \varphi_{nlm}(\boldsymbol{r}_2) f_{nlm}(O_1)\, e^{ik_n r_1}$$
$$+ \frac{1}{r_1} \int_{v \leq \sqrt{2E_t}} \varphi(\boldsymbol{v}, \boldsymbol{r}_2) f_v(O_1) \exp[i k r_1 + i\eta(\boldsymbol{v}, \boldsymbol{r}_1)]\, d\boldsymbol{v}\,, \quad (5.6)$$

$$\Psi(\boldsymbol{r}_1, \boldsymbol{r}_2) \underset{r_2 \to \infty}{\sim} \frac{1}{r_2} \sum_{nlm} \varphi_{nlm}(\boldsymbol{r}_1)\, g_{nlm}(O_2)\, e^{ik_n r_2}$$
$$+ \frac{1}{r_2} \int_{v \leq \sqrt{2E_t}} \varphi(\boldsymbol{v}, \boldsymbol{r}_1)\, g_v(O_2) \exp[i k r_2 + i\eta(\boldsymbol{v}, \boldsymbol{r}_2)]\, d\boldsymbol{v}\,, \quad (5.7)$$

where O_1 and O_2 are the directions of the vectors $\boldsymbol{r}_1$ and $\boldsymbol{r}_2$, respectively. The unperturbed wave function $\psi^0(\boldsymbol{r}_1\, \boldsymbol{r}_2)$ is the solution of (5.5) without the term r_{12}^{-1} (the coordinate $\boldsymbol{r}_2$ relates to the bound ionic electron),

$$\Psi^0(\boldsymbol{r}_1\, \boldsymbol{r}_2) = \varphi^+(\boldsymbol{k}_0, \boldsymbol{r}_1)\, \varphi_0(\boldsymbol{r}_2) \quad (5.8)$$

and φ^+ is the Coulomb wave function for the continuous spectrum, defined by (3.4) in Sect. 3.1. In the final channel, the asymptotic wave function of the scattered electron has the form of a plane wave + an incoming wave, see (3.5),

$$\varphi(\boldsymbol{v}, \boldsymbol{r}) = \varphi^{(-)}(\boldsymbol{v}, \boldsymbol{r}) = (2\pi)^{-3/2}\, e^{\pi Z_n/2v}\, \Gamma\left(1 + i\frac{Z_n}{v}\right)$$
$$\times e^{i\boldsymbol{v}\cdot\boldsymbol{r}} F\left[-i\frac{Z_n}{v}, 1; -i(v r + \boldsymbol{v}\cdot\boldsymbol{r})\right], \quad (5.9)$$

where $\Gamma(x)$ is the gamma function and $F(a, b; x)$ is the confluent hypergeometric function. In (5.6 and 7), the numbers n, l and m denote the principal, orbital and magnetic quantum numbers of discrete states, respectively, and f_{nlm} and g_{nlm} are the amplitudes of direct and exchange excitation of discrete levels, which include the Coulomb scattering phases (4.8). The electron momenta are defined by the energy conservation conditions:

$$\frac{1}{2} k_n^2 = \frac{k_0^2}{2} - I_n = \frac{k_0^2}{2} - \frac{Z_n^2}{2n^2}\,, \quad (5.10)$$

where I_n is the ionization potential for the nth level, i.e. $I_1 = I_Z$. From (5.4), it follows that the integration region in (5.6 and 7) is given by

$$0 \leqq v \leqq \sqrt{2E_t} . \tag{5.11}$$

The phases $\eta(v, \boldsymbol{r})$ in (5.6 and 7) differ from the Coulomb phases and exist only in the asymptotic part of the wave function, which corresponds to ionization. This statement can be understood as follows. In the case of electron-impact excitation, the electron is effectively located at large distances both from the nucleus and the bound ionic electron. For this case the phase shift is due to the long-range Coulomb field, and corrections due to short-range interactions are small. This phase shift is already taken into account in the amplitudes of direct scattering f_{nlm}, f_v, and of exchange scattering, g_{nlm}, g_v. When we consider the ionization problem, we have to take into account the existence of regions in configuration space in which, although both vectors $\boldsymbol{r}_1$ and $\boldsymbol{r}_2$ tend to infinity, the interelectron distance can still be small, $r_{12} = |\boldsymbol{r}_1 - \boldsymbol{r}_2| \to 0$. In these regions, the interelectron interaction $1/r_{12}$ can be comparable with or even greater than the electron interaction with the nucleus, $Z_n/r_1 + Z_n/r_2$. One can say that in the asymptotic region the excitation problem is a two-particle one, whereas the ionization problem is still a three-particle problem.

In analogy with the excitation problem, the amplitudes f_v and g_v may be called "direct" and "exchange" ionization amplitudes. However, when both electrons are far from the nucleus, the definitions of direct and exchange processes need to be refined. The amplitude $g_v(O_2)$ corresponds to the process in which the incident electron moves with a momentum $\boldsymbol{v}$ after the collision, whereas the initially bound electron moves with the momentum $\boldsymbol{k}$. The modulus of $\boldsymbol{k}$ is given by (5.4) and its direction coincides with the O_2 direction. The same situation is, however, described by the amplitude $f_k(O_1)$ if O_1 coincides with the direction of $\boldsymbol{v}$. Therefore, for the ionization process, the amplitudes f and g are not independent. In order to reveal the relation between them, it is convenient to introduce a more symmetric form:

$$f_v(O_1) = f(\boldsymbol{k}, \boldsymbol{v}) , \tag{5.12}$$

$$g_v(O_2) = g(\boldsymbol{k}, \boldsymbol{v}) , \tag{5.13}$$

where both of the vectors $\boldsymbol{k}$ and $\boldsymbol{v}$ have independent directions and their absolute values are related by (5.4). In (5.12) the $\boldsymbol{k}$ direction coincides with O_1, $\boldsymbol{k} \parallel O_1$, and in (5.13) $\boldsymbol{k} \parallel O_2$. Now, it follows that [5.6, 7]

$$|g(\boldsymbol{k}, \boldsymbol{v})| = |f(\boldsymbol{v}, \boldsymbol{k})| , \tag{5.14}$$

or, after designating $\boldsymbol{k} = \boldsymbol{k}_1$ and $\boldsymbol{v} = \boldsymbol{k}_2$, one obtains

$$|g(\boldsymbol{k}_1, \boldsymbol{k}_2)| = |f(\boldsymbol{k}_2, \boldsymbol{k}_1)|. \tag{5.15}$$

In the calculations of ionization cross sections, one has to integrate over the final momenta. It is, however, more convenient to integrate over energies:

$$d\boldsymbol{k}_i = k_i\, d\varepsilon_i\, dO_i, \quad \varepsilon_i = \tfrac{1}{2} k_i^2, \quad i = 1, 2. \tag{5.16}$$

The expression for the cross section can now be given in the form [5.6, 7]

$$\sigma_0 = \int_0^{E_t} \frac{k_1 k_2}{k_0} d\varepsilon_2 \int |f(\boldsymbol{k}_1, \boldsymbol{k}_2)|^2 dO_1\, dO_2, \tag{5.17}$$

or

$$\sigma_0 = \int_0^{E_t} \frac{k_1 k_2}{k_0} d\varepsilon_1 \int |g(\boldsymbol{k}_2, \boldsymbol{k}_1)|^2 dO_1\, dO_2, \tag{5.18}$$

where $\varepsilon_1 + \varepsilon_2 = E_t$, in accordance with (5.4). Equation (5.15) enables one to write down the ionization cross section in a more symmetric form:

$$\sigma_0 = \int_0^{E_t/2} \frac{k_1 k_2}{k_0} d\varepsilon_2 \int [|f(\boldsymbol{k}_1, \boldsymbol{k}_2)|^2 + |g(\boldsymbol{k}_1, \boldsymbol{k}_2)|^2]\, dO_1\, dO_2\,, \tag{5.19}$$

in which the integration over ε_2 $(0 \leq \varepsilon_2 \leq E_t/2)$ includes all the cases when one of the electrons has an energy equal to ε_2. This result reveals the significant effect of the exchange, even in the case when the exchange amplitude $g(\boldsymbol{k}_1, \boldsymbol{k}_2)$ is much smaller than the direct one, $f(\boldsymbol{k}_1, \boldsymbol{k}_2)$, as is the case in the region $0 \leq \varepsilon_2 \leq E_t/2$. In the region $E_t/2 \leq \varepsilon_2 \leq E_t$, the direct amplitude, $f(\boldsymbol{k}_1, \boldsymbol{k}_2)$, is much smaller than the exchange one, in accordance with (5.15). It is worth noting that in the approximate calculations which exclude the exchange $[g(\boldsymbol{k}_1, \boldsymbol{k}_2) = 0]$, (5.19) gives more accurate results than (5.17), which can be confirmed by direct comparison with the experimental data [5.6].

In obtaining (5.19) the indistinguishability of the electrons has been accounted for in the final channel, but not in the initial one. From a mathematical point of view, this fact is reflected in the form of the initial unperturbed wave function ψ^0, (5.8), and in the absence of interference between f and g in (5.19).

A collision process involving two identical electrons should be described by symmetric and antisymmetric wave functions of the form

$$\Psi^{\pm} = \Psi(\boldsymbol{r}_1, \boldsymbol{r}_2) \pm \Psi(\boldsymbol{r}_2, \boldsymbol{r}_1)\,, \tag{5.20}$$

and the function Ψ^0 (5.8) has also to be written in a similar form. Under these conditions, the initial wave exists both at $\boldsymbol{r}_1 \to \infty$ and at $\boldsymbol{r}_2 \to \infty$, and the asymmetry of expressions (5.6 and 7) disappears. The asymptotic form for $\Psi^\pm$ at $r_1 \to \infty$ can be obtained by replacement of the amplitude f in (5.6) by

$$f^\pm = f \pm g\,. \tag{5.21}$$

The relation (5.15) now takes the form

$$f^\pm(\boldsymbol{k}_1, \boldsymbol{k}_2) = \exp[\mathrm{i}\,\tau(\boldsymbol{k}_1, \boldsymbol{k}_2)]\, f^\pm(\boldsymbol{k}_2, \boldsymbol{k}_1) \tag{5.22}$$

where τ has the physical meaning of the phase difference between the direct and the exchange amplitudes. The cross section is now equal to [5.6]

$$\sigma^\pm = \int_0^{E_t/2} \frac{k_1 k_2}{k_0}\, d\varepsilon_2 \int |f^\pm(\boldsymbol{k}_1, \boldsymbol{k}_2)|^2\, dO_1\, dO_2 \tag{5.23}$$

$$= \int_0^{E_t/2} \frac{k_1 k_2}{k_0}\, d\varepsilon_2 \int |f(\boldsymbol{k}_1, \boldsymbol{k}_2) \pm g(\boldsymbol{k}_1, \boldsymbol{k}_2)|^2\, dO_1\, dO_2\,. \tag{5.24}$$

By using (5.22), one can rewrite (5.24) in the form

$$\sigma^\pm = \sigma_0 \pm 2\sigma_{\mathrm{int}}\,, \tag{5.25}$$

where σ_0 is defined by (5.19), and σ_{int} is the interference term

$$\sigma_{\mathrm{int}} = \int_0^{E_t/2} \frac{k_1 k_2}{k_0}\, d\varepsilon_2 \int \mathrm{Re}\{f(\boldsymbol{k}_1, \boldsymbol{k}_2) \times \exp[-\mathrm{i}\,\tau(\boldsymbol{k}_1, \boldsymbol{k}_2)]\, f^*(\boldsymbol{k}_2, \boldsymbol{k}_1)\}\, dO_1\, dO_2\,. \tag{5.26}$$

For hydrogen-like ions, the ionization cross section, averaged over the spins, has the form

$$\sigma = \tfrac{1}{4}\sigma^+ + \tfrac{3}{4}\sigma^- = \sigma_0 - \sigma_{\mathrm{int}}\,. \tag{5.27}$$

All the above formulae are derived from first principles without employing any approximations and, therefore, they are exact. However, the value of σ_{int} and its contribution to the ionization cross section depends on the phase $\tau(\boldsymbol{k}_1, \boldsymbol{k}_2)$, whose calculation involves the solution of a dynamic scattering problem.

Concluding this section, we underline that the most significant difference between the theory presented here and that of *Massey* and *Moore* [5.8] is the integration over ejected electron energies in the interval $0 \le \varepsilon_2 \le E_t/2$. This difference, as we have already pointed out, has a significant influence on the accuracy of practical calculations.

5.3 Threshold Behaviour of the Cross Section

Unlike the electron-ion excitation cross sections the ionization cross section is equal to zero at the threshold: $\sigma(E_t=0)=0$. The threshold behaviour of the ionization cross section, $\sigma(E_t)$ as $E_t \to 0$, is a very complicated problem, since in the region of low energies the interelectron repulsion is a comparatively strong interaction and the three-body character of the collision problem is strongly pronounced. As is well known, the treatment of the three-body problem is extremely difficult both in classical and quantum mechanics. *Wannier* [5.9] took into account the applicability of the classical description for the motion of scattered electrons in a significant part of configuration space, and derived the following well-known expression for the threshold behaviour of the ionization cross section:

$$\sigma \underset{E_t \to 0}{\sim} E_t^{\alpha}, \qquad \alpha = \alpha(Z_n), \tag{5.28}$$

where Z_n is the nuclear charge. The Wannier law has subsequently been confirmed within the semi-classical approximation [5.10–12]. For the ionization cross section of a neutral hydrogen atom ($Z_n = 1$) the Wannier law gives

$$\sigma(E_t) \sim E_t^{1.127}, \tag{5.29}$$

whereas the Born approximation predicts

$$\sigma^B(E_t) \sim E_t^{3/2}. \tag{5.30}$$

It is worth noting that the Coulomb-Born approximation, which may be used for neutral atom ionization estimations, leads to a linear law:

$$\sigma^{CB} \sim E_t. \tag{5.31}$$

In the case of neutral atoms, the experimental results on the threshold-cross-section behaviour, obtained by using high-resolution crossed beam techniques, can serve as practical criteria for the applicability of different theoretical methods in the region $E_t \to 0$, which is the most difficult one for theoretical description. A detailed analysis of theoretical methods and results is given, for example, in [5.6].

The situation in the case of ionization of positive ions is different from that for atoms. For a hydrogen-like helium ion ($Z_n = 2$) the *Wannier* formula (5.28) gives

$$\sigma \sim E_t^{1.056}, \qquad Z_n = 2, \tag{5.32}$$

and for $Z_n = 5$ the result approaches the linear law (5.31): $\sigma \sim E_t^{1.007}$. From a physical point of view this is a consequence of the smallness of the inter-electron interaction in comparison with electron-nucleus interactions, i.e. it corresponds to the applicability of the expansion in powers of $(1/Z)^n$, described in Sect. 4.5 for the problem of electron-impact excitation of ions. Thus, for electron-impact ionization of ions with $Z \geq 5$ the first correct term of the perturbation series is the Coulomb-Born approximation with exchange (CBE). Practical calculations of $\sigma(E_t)$ in the threshold region are, however, rather complicated even within the CBE approximation.

5.4 Quantum-Mechanical Methods for Ionization Cross Section Calculations

In the case of ionization of hydrogen-like ions, no resonance structures appear in the cross section, since all the autoionization states of a system involving a nucleus + two electrons lie below the ionization threshold. We have already pointed out in Sect. 4.5 that for a multiply charged ion with $Z \geq 5$ the distorted-wave exchange (DWE) approximation practically coincides with the CBE approximation. For a hydrogen-like ion, the CBE approximation gives the following expressions for the direct (f) and exchange (g) amplitudes:

$$f(\boldsymbol{k}_1, \boldsymbol{k}_2) = \frac{1}{2\pi} \int \varphi^+(\boldsymbol{k}_0, \boldsymbol{r}_1)\, \varphi_0(\boldsymbol{r}_2)\, \frac{1}{r_{12}}\, \varphi^*(\boldsymbol{k}_1, \boldsymbol{r}_1)\, \varphi^*(\boldsymbol{k}_2, \boldsymbol{r}_2)\, d\boldsymbol{r}_1\, d\boldsymbol{r}_2\,, \tag{5.33}$$

$$g(\boldsymbol{k}_1, \boldsymbol{k}_2) = \frac{1}{2\pi} \int \varphi^+(\boldsymbol{k}_0, \boldsymbol{r}_1)\, \varphi_0(\boldsymbol{r}_2)\, \frac{1}{r_{12}}\, \varphi^*(\boldsymbol{k}_1, \boldsymbol{r}_2)\, \varphi^*(\boldsymbol{k}_2, \boldsymbol{r}_1)\, d\boldsymbol{r}_1\, d\boldsymbol{r}_2 \tag{5.34}$$

where φ_0 is the wave function of the bound state, φ^+ is the Coulomb wave function with the outgoing wave asymptotic behaviour (3.4), $\varphi = \varphi^-$ is the Coulomb wave function with the incoming wave asymptotic form (5.9). Cross section calculations can be performed by using (5.24–27). In practice, the partial wave expansion is used:

$$\varphi^{\pm}(\boldsymbol{k}, \boldsymbol{r}) = \frac{1}{2k} \sum_{l=0}^{\infty} \mathrm{i}^l (2l+1)\, \mathrm{e}^{\pm \mathrm{i}\delta_l} F_l(kr)\, P_l\left(\frac{\boldsymbol{k}\cdot\boldsymbol{r}}{kr}\right). \tag{5.35}$$

Here $F_l(kr)$ are the radial Coulomb wave functions satisfying the equation

$$\left(\frac{d^2}{dr^2} - \frac{l(l+1)}{r^2} + \frac{2Z_n}{r} + k_n^2\right) F_l(kr) = 0 \tag{5.36}$$

with boundary conditions $F_l(k\,r) \sim \sin(k\,r - \pi\,l/2 + \delta_l)$, i.e. solutions of the system (4.38) without the right-hand side; P_l are the Legendre polynomials and the Coulomb phases δ_l are defined by

$$\exp(\pm\,\mathrm{i}\,2\,\delta_l) = \frac{\Gamma(l+1 \pm \mathrm{i}\,Z_\mathrm{n}/k)}{\Gamma(l+1 \mp \mathrm{i}\,Z_\mathrm{n}/k)}\,. \tag{5.37}$$

The use of the DWE approximation instead of the CBE leads to replacing the functions $F_l(k\,r)$ in (5.35) by the functions $\mathscr{F}$, which are solutions of (4.10) without the right-hand side. From a physical point of view this means that (5.36) should also contain on the left-hand side, in addition to the Coulomb potential $2\,Z_\mathrm{n}/r$, a short-range potential, a polarization potential and an exchange potential, calculated as diagonal matrix elements.

The formulae given in this section are relatively simple and they demonstrate the general structure of the expressions which define the ionization cross sections. Only the partial wave expansion technique can be used for practical quantum calculations in the more complicated cases, i.e. for ions with two (or more) equivalent electrons, for analysis of relativistic corrections in the unperturbed ion wave functions (intermediate coupling), and so on. Corresponding formulae should obviously include summation over all the quantum numbers of both scattered electrons and integration over the final-state energies; they are a natural extension of the formulae given in Sects. 4.2–5 for ion excitation. Since the calculations in the CBE approximation require computation of the ion eigenfunctions, and the DWE approximation requires numerical solution of the integro-differential equations, we do not give more complicated general formulae here. Instead, we shall present some results of numerical calculations.

5.5 Classical and Semi-Empirical Formulae

The classical character of the motion of scattered electrons in some regions of configuration space permits one to use the methods of classical mechanics to derive simple formulae for ionization, which provide order-of-magnitude estimates of the cross section and its dependence on the main parameters of the problem. The first such formula was obtained by *Thomson* [5.1]:

$$\sigma = 4q\left(\frac{\mathrm{Ry}}{I_Z}\right)^2 \frac{u}{(u+1)^2}\,(\pi\,a_0^2)\,, \tag{5.38}$$

$$u = (E - I_Z)/I_Z$$

where q is the number of equivalent electrons in the shell $n\,l^q$ with binding energy I_Z, $E = k_0^2/2$ is the energy of the incident electron, 1 Ry = 13.6 eV.

Quantum mechanics has later shown that this formula incorrectly predicts the dependence of σ on u, but the Z dependence of σ is predicted correctly. Since $I_Z \sim Z^2$, $\sigma \sim Z^{-4}$.

A detailed list and analysis of semi-empirical formulae for electron-ion impact ionization is given in [5.13]. Comparison with the available experimental data shows that the most suitable semi-empirical formula for the case of multiply charged ion ionization is the one proposed by *Lotz* [5.14],

$$\sigma(u) = 2.76 \left(\frac{\text{Ry}}{I_Z}\right)^2 \frac{\ln(u+1)}{u+1} \quad [\pi a_0^2]\,. \tag{5.39}$$

This formula is based on the Coulomb-Born approximation for H-like ions and is widely used because of its simplicity. It gives acceptable results for ions with $Z = 4-6$, but overestimates the available experimental data for $Z > 6$.

Among widely used formulae one has to mention the Burgess formula (see, e.g. [5.15]), obtained on the basis of the semi-classical electron collision impact parameter (ECIP) approach:

$$\begin{aligned}\sigma(E) = {} & \frac{4\pi a_0^2\,\text{Ry}^2 q}{(E+I_Z+E_Z)}\left[\left(\frac{1}{I_Z}-\frac{1}{E}\right)+\frac{2}{3}E_Z\left(\frac{1}{I_Z^2}-\frac{1}{E^2}\right)-\frac{\ln(E/I_Z)}{E+I_Z}\right] \\ & + \sigma^{\text{IP}}(E) \equiv \sigma^{\text{EC}} + \sigma^{\text{IP}}\,,\end{aligned} \tag{5.40}$$

where E_Z is the kinetic energy of the bound electron and σ^{IP} is the so-called collisional part of the cross section, calculated using the photoionization cross section and, as a rule, contributing about 20% to the cross section. The need for calculations of E_Z and σ^{IP} introduces certain difficulties in the use of (5.40). As a rule, the ECIP approximation underestimates the ionization cross sections for multiply charged ions by a factor of 1.5 to 2.

We shall not discuss in detail the advantages and disadvantages of the formulae presented here. We note, however, that the Lotz formula (5.39) is the most used despite its overestimation of the cross sections for highly charged ions. It is clear that the CBE approximation should give more accurate results than the Lotz formula.

5.6 Excitation-Autoionization

In the case of many-electron ions, one has also to consider, in addition to direct ionization, the process of excitation of inner-shell electrons which leads to the formation of autoionizing states above the first ionization threshold. Their nonradiative decay (i.e. autoionization) gives an additional contribution to the observed ionization cross section. This process is called excitation-autoionization. For example, in the case of the Fe XVI ion, with ground state configuration ($1s^2\,2s^2\,2p^6\,3s$), the binding energy for the $3s$ electron is equal to 489 eV, whereas the excitation energy for the $2p^6\,3s \rightarrow 2p^5\,3s\,3d$ transition is equal to 803 eV, larger than the ionization threshold.

Let us denote by j the state of the ion $(X^{Z+})^{**}$ with an inner electron excited, i.e. the autoionization state. Its decay may take place via two channels: radiation and autoionization. The total ionization cross section, therefore, can be written in a form similar to (4.88):

$$\sigma_t = \sigma^d + \sum_j B_j\, \sigma_j^{exc}\,, \tag{5.41}$$

$$B_j = \frac{\sum_k A_{jk}^a}{\sum_k A_{jk}^a + \sum_l A_{jl}^r}\,, \tag{5.42}$$

where B_j is the branching ratio for the level j, A^r and A^a are the radiative and autoionization rates, respectively, k denotes all the states of the product $X^{(Z+1)+}$ ion, l denotes the states of the X^{Z+} ion formed by radiative decay, σ^d is the cross section for direct ionization and σ_j^{exc} is the cross section for excitation to the autoionizing state j.

The importance of the excitation-autoionization process was first pointed out by *Fox* et al. [5.16] and *Goldberg* et al. [5.17]. It has been experimentally confirmed in the crossed beam experiments by *Peart* and *Dolder* [5.18], who measured the ionization cross section of the Ba^+ ion. For Li-like ions (C^{3+}, N^{4+}, O^{5+}), the excitation-autoionization process gives an additional maximum in the cross section, and the ratio $\sigma_{t,\max}/\sigma^d_{\max}$ increases from a value of 0.92 up to 1.11 with increasing ion charge. The theoretical estimate [5.19] of this ratio gives a value 1.28 for $Z \rightarrow \infty$. The contribution of excitation-autoionization to the total ionization cross section has also been observed for Na-like ions with $Z \leq 4$ [5.20]. However, the role of the excitation-autoionization process in the electron-highly charged ion ionization has not yet been investigated in detail. It is worth remembering that with increasing ionic charge, the radiation decay rate A^r increases as Z^4, whereas the autoionization rate A^a almost does not depend on Z.

5.7 Experimental Data and Comparison with Theoretical Calculations

There are three sources which are presently used for experimental investigations of the ionization cross sections σ and the ionization rates $\langle v\,\sigma\rangle$: ion beams, ion traps, and laboratory plasmas (θ – pinch, stellarator, tokamak, etc.). With laboratory plasma sources only the ionization rates can be measured. The crossed beam experiments provide the most precise mea-

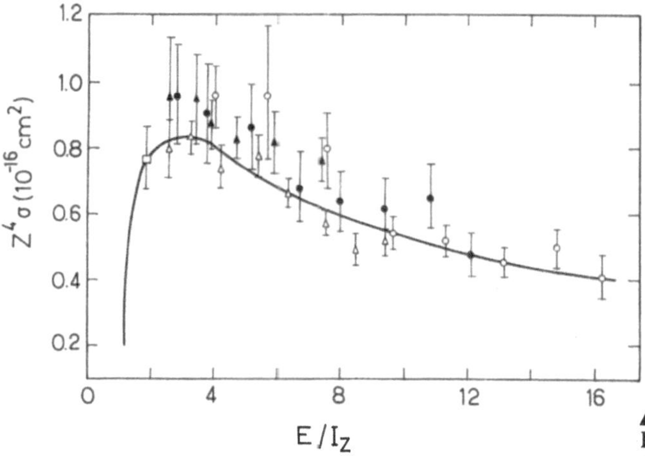

▲ **Fig. 5.1.** Cross sections for ionization of hydrogen-like ions from the ground state vs. electron impact energy (I_Z is the ion ionization potential). Experiment [5.22]: (○) C^{5+}, (●) N^{6+}, (△) O^{7-}, (▲) Ne^{9+}, (□) Ar^{17+}. Solid curve represents theoretical calculations [5.23] for $Z_n = 128$

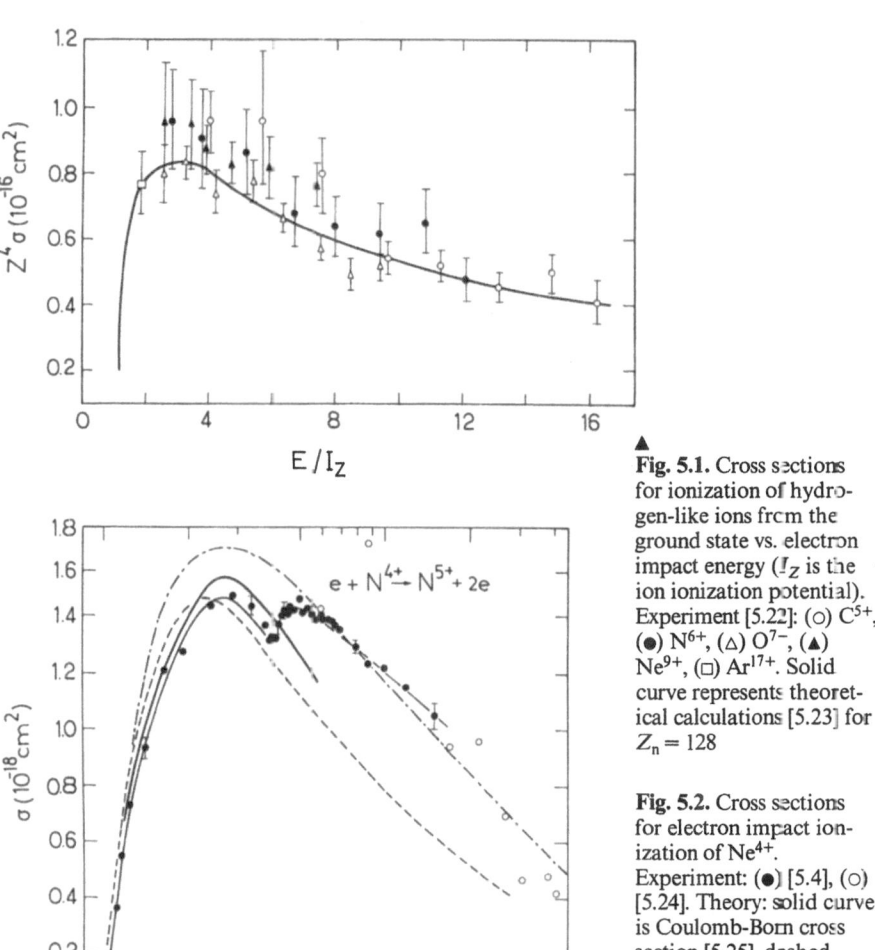

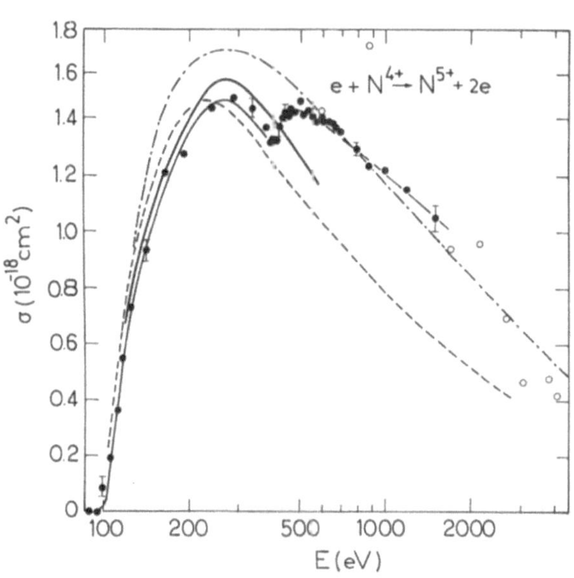

Fig. 5.2. Cross sections for electron impact ionization of Ne^{4+}. Experiment: (●) [5.4], (○) [5.24]. Theory: solid curve is Coulomb-Born cross section [5.25], dashed curve is scaled Coulomb-Born cross section [5.26], dot-dashed curve is result of Lotz formula [5.14]

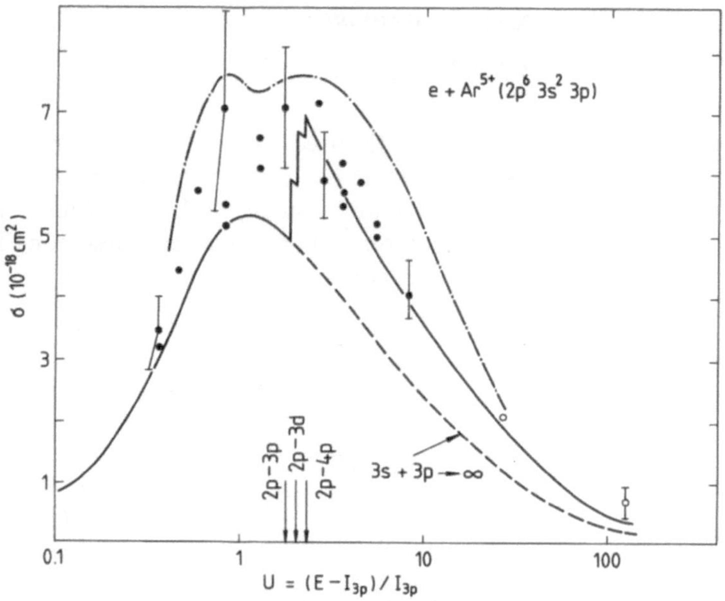

Fig. 5.3

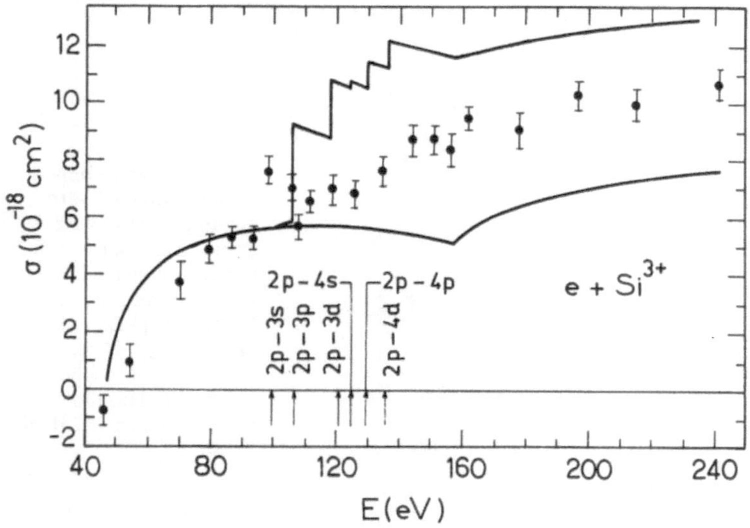

FFig. 5.4

Fig. 5.3. Ionization cross sections of Ar^{5+} vs. electron impact energy in threshold units. Experiment: (●) [5.32], (○) [5.22]. Theory: dot-dashed curve is result of classical calculations [5.28], solid curve is the CBE result with contribution from the excitation-auotionization of $2p$ shell [5.29], dashed curve is result of CBE calculations for direct ionization from $3s$ and $3p$ shells [5.29]

Fig. 5.4. Ionization cross section of Si^{3+} near threshold vs. electron impact energy. Points are experimental data [5.20]. Theory: solid curve is the result of DWE approximation for direct ionization [5.30], the curve added to the DWE is the excitation-autoionization contribution [5.31]

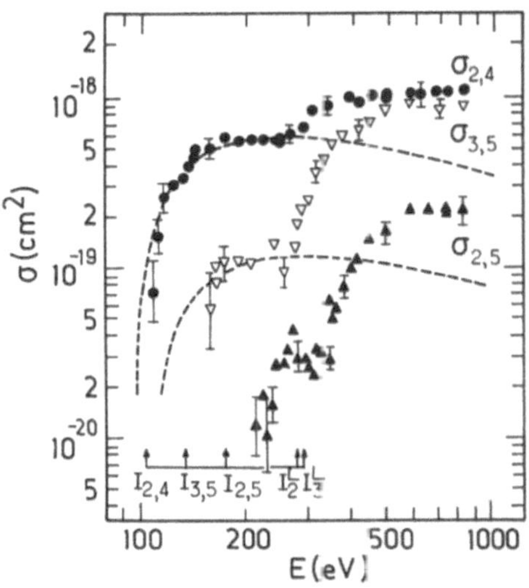

Fig. 5.5. Multiple ionization of Ar^{2+} and Ar^{3+} by electron impact [5.27]. Experimental data points are labelled by the initial and final charge states. Ionization potentials for initial and final charge states are indicated as $I_{i,f}$ and direct ionization of an L-shell electron by I_i^L. Dashed curves are normalized calculations for direct double-electron ionization in the Bethe approximation

surements of σ; however, the absence of suitable ion sources has restricted (up to now) the application of crossed beam methods to ions with $Z \leq 6$. The use of magnetic traps can provide more accurate results than the plasma sources; however, model calculations are required in order to interprete the results. Details of the experimental techniques are given in some recent review papers [5.4, 21]. An advantage of the use of magnetic traps is the possibility they provide to investigate the electron-impact ionization of ions with $Z \geq 10$. However, this method does not allow the investigation of the effects of excitation-autoionization, the presence of metastables, etc., on the ionization cross section.

The experimental results for the ionization cross sections of H-like ions, obtained by different methods, are given in Fig. 5.1. In Fig. 5.2, the crossed beam experimental data for electron-impact ionization of Ne^{4+} [5.24] are shown, together with different theoretical calculations. In Fig. 5.3, the contribution of the autoionizing transitions from the $2p^6$ shell to the ionization cross section of Ar^{5+} is shown. The effect of the excitation-autoionization process on the total ionization cross section of Si^{3+} [5.20] is demonstrated in Fig. 5.4.

The above discussion concerns only one-electron ionization. Multiple-electron impact ionization of highly charged ions is also a possible process, investigations of which have recently begun [5.27, 32, 33]. Some experimental results for Ar^{2+} and Ar^{3+} are shown in Fig. 5.5 and compared with the predictions of Bethe theory for double ionization.

5.8 Analytic Approximations for the Quantum-Mechanical Cross Sections

For the purposes of many applied problems (such as the calculations of the ionization balance in a plasma, for example), one needs to know the ionization cross section σ in a rather simple analytic form and with a better accuracy than provided by the formulae given in Sect. 5.6. For such purposes, analytic approximations to the numerical results, obtained by more sophisticated quantum-mechanical methods may be very useful. An example of such an analytic approximation is the four-parameter analytic fit of the CBE numerical results obtained with hydrogen-like wave functions, which was proposed by *Golden* and *Sampson* [5.26],

$$\sigma = \pi a_0^2 \left(\frac{n}{Z_{\text{eff}}}\right)^2 \left(\frac{\text{Ry}}{I_Z}\right) \frac{q}{u+1}$$

$$\cdot \left[A \ln(u+1) + D\left(\frac{u}{u+1}\right)^2 + \left(\frac{c}{u+1} + \frac{d}{(u+1)^2}\right)\frac{u}{u+1}\right], \tag{5.43}$$

$$u = \frac{E - I_Z}{I_Z}$$

where A, D, c, d are the fitting parameters, tabulated in [5.26], and Z_{eff} is the effective charge for a given electron shell $n\,l$.

Vainshtein and *Shevelko* [5.29, 34] also used the CBE approximation, employing, however, more accurate ion eigenfunctions obtained from the numerical solution of the radial Schrödinger equation. The CBE ionization cross sections can be approximated (within 10%) by the following two-

Table 5.1. Fitting parameters C and φ (5.44) for ionization of outer- and inner-shell electrons of multiply charged ions [5.34]

Outer shells				Inner shells		
$n\,l^q$	Isoelectronic sequence	C	φ	$n\,l^q$	C	φ
$1\,s^q$	H, He	8.18	3.25	$1\,s^2$	9.10	3.86
$2\,s^q$	Li, Be	5.65	1.98	$2\,s^2$	6.03	2.20
$2\,p^q$	B ÷ Ne	6.63	1.34	$2\,p^6$	6.00	1.40
$3\,s^q$	Na, Mg	4.40	1.78	$3\,s^2$	4.42	1.72
$3\,p^q$	Al ÷ Ar	4.26	1.12	$3\,p^6$	4.33	1.14
$4\,s^q$	K, Ca	3.29	1.71			

parameter fitting formula:

$$\sigma = \pi a_0^2 q \left(\frac{\mathrm{Ry}}{I_Z}\right)^2 \frac{u}{u+1} \frac{C}{u+\varphi}, \quad u = \frac{E - I_Z}{I_Z}. \tag{5.44}$$

The fitting parameters C and φ are given in Table 5.1 for the outer and inner shells of the isoelectronic sequences from H to Ca, calculated for ions with nuclear charge $Z_n = 99$. Calculations show that C and φ, for given quantum numbers $n\,l$, have a weak dependence on Z_n and the number of equivalent electrons q if the ion charge $Z \geq Z_n/2$ (see [5.29, 34] for details).

6. Collisions of Atoms with Highly Charged Ions: General Theoretical Description

The theoretical description of ion-atom or ion-ion collision processes is more complex than that of radiative and electron-ion collision processes due to the fact that, besides the electronic motion, the relative motion of the nuclei should also be adequately described. Moreover, during the collision, the electronic and nuclear motions are mutually coupled, and this introduces new aspects in the collision dynamics. Since the colliding system is, in general, a multielectron system, the number of inelastic processes that can occur is fairly large, and includes target excitation or ionization, electron capture, simultaneous capture and excitation (or ionization), and other processes involving one- or many-electron transitions. The theoretical descriptions of most of these processes differ and depend on the collision energy. However, many of them can be described by unifying theoretical approaches, and the specific features of each particular process appear only in the latter stages of the treatment.

In this chapter we shall consider those general theoretical methods which constitute a conceptual framework for the description of various inelastic ion-atom collision processes. The expansion (or close-coupling) methods, perturbation theory and different distorted-wave formalisms are certainly the most powerful among them. Classical mechanics may also be successfully used for a description of some classes of inelastic ion-atom processes. However, before going into the presentation of these methods, it is convenient to give some basic information about the one-electron two-Coulomb-centre system. Apart from its direct relevance to the processes in hydrogen atom-fully stripped ion collisions, this system often serves as a suitable model for describing the physical situation in more complex ion-atom systems ("one-electron approximation"). Its consideration will also allow us to introduce some basic notions used in the analysis of low-energy ion-atom collisions.

6.1 Two-Coulomb-Centre Problem

The determination of the eigenfunctions and eigenenergies of an electron in the field of two fixed Coulomb centres 1 and 2, having charges Z_1 and Z_2, respectively, is one of the basic problems of quantum mechanics. The significance of this so-called two-Coulomb-centre problem for molecular physics is analogous to that of the hydrogen-like atom in atomic physics. Unfortunately, the two-Coulomb-centre problem cannot be solved exactly. However, by using the approximate methods of quantum mechanics, many analytical results can be obtained for this problem for small and large distances between the Coulomb centres. Also, at present, there exist well-developed numerical algorithms for solving the two-Coulomb-centre eigenvalue problem. All these results, as well as an extensive bibliography, are collected in the monograph by *Komarov* et al. [6.1].

6.1.1 The Eigenvalue Problem

In the non-relativistic approximation, the Schrödinger equation for the electronic motion in the (Z_1, e, Z_2) system at a fixed distance R between the Coulomb centres, has the form

$$[\nabla_r^2/2 + Z_1/r_1 + Z_2/r_2 + E(R)]\,\Phi(\boldsymbol{r}, \boldsymbol{R}) = 0\,, \tag{6.1}$$

where $\boldsymbol{r}_1$, $\boldsymbol{r}_2$ and $\boldsymbol{r}$ are the position vectors of the electron with respect to the centres "1", "2" and the mid-point of the inter-nuclear separation, and $\Phi(\boldsymbol{r}, \boldsymbol{R})$ and $E(R)$ are, respectively, the electron wave function and energy. Equation (6.1) is separable in prolate spheroidal (confocal elliptic) coordinates (ξ, η, ϕ) [6.2],

$$\begin{aligned} \xi &= (r_1 + r_2)/R, & 1 \leqq \xi < \infty\,, \\ \eta &= (r_1 - r_2)/R, & -1 \leqq \eta \leqq 1\,, \\ \phi &= \arctan(y/x)\,, & 0 \leqq \phi < 2\pi\,, \end{aligned} \tag{6.2}$$

in which the positions of the centres Z_1 and Z_2 are defined by $(\xi = 1, \eta = -1)$ and $(\xi = 1, \eta = +1)$, respectively. Representing the wave function $\Phi(\boldsymbol{r}, \boldsymbol{R})$ in the form

$$\Phi(\boldsymbol{r}, \boldsymbol{R}) = X(\xi)\, Y(\eta)\, \mathrm{e}^{\pm im\phi}\,, \tag{6.3}$$

one obtains the following equations for the functions $X(\xi)$ and $Y(\eta)$:

$$\frac{d}{d\xi}\left[(\xi^2 - 1)\frac{dX}{d\xi}\right] + \left(\lambda - p^2\,\xi + a\,\xi - \frac{m^2}{\xi^2 - 1}\right) X(\xi) = 0\,, \tag{6.4}$$

$$\frac{d}{d\eta}\left[(1 - \eta^2)\frac{dY}{d\eta}\right] + \left(-\lambda + p^2\,\eta + b\,\eta - \frac{m^2}{1 - \eta^2}\right) Y(\eta) = 0\,, \tag{6.5}$$

where

$$p^2 = -R^2 E/2, \quad a = R(Z_1 + Z_2), \quad b = R(Z_2 - Z_1) \tag{6.6}$$

and λ is the separation constant. For the sake of definiteness, we assume $Z_2 \geqq Z_1$. In the case of the discrete spectrum ($E < 0$) the functions $X(\xi)$ and $Y(\eta)$ satisfy the boundary conditions

$$|X(1)| < \infty, \quad X(\xi) \xrightarrow[\xi \to \infty]{} 0, \tag{6.7}$$

$$|Y(\pm 1)| < \infty. \tag{6.8}$$

With these boundary conditions, (6.4, 5) each define a Sturm-Liouville problem, in which λ is the spectral parameter. The eigenfunctions and eigenvalues of these boundary-value problems are characterized by two quantum numbers n_ξ and m for $X(\xi)$ and $\lambda^{(\xi)}$ from (6.4), and n_η and m for $Y(\eta)$ and $\lambda^{(\eta)}$ from (6.5). Each of the quantum numbers n_ξ, n_η and m may take values 0, 1, 2, ...; n_ξ and n_η are equal to the number of nodes of the functions $X_{n_\xi m}(\xi)$ and $Y_{n_\eta m}(\eta)$, respectively. From the general theory of the Sturm-Liouville problems [6.2], it follows that the quantum numbers n_ξ, n_η and m are conserved when the parameter R varies and that none of the eigenvalues $\lambda^{(\xi)}_{n_\xi m}(p, a)$ or $\lambda^{(\eta)}_{n_\eta m}(p, b)$ can be degenerate.

In order for the function (6.3) to be a solution of the initial equation (6.1), the separation constants $\lambda^{(\xi)}_{n_\xi m}(p, a)$ and $\lambda^{(\eta)}_{n_\eta m}(p, b)$ have to be equal:

$$\lambda^{(\xi)}_{n_\eta m}(p, a) = \lambda^{(\eta)}_{n_\eta m}(p, b). \tag{6.9}$$

For fixed parameters a and b, and fixed quantum numbers n_ξ, n_η, m, (6.9) has a unique solution (if it exists):

$$p = p_{n_\xi n_\eta m}(a, b). \tag{6.10}$$

Using the relation (6.6) between p and E, one can derive the discrete energy spectrum of the two-Coulomb-centre problem,

$$E_j(R) = E_{n_\xi n_\eta m}(R, Z_1, Z_2), \tag{6.11}$$

where j designates the set of quantum numbers $\{n_\xi, n_\eta, m\}$. For an arbitrary, finite value of R, the boundary-value problems (6.4, 7) and (6.5, 8) are usually solved numerically, and the energy is obtained from (6.9–11). Let us now consider the behaviour of the molecular energies $E_j(R)$ when the internuclear distance R varies.

6.1.2 Correlation Rules and Pseudocrossings

It is well known that prolate spheroidal coordinates go over into spherical ones when $R \to 0$ (united atom limit), and into two parabolic systems of coordinates when $R \to \infty$ (separated atom limit), connected with each of the Coulomb centres. In the united atom limit, it is natural to chracterize the states of the system by the set of spherical quantum numbers $(N\, l\, m)$. In the separated atom limit the states can be characterized by the set of parabolic quantum numbers $[n\, n_1\, n_2\, m]$ if the electron is localized around Z_2, or by $[n'\, n_1'\, n_2'\, m]$ when the electron is localized on Z_1. The azimuthal quantum number m remains the same in all three systems of coordinates, since the projection L_z of the electron angular momentum $\boldsymbol{L}$ on the internuclear axis is an integral of motion. Since the nodes n_ξ and n_η of the functions $X_{n_\xi m}(\xi; R)$ and $Y_{n_\eta m}(\eta; R)$ are conserved when R varies, it is obvious that the set of quantum numbers $\{n_\xi\, n_\eta\, m\}$ can be related to the sets $(N\, l\, m)$ and $[n\, n_1\, n_2\, m]$ (or $[n'\, n_1'\, n_2'\, m]$) in the $R \to 0$ and $R \to \infty$ limits, respectively. The connection between $\{n_\xi\, n_\eta\, m\}$ and $(N\, l\, m)$ can be derived easily. If (r, θ, ϕ) is a spherical coordinate system with the origin at the midpoint of the internuclear separation, and the z axis (from which the angle θ is measured) oriented from centre "1" towards centre "2", then for $R \to 0$ and finite r, one obtains from (6.2)

$$\xi \to 2\, r/R, \quad \eta \to \cos\theta. \tag{6.12}$$

Inserting $\eta \to \cos\theta$ in (6.5) and neglecting terms which tend to zero when $R \to 0$, (6.5) becomes the equation for the associated Legendre polynomials. This immediately gives

$$\lambda|_{R=0} = -l(l+1), \quad l = 0, 1, 2, \ldots \tag{6.13}$$

$$Y_{n_\eta m}(\eta)|_{R=0} = P_l^m(\cos\theta), \quad 0 \leq m \leq l, \tag{6.14}$$

where the superscript η in λ has been omitted. Substituting (6.13) for λ in (6.4) and retaining only the leading terms in R, we obtain the equation for the radial functions of a hydrogen-like system with charge $Z_1 + Z_2$. It then follows that

$$X_{n_\xi m}(\xi)|_{R=0} = R_{Nl}(r), \quad N = 1, 2, 3, \ldots, \quad l = 0, 1, 2, \ldots, N-1, \tag{6.15}$$

$$E_j(R)|_{R=0} = -(Z_1 + Z_2)^2/(2\, N^2). \tag{6.16}$$

Since the functions $R_{Nl}(r)$ and $P_l^m(\cos\theta)$ have, respectively, $N - l - 1$ and $l - m$ nodes [6.2] the conservation of the elliptic quantum numbers n_ξ

and n_η requires

$$\eta_\xi = N - l - 1, \quad 0 \leqq l \leqq N-1, \tag{6.17}$$

$$n_\eta = l - m, \quad 0 \leqq m \leqq l. \tag{6.18}$$

The molecular states are usually designated by the quantum numbers N, l, m. States with $l = 0, 1, 2, \ldots$ are termed $s, p, d, \ldots$ states, whereas those with $m = 0, 1, 2, \ldots$ are $\sigma, \pi, \delta, \ldots$ states. However, the molecular states can alternatively be designated by the set of parabolic quantum numbers $[n\, n_1\, n_2\, m]$ which, at large distances, may be more appropriate.

Due to the fact that, at $R \to 0$, (6.4, 5) go over respectively into the radial and angular equations for the hydrogen-like system, (6.4) is called "quasi-radial", and (6.5) is called "quasi-angular".

The correspondence between n_ξ, n_η and the parabolic quantum numbers in the separated atom limit has been established by *Gershtein* and *Krivchenkov* [6.3]. Introducing new variables

$$v = (\xi - 1)\, R, \quad \mu_2 = (1 - \eta)\, R, \quad v, \mu_2 \ll R, \tag{6.19}$$

(6.4, 5) go over at $R \to \infty$ into the separated equations of the hydrogen-like atom in parabolic coordinates, located at centre "2". Thus, the correspondence between $X_{n_\xi m}(\xi; R)$ and $Y_{n_\eta m}(\eta; R)$ and the solutions of the one-centre Coulomb problem is

$$X_{n_\xi m}(\xi)\,|_{R\to\infty} = f_{n_1 m}(v), \qquad n_1 = 0, 1, 2, \ldots, \tag{6.20}$$

$$Y_{n_\eta m}(\eta)\,|_{R\to\infty} = f_{n_2 m}(\mu_2), \quad n_2 = 0, 1, 2, \ldots. \tag{6.21}$$

Consequently, for the energy $E_j(R)$ one has

$$E_j(R)\,|_{R\to\infty} = -Z_2/(2n^2), \quad n = n_1 + n_2 + m + 1. \tag{6.22}$$

With the transformation

$$v = (\xi - 1)\, R, \quad \mu_1 = (1 + \eta)\, R, \quad v, \mu_1 \ll R, \tag{6.23}$$

one obtains analogous results for the left centre "1" at $R \to \infty$:

$$X_{n_\xi m}(\xi)\,|_{R\to\infty} = f_{n'_1 m}(v), \qquad n'_1 = 0, 1, 2, \ldots, \tag{6.24}$$

$$Y_{n_\eta m}(\eta)\,|_{R\to\infty} = f_{n'_2 m}(\mu_1), \qquad n'_2 = 0, 1, 2, \ldots, \tag{6.25}$$

$$E_j(R)\,|_{R\to\infty} = -Z_1^2/(2n'^2), \quad n' = n'_1 + n'_2 + m + 1. \tag{6.26}$$

The relation between n_ξ and n_1, n_1' follows from the conservation theorem for the nodes of the solution of the Sturm-Liouville problem and has the form

$$n_\xi = n_1 = n_1' . \tag{6.27}$$

However, the relation between n_η and n_2, n_2' is somewhat more complex due to the fact that when $R \to \infty$, η may tend either to $+1$ or -1. Therefore, at large values of R, the function $Y_{n_\eta m}(\eta; R)$ is a linear combination of $f_{n_2 m}(\mu_2)$ and $f_{n_2' m}(\mu_1)$, the coefficients of which depend on the charges Z_1 and Z_2. Such a linear combination for $X_{n_\xi m}(\xi; R)$ is not needed. The analysis of the nodes of the linear combination for $Y_{n_\eta m}(\eta; R)$ at $R \to \infty$ leads to the following relations between n_η, n_2, n_2' $(Z_2 > Z_1)$ [6.3]:

– for the $(e\, Z_2)$ levels,

$$\begin{aligned} n_\eta &= n_2, \quad n_2 < n(Z_2 - Z_1)/Z_2 \\ &= n_2 + 1 + \mathrm{Ent}\,\{n_2 - n(Z_2 - Z_1)/Z_2\}, \quad n_2 \geqq n(Z_2 - Z_1)/Z_1 \end{aligned} \tag{6.28}$$

– for the $(e\, Z_1)$ levels,

$$\begin{aligned} n_\eta &= 2\,n_2' + n'(Z_2 - Z_1)/Z_1, \quad n'\, Z_2/Z_1 = \text{integer}, \\ &= 2\,n_2' + 1 + \mathrm{Ent}\,\{n'(Z_2 - Z_1)/Z_1\}, \quad n'\, Z_2/Z_1 \neq \text{integer}, \end{aligned} \tag{6.29}$$

where Ent $\{x\}$ is the entire part of x (the Gauss symbol). The different relations between n_η and $n_2(n_2')$ for the levels around different centres (when $Z_2 \neq Z_1$) seem, at first glance, to violate the node conservation theorem for the Sturm-Liouville problems. This is not so; the zeroes of $Y_{n_\eta m}(\eta; R)$ do not disappear, but part of them lie in the exponential tail of the wave function and are not essential for its determination.

The sets of relations (6.17, 18) and (6.27–29) constitute the rules for the correlation of the states of the (Z_1, e, Z_2) system in the united atom and separated atom limits. They can be written as

– correlation of $(N\, l\, m)$ levels with the levels of $(e\, Z_2)$:

$$\begin{aligned} N &= n \\ l &= n_2 + m \end{aligned}$$

for $n_2 < n(Z_2 - Z_1)/Z_2$, (6.30)

$$\begin{aligned} N &= n + 1 + \mathrm{Ent}\,\{n_2 - n(Z_2 - Z_1)/Z_2\} \\ l &= n_2 + m + 1 + \mathrm{Ent}\,\{n_2 - n(Z_2 - Z_1)/Z_2\} \end{aligned}$$

for $n_2 \geqq n(Z_2 - Z_1)/Z_2$,

– correlation of $(N\,l\,m)$ levels with the levels of $(e\,Z_1)$:

$$N = n' + n_2' + n'(Z_2 - Z_1)/Z_1$$
$$l\ = m + 2\,n_2' + n'(Z_2 - Z_1)/Z_1$$

for $\quad n'(Z_2 - Z_1)/Z_1 = \text{integer},$ (6.31)

$$N = n' + n_2' + 1 + \text{Ent}\,\{n'(Z_2 - Z_1)/Z_1\}$$
$$l\ = m + 2\,n_2' + 1 + \text{Ent}\,\{n'(Z_2 - Z_1)/Z_1\}$$

for $\quad n'(Z_2 - Z_1)/Z_1 \neq \text{integer}.$

In the case of a homonuclear system $(Z_1 = Z_2)$, the correlation rules (6.30, 31) have a simpler form. Besides commuting with the operator L_z (the projection of the electron angular momentum on the internuclear axis), the Hamiltonian in this case also commutes with the operator w for the reflection of the electronic coordinates through the $z = 0$ plane $(\xi \to \xi$, $\eta \to -\eta$, $\phi \to \phi)$, and with the operator I for the inversion of the electronic coordinates with respect to the mid-point of the internuclear axis $(\boldsymbol{r} \to -\boldsymbol{r}$, or $\xi \to \xi$, $\eta \to -\eta$, $\phi \to \phi + \pi)$. The eigenvalues of these operators respectively are [6.2] $w = (-1)^{l-m}$ and $I = (-1)^l$. With respect to the reflection symmetry, the states are classified as symmetrical $(w = +1)$ or antisymmetrical $(w = -1)$ whereas with respect to the inversion symmetry, they can have either an even parity ($I = 1$; gerade states) or an odd parity ($I = 1$; ungerade states). Since in the $Z_1 = Z_2$ case the zeroes of the function $Y_{n_\eta m}(\eta; R)$ when $R \to \infty$ are equal at both centres $(n_2 = n_2')$, and since the antisymmetric function $Y^{(a)}_{n_\eta m} \sim [f_{n_2 m}(\mu_1) - f_{n_2 m}(\mu_2)]$ has an additional zero, from the node conservation theorem, one gets

$$n_\eta^{\mathrm{s}} = 2\,n_2, \quad n_\eta^{\mathrm{a}} = 2\,n_2 + 1\,. \tag{6.32}$$

Analogous considerations lead to

$$n_\eta^{\mathrm{g}} = 2\,n_2 + [1 - (-1)^m]/2, \quad n_\eta^{\mathrm{u}} = 2\,n_2 + [1 + (-1)^m]/2\,. \tag{6.33}$$

Taking these relations into account together with (6.17, 18), one obtains the following correlation rules for the $(Z,\ \mathrm{e},\ Z)$ system:

$$\left.\begin{aligned} N &= n + n_2 \\ l &= m + 2\,n_2 \end{aligned}\right\} \quad \text{if } l \text{ and } m \text{ have the same parity,}$$

(6.34)

$$\left.\begin{aligned} N &= n + n_2 + 1 \\ l &= m + 2\,n_2 + 1 \end{aligned}\right\} \quad \text{if } l \text{ and } m \text{ have different parity.}$$

In the context of this and following chapters, we shall be, however, mostly interested in heteronuclear systems.

As can be seen from (6.30, 31, 34), except for the case $n_2 < n(Z_2 - Z_1)/Z_2$, $(Z_2 \neq Z_1)$, the principal quantum number N in the united atom limit is always higher than the principal quantum numbers n and n' in the separated atom limit. The effect of increasing the principal quantum number when going to the united atom limit is called the "state-promotion" effect. The higher the value of the parabolic quantum number n_2, the stronger the state-promotion effect is.

The molecular energies of the $(Z_1, e\, Z_2)$ system as $R \to 0$ and $R \to \infty$ can be obtained using ordinary perturbation theory [6.1–3]:

$R \to 0$,

$$E_{Nlm}(Z_1, Z_2, R) = -\frac{(Z_1 + Z_2)^2}{2N^2} - \frac{2 Z_1 Z_2 [l(l+1) - 3m^2](Z_1 + Z_2)^2}{N^3 l(l+1)(2l-1)(2l+1)(2l+3)} R^2 + O(R^3). \tag{6.35}$$

$R \to \infty$,

$$E_{nn_1n_2m}(Z_1, Z_2, R) = -\frac{Z_2^2}{2n^2} - \frac{Z_1}{R} + \frac{3 n Z_1 \Delta}{2 Z_2 R^2} + O(R^{-3}),$$

$$\Delta = n_1 - n_2, \tag{6.36a}$$

$$E_{n'n_1'n_2'm}(Z_1, Z_2, R) = E_{nn_1n_2m}(Z_1 \leftrightarrow Z_2, n \to n', \Delta \to \Delta'). \tag{6.36b}$$

For the quantity $p = R(-2E)^{1/2}/2$ at $R \to \infty$ one gets

$$p_{nn_1n_2m}(R) = \frac{Z_2 R}{2n} + \frac{n Z_1}{2 Z_2} - \frac{n^2 Z_1}{4 Z_2^3 R}(3 \Delta Z_2 + n Z_1) + O(R^{-2}), \tag{6.37a}$$

$$p_{n'n_1'n_2'm}(R) = p_{nn_1n_2m}(Z_1 \leftrightarrow Z_2, n \to n', \Delta \to \Delta'). \tag{6.37b}$$

The higher-order terms in the asymptotic expansions (6.36, 37) can be calculated using any of three methods: a modification of perturbation theory [6.4], use of the "comparison-equation" method [6.1, 5] and the continued-fraction expansion method [6.6]. These expansions are useful, since the numerical integration of the boundary-value problem (6.4–11) in the asymptotic region is rather difficult.

Besides the correlation rules (6.30, 31), the energies $E_j(R)$ as a function of R should also obey the von Neumann-Wigner non-crossing theorem [6.2, 7]. According to this theorem, the eigenenergies of states with the same symmetry cannot be degenerate at any finite value of the internuclear distance. At some particular distance R_c, the eigenenergies of such states can, however, closely approach each other. If the symmetry of these states is broken for any reason, then the eigenenergies can cross at R_c. Such crossing is called pseudocrossing, since by removing the cause of the symmetry-breaking, the crossing (or degeneracy) is lifted. The symmetry of diatomic molecular states is, in general, determined by the geometrical group of symmetries. In addition to this group of symmetries, the one-electron two-Coulomb-centre system also possesses a dynamical symmetry, connected with the possibility of separating the variables in prolate spheroidal coordinates [6.8, 9]. The operator associated with this symmetry is the so-called separation-constant operator, and is given by [6.9]

$$\Lambda = \boldsymbol{L}^2 + \frac{1}{4} R^2 \nabla^2 - \left[\left(\frac{1}{2} \boldsymbol{R} \nabla \right)^2 - \frac{Z_1}{r_1} \boldsymbol{R} \cdot \boldsymbol{r}_1 - \frac{Z_2}{r_2} \boldsymbol{R} \cdot \boldsymbol{r}_2 \right] . \tag{6.38}$$

Its eigenvalues are

$$\lambda^{(\xi)}_{n_\xi m}(p, a) = \lambda^{(\eta)}_{n_\eta m}(p, b) . \tag{6.39}$$

In (6.38) $\boldsymbol{L}$ is the electron angular momentum operator.

The operator Λ commutes both with H and L_z; therefore these three operators can be diagonalized simultaneously, and the wave functions in this representation are given by Eq. (6.3).

The existence of dynamical symmetry in the (Z_1, e, Z_2) system makes the von Neumann-Wigner theorem more restrictive. Namely, it now requires that states with the same eigenvalues of the operators L_z and Λ be non-degenerate for finite R. Two energies, $E_{n_\xi n_\eta m}(R)$ and $E_{n'_\xi n'_\eta m'}(R)$, of the (Z_1, e, Z_2) system may always cross if two quantum numbers of the sets $\{n_\xi\, n_\eta\, m\}$ and $\{n'_\xi\, n'_\eta\, m'\}$ are different. In particular, energies with the same values of m ($m = m'$, but $n_\xi \neq n'_\xi$, $n_\eta \neq n'_\eta$) may always cross, in contrast to the case of many-electron diatomic systems, where the geometrical symmetry is the only symmetry. A pseudocrossing (or avoided crossing) in the system (Z_1, e, Z_2) for real values of R occurs only when the following relations hold [6.5, 6]:

$$n_\xi = \eta'_\xi , \quad n_\eta = n'_\eta + 1 , \quad m = m' . \tag{6.40}$$

The pseudocrossings defined by (6.40) are connected with the one-dimensional motion along the η coordinate. A pseudocrossing of two molecular energies at the point R_c on the real R axis corresponds to a real crossing of these energies in the complex R-plane at some point R_p [6.10], such that $R_c = \mathrm{Re}\{R_p\}$. The singular point R_p is a branching point and for the pseudocrossings (6.40) it lies in the region of internuclear distances for which the electron energy is below the top of the barrier separating the Coulomb potential wells (underbarrier region).

The correlation rules (6.30, 31), the non-crossing theorem and (6.35, 36) which give $E_j(R)$ at $R \to 0$ and $R \to \infty$ allow one to construct qualitatively the behaviour of the molecular energies at all R. These so-called correlation

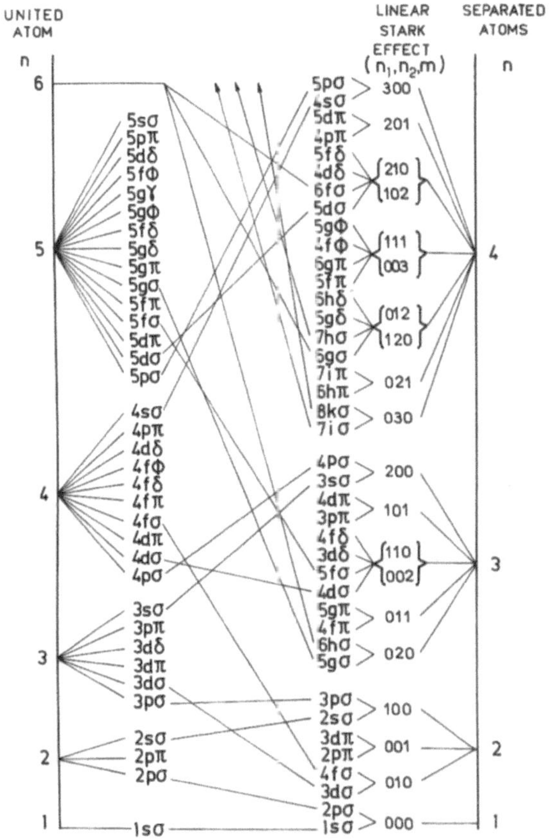

Fig. 6.1. Correlation diagram for the σ states of a homonuclear one-electron system

diagrams are frequently useful for giving a direct (albeit only qualitative) insight into the collision dynamics of the (Z_1, e, Z_2) system. The correlation diagram for the lower σ states of the H_2^+ system is shown in Fig. 6.1. The order of the energy levels at small and large R is determined by the linear Stark effect. The promotion effect is clearly seen in the figure.

If (6.4, 5) with the boundary conditions (6.7, 8) are solved in the complex R-plane, then besides the pseudocrossings defined by (6.40), one also finds pseudocrossings $\tilde{R}_p$ defined by the conditions [6.11]

$$n_\xi = n'_\xi + 1, \quad n_\eta = n'_\eta, \quad m = m'. \tag{6.41}$$

These pseudocrossings are associated with the electronic motion along the ξ coordinate. For small values of m, the pseudocrossings $\tilde{R}_p$ lie close to the real axis and form an infinite Rydberg series. The point $\tilde{R}_c = \mathrm{Re}\{\tilde{R}_p\}$ is connected with the radius of convergence of the energy expansion in the united atom limit (6.35). The analytical properties of the complex energies in the (Z_1, e, Z_2) system are still not well known. Therefore, in the rest of this chapter we shall be concerned only with real energies and pseudocrossings of the type (6.40).

6.1.3 Energy Splitting at a Pseudocrossing

The pseudocrossings of molecular energies have important implications for the collision dynamics (see below). Of special importance are the crossings at (relatively) large internuclear distances. If one uses the expansions (6.36) for the molecular energies, then independently of (6.40), one finds that the energies may always cross. This apparent violation of the von Neumann-Wigner theorem is a consequence of the fact that the asymptotic expansions (6.36) are obtained with an exponential accuracy, i.e. by assuming that the electron is dominantly localized around one of the centres and the influence of the other centre is accounted for by perturbation theory. If the electron is located predominantly around one of the centres, its wave function in the region of the other centre is exponentially small; its contribution to the energy is also exponentially small and can be neglected with respect to the multipole terms of the expansion. However, if there is a state around the other centre which is nearly degenerate with the first one, then the electron is described by a superposition of these two states and is shared between the centres. In that case the perturbational approach for the calculation of the molecular energy becomes incorrect, i.e. exponential corrections to the energy (having their origin in the delocalized character of the electronic motion) have to be retained. This is exactly the situation when two molecular states with the same symmetry have a pseudocrossing. The exponentially small terms (known as exchange terms) in the energy give rise to a splitting of the corresponding molecular energies and the crossing is avoided. The energy splitting in the pseudocrossing region is thus a measure of the electron exchange effects and is frequently called the exchange interaction. The delocalized character of the electronic motion in the pseudocrossing region is the physical basis for transitions between the pseudocrossing states.

The pseudocrossing point R_c of the molecular energies $E_{nn_1n_2m}(R)$ and $E_{n'n_1n_2'm}(R)$ [note that $n_1' = n_1$, see (6.27)], is defined by

$$E_{nn_1n_2m}(R_c) = E_{n'n_1n_2'm}(R_c) . \tag{6.42}$$

One can show that the energy at $R = R_c$ is given by [6.6]

$$E_c = \frac{1}{2}[E_{nn_1n_2m}(R_c) + E_{n'n_1n_2'm}(R_c)] = -\frac{1}{2}\left(\frac{Z_2 - Z_1}{n_2 - n_2'}\right)^2 . \tag{6.43}$$

Using in this equation only the first two terms of the expansions (6.36 a, b), one obtains the following expression for R_c:

$$R_c = (Z_1 + Z_2)\left[\left(\frac{Z_2 - Z_1}{n_2 - n_2'}\right)^2 - \frac{1}{2}\left(\frac{Z_2^2}{n^2} + \frac{Z_1^2}{n'^2}\right)\right]^{-1}. \tag{6.44}$$

The appearance of an exponentially small energy splitting in the region δR_c around R_c results from the existence of exponentially small terms in the eigenvalues $\lambda_{n_\eta m}^{(\eta)}(p, b)$. Physically, these terms reflect the tunnel transitions of the electron through the potential barrier in (6.5). The eigenvalues $\lambda_{n_\eta m}^{(\eta)}(p, b)$ and $\lambda_{n_\eta+1,m}^{(\eta)}(p, b)$ in the pseudocrossing region coincide in the zero-order approximation. In the first-order approximation they are split and have the values [6.5]

$$\lambda_{\substack{n_\eta m \\ \eta_\eta+1,m}}^{(\eta)} = \lambda_0^{(\eta)} \pm \frac{(4p)^{n_2+n_2'+m+2}\,e^{-2p}}{[n_2!\,(n_2+m)!\,n_2'(n_2'+m)!]^{1/2}}[1 + O(p^{-1})]. \tag{6.45}$$

On the basis of the symmetry properties of (6.4–6), it can be shown that the eigenvalues $\lambda_{n_\xi m}^{(\xi)}$ are connected with $\lambda_{n_\eta m}^{(\eta)}$ by the relation

$$\lambda_{n_\xi m}^{(\xi)}(p, a) = \lambda_{n_\xi m}^{(\eta)}(-p, -a). \tag{6.46}$$

Using (6.45, 46) in the equation $\lambda_{n_\xi m}^{(\xi)}(p, a) = \lambda_{n_\eta m}^{(\eta)}(p, b)$, one obtains the following expression for the energy splitting at the pseudocrossing point R_c [6.5]:

$$\Delta(R_c) = 4E_c\frac{(4p)^{n_2+n_2'+m+2}\,e^{-2p}}{[n_2!\,(n_2-m)!\,n_2'!\,(n_2'+m)!]^{1/2}}[1 + O(p^{-1})], \tag{6.47}$$

where E_c is given by (6.43) and

$$p = p_c = R_c(-E_c/2)^{1/2}. \tag{6.48}$$

The higher-order terms in the asymptotic expansion of the energy splitting (6.47) can also be calculated, and for some specific values of Z_1 and Z_2 are known [6.1]. In the symmetrical case ($Z_1 = Z_2$) the expression $\Delta(R)$ gives the energy splitting of the *gerade* and *ungerade* states, which become degenerate as $R \to \infty$.

The expression (6.47) is valid for the low-lying states of the (Z_1, e, Z_2) system. For the highly excited states, two large parameters exist in the problem (p and the principal quantum number) which implies that different expansions must be made for the solution of (6.5). The energy splitting in the pseudocrossing region has the form [6.12]

$$\Delta(R_c) = \left[\frac{Z_1^2}{n'^3}\left(1 + \frac{Z_1^2(n-n')^3}{n'^3(Z_2-Z_1)^2}\right)^{-1/2} + \frac{Z_2^2}{n^3}\left(1 - \frac{Z_2^2(n-n')^3}{n^3(Z_2-Z_1)^2}\right)^{-1/2}\right]\delta(R_c) \tag{6.49}$$

where

$$\delta(R_c) = \frac{1}{2\pi}\left(\frac{4pe}{n_2'+(m+1)/2}\right)^{n_2'+(m+1)/2}\left(\frac{4pe}{n_2+(m+1)/2}\right)^{n_2+(m+1)/2}$$
$$\times e^{-2p}[1+O(1/n_2')+O(1/n_2)+O(1/p)], \tag{6.50}$$

with $p = p_c$.

In the case $Z_2 \gg Z_1$ (and $n_2 \gg 1$) there are again two large parameters in the problem. The energy splitting can be obtained in the form of (6.49) with $\delta(R_c)$ given by [6.13]

$$\delta(R_c) = \frac{(2p)^{n_2+(m+1)/2}}{[2\pi n_2'!(n_2'+m)!]^{1/2}}\exp\{-[2^{1/2}-\ln(2^{1/2}-1)]p+2B\ln(2^{1/2}+1)$$
$$+5[n_2'+(m+1)/2](\ln 2)/2\}[1+O(1/n_2)+O(1/p)] \tag{6.51}$$

with

$$B = 2^{1/2}[7(n_1'-n_2')/8+1]$$

and $p = p_c$.

In connection with the collision processes of highly charged ions with atoms, the system $Z_1 = 1$, $Z_2 = Z \gg 1$ is of special importance. If the electron around the centre Z_1 (the proton) is in the ground state $[n'\,n_1'\,n_2'\,m] = [1000]$, then the non-crossing rules (6.40) imply that from all the $[n\,n_1\,n_2\,m]$ electron states around the centre Z_2 (which at $R \to \infty$ are degenerate), only the state $[n, 0, n-1, 0]$ will exhibit a pseudocrossing. Hence, each of the σ-molecular states, which at $R \to \infty$ correlate to different n states of the (e Z_2) system, exhibits a pseudocrossing with the molecular state which, at $R \to \infty$, correlates with the hydrogen-atom ground state. Using the asymptotic expansions (6.36) in (6.42) and retaining the linear Stark terms in them, one obtains the following value for the pseudocrossing point of the [1000] and $[n, 0, n-1, 0]$ molecular states:

$$R_{cn} = \frac{n^2(Z_1-1)}{Z^2-n^2}\left\{1+\left[1-\frac{3n(n-1)}{Z(Z-1)^2}\left(\frac{Z^2}{n^2}-1\right)\right]^{1/2}\right\}, \quad n < Z \tag{6.52}$$

$$\approx \frac{2n^2(Z-1)}{Z^2-n^2}, \quad Z \gg 1, \quad n < Z. \tag{6.52a}$$

An example of pseudocrossing potential energy curves is shown in Fig. 6.2 for the $(OH)^{8+}$ system.

For the system $(1, e, Z)$, the energy splitting at $R = R_{cn}$ of the σ-molecular energies, obtained from the numerical calculations of the eigenvalue problem for $4 \le Z \le 54$, can be fitted to the following analytical

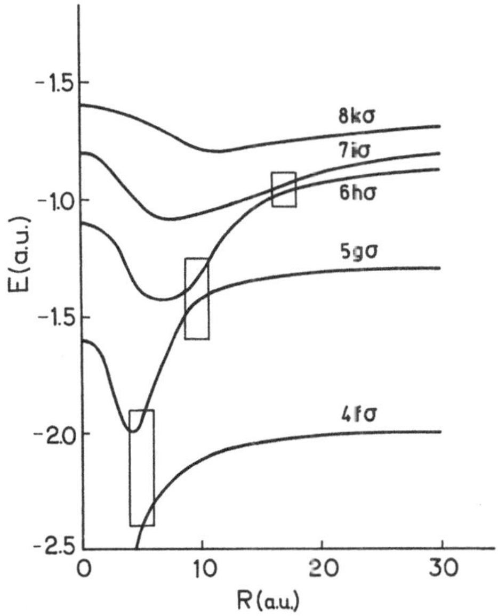

Fig. 6.2. Part of the potential energy diagram for the $(OH)^{8+}$ system. The pseudocrossing regions are indicated by boxes

expression [6.14] (within a 17% accuracy):

$$\Delta(R_{cn}) = 18.26\, Z^{-1/2} \exp(-1.324\, Z^{-1/2} R_{cn}) . \tag{6.53}$$

All the above expressions for $\Delta(R_c)$ are meaningful only for R_c which satisfy the condition $(Z_1 = 1, Z_2 = Z)$

$$R_c > R_0 = 2n' [(2n_1' + m + 1)(2n'Z - 2n_2' - m - 1)^{1/2} - n_1' + n_2'] , \tag{6.54}$$

where R_0 is the internuclear distance at which the energy level in (6.5) becomes equal to the top of the potential barrier in the η direction. For the ground [1000] state, $R_0 = 2(2Z - 1)^{1/2}$.

The number of pseudocrossings in the region $R > R_0$ is approximately given by [6.15]

$$r_0 = \mathrm{Ent}\,\{n'(x-1)[x + 1 - (1 + 2x)^{1/2}]\} , \quad x = (Z_2/Z_1)^{1/2} . \tag{6.55}$$

According to this estimate, the first pseudocrossing appears for $Z_2/Z_1 = 5$. With increasing x^2, the number of crossings increases rapidly.

In the case $Z_2 \gg Z_1$, the electronic motion under the potential barrier $(R > R_0)$ can be described within the quasi-classical approximation. In this case, the energy splitting can be expressed in terms of the quantum-mechanical penetrability of the potential barrier [6.16]. The calculation of the penetrability integral can be carried out analytically by expanding the quasi-classical momentum in terms of the small quantity R_0/R, where $R_0 = [2(Z_1 Z_2)^{1/2} + Z_1]/E_1$ and E_1 is the initial-state electron energy. R_0 has the same definition as in (6.54). However, it is now derived from the three-dimensional two-Coulomb-centre potential along the z direction. For the (Z_1, e, Z_2) system, the result of the quasi-classical method for the crossing of the initial $|100\rangle$ and final $|nlm\rangle$ states is [6.17]

$$\Delta_{nlm} = \delta_{m0}\, G_l(Z_1, Z_2)\, \Delta_{n00}(Z_1, Z_2, R) \tag{6.56}$$

where

$$\Delta_{n00}(Z_1, Z_2, R) = 2(2/\pi)^{1/2} Z_1^3 R\, n^{-3/2} \tag{6.57}$$

$$\times \exp\left\{- Z_1 R + \frac{2(Z_2/Z_1 - 1)}{(1 + 2Z_2/Z_1^2 R)^{1/2}} \ln\left[\left(\frac{Z_1^2 R}{2Z_2}\right)^{1/2} + \left(1 + \frac{Z_1^2 R}{2Z_2}\right)^{1/2}\right]\right\},$$

$$G_l(Z_1, Z_2) = (2l+1)^{1/2} \exp[-l(l+1)Z_1/2Z_2]\,. \tag{6.58}$$

In two limiting situations (6.57) can be simplified. For $(2Z_2/Z_1)^{1/2} \ll Z_1 R \ll 2Z_2/Z_1$, one has

$$\Delta_{n00}(Z_1, Z_2, R) = 2\left(\frac{2}{\pi}\right)^{1/2} \frac{Z_1 R}{n^{3/2}} \exp\left(-\frac{Z_1^3 R^2}{3Z_2}\right), \tag{6.59}$$

and for $Z_1 R \gg 2Z_2/Z_1$, the result is

$$\Delta_{n00}(Z_1, Z_2, R) = \frac{2^{Z_1/Z_2+1}}{(Z_2/Z_1)\,\Gamma(Z_2/Z_1)}\, e^{-Z_2/Z_1} R^{Z_2/Z_1} \exp(-Z_1 R)\,. \tag{6.60}$$

The above result can be generalized also to the case of an arbitrary atom – partially (or fully) stripped ion system [6.17, 18].

6.1.4 Algorithms for Solving the Eigenvalue Problem for the (Z_1, e, Z_2) System

The main feature of the eigenvalue problem for the (Z_1, e, Z_2) system is that the Sturm-Liouville equations, (6.4, 7) and (6.5, 8), have to be solved simultaneously. In the numerical treatment of the boundary-value problems, the following expansions are used [6.1]:

$$X_{n_\xi m}(p, a; \xi) = (\xi^2 - 1)^{m/2}\, e^{-p(\xi-1)} f(\xi)\,, \tag{6.61}$$

$$f(\xi) = \sum_{k=0}^{\infty} s_k\, L_{k+m}^m [2p(\xi - 1)]\,, \tag{6.61 a}$$

$$Y_{n_\eta m}(p, b; \eta) = e^{-p(1+\eta)}\, g^{(\pm)}(\eta)\,, \tag{6.62}$$

$$g^{(\pm)}(\eta) = \sum_{k=0}^{\infty} t_k^{(\pm)}\, P_{k+m}^m(\eta)\,, \tag{6.62 a}$$

where $\mathrm{L}^m_{k+m}(x)$ and $\mathrm{P}^m_{k+m}(x)$ are, respectively, the associated Laguerre and Legendre polynomials. The signs $\pm$ in (6.62, 62a) correspond to the expansions around the centres $\eta = \mp 1$. By inserting (6.61) into (6.4) and (6.62) into (6.5), one obtains a three-term recurrence relation for the coefficients s_k, $t_k^{(+)}$ and $t_k^{(-)}$:

$$\alpha_k r_{k+1} - \beta_k r_k + \gamma_k r_{k-1} = 0 \tag{6.63}$$

where r_k denotes any of s_k and $t_k^{(\pm)}$, and the constants α_k, β_k and γ_k are expressed in terms of p and λ. In matrix form, (6.63) can be expressed as

$$\underline{A}\,\underline{r} = \begin{Vmatrix} -\beta_0 & \alpha_0 & 0 & 0 & \dots \\ \gamma_1 & -\beta_1 & \alpha_1 & 0 & \dots \\ 0 & \gamma_2 & -\beta_2 & \alpha_2 & \dots \\ \cdot & \cdot & \cdot & \cdot & \dots \end{Vmatrix} \cdot \begin{Vmatrix} r_0 \\ r_1 \\ r_2 \\ \vdots \end{Vmatrix} = 0\,. \tag{6.64}$$

In order to terminate the expansions (6.61, 62) for negative k values, one has to set

$$r_{-1} = 0\,. \tag{6.65}$$

It can be shown [6.1, 6] that the eigenvalues of the boundary-value problems (6.4, 7) and (6.5, 8) can be determined from the condition that a solution of (6.64) exists,

$$\det \underline{A} = \mathscr{A}(p, \lambda) = 0\,. \tag{6.66}$$

An important practical feature of this approach to the eigenvalue problems (6.4–7) and (6.5–8) is that for a three-term recurrence relation, the determinant $\mathscr{A}(p, \lambda)$ can be represented as an equivalent continued fraction [6.19],

$$\mathscr{A}(p, \lambda) = \beta_0 - \cfrac{\alpha_0 \gamma_1}{\beta_1 - \cfrac{\alpha_1 \gamma_2}{\beta_2 - \dots}}\,. \tag{6.67}$$

$\mathscr{A}(p, \lambda)$ is convergent if the condition $|\alpha_{k-1} \gamma_k / \beta_{k-1} \beta_k| < 1/4$ is satisfied.

If $\mathscr{A}^{(\xi)}_{N_1}(p, \lambda)$ and $\mathscr{A}^{(\eta)}_{N_2}(p, \lambda)$ are the continued fractions (containing $N_1 - 1$ and $N_2 - 1$ terms, respectively) for the boundary-value problems (6.4–7) and (6.5–8), then the common roots of the equations

$$\mathscr{A}^{(\xi)}_{N_1}(p, \lambda) = 0\,, \quad \mathscr{A}^{(\eta)}_{N_2}(p, \lambda) = 0 \tag{6.68}$$

determine the values $p(R)$ and $\lambda(R)$ of the spectral parameters p and λ for any value of R and any set of the quantum numbers $\{n_\xi, n_\eta, m\}$. The numbers N_1 and N_2 are chosen so that the required accuracy in the energy $E_j(R)$ is attained.

Instead of using the expansion (6.61 a) for $f(\xi)$, one can take the Jaffé expansion [6.20]

$$f(\xi) = (\xi+1)^\sigma \sum_{k=0}^{\infty} s_k \left(\frac{\xi-1}{\xi+1}\right)^k, \quad \sigma = \frac{a}{2p} - m - 1\,, \tag{6.69 a}$$

which also leads to a three-term recurrence relation and provides better stability in the computational procedure.

The constants α_k, β_k, γ_k, in the three-term recurrence relation (6.63) have, for the expansions (6.61, 69 a) and (6.62, 62 a), respectively, the form

$$\begin{aligned} \alpha_k^{(\xi)} &= (k+1)(k+m+1)\,, \\ \beta_k^{(\xi)} &= 2k(k+2p-\sigma) - (m+\sigma)(m+1) - 2p\sigma + \lambda\,, \\ \gamma_k^{(\xi)} &= (k-1-\sigma)(k-1-m-\sigma)\,, \end{aligned} \tag{6.69 b}$$

$$\begin{aligned} \alpha_k^{(\eta)} &= (k+2m+1)[b-2p(k+m+1)]/(2k+2m+3), \\ \beta_k^{(\eta)} &= (k+m)(k+m+1) - \lambda\,, \\ \gamma_k^{(\eta)} &= k[b+2p(k+m)]/(2k+2m-1)\,. \end{aligned} \tag{6.69 c}$$

Details of numerical algorithms for solving the eigenvalue problem for the discrete spectrum of the two-Coulomb-centre problem can be found elsewhere [6.21–24].

6.1.5 Application of the Two-Centre Problem to Many-Electron Diatomic Molecules: Molecular Orbital Model

The one-electron two-Coulomb-centre problem serves as a basis for the independent-electron molecular orbital (MO) model of many-electron diatomic systems [6.25]. The MO model is particularly useful in describing the motion of inner-shell electrons, which is dominated by the nuclear Coulomb fields. Within the MO model, the presence of the other electrons is taken into account through their screening effects. These screening effects destroy the dynamical symmetry of the two-Coulomb-centre problem, as well as the degeneracy of the single-electron states in the separated atom limit.

For a fixed internuclear distance R, a molecular orbital is determined by the single-electron Schrödinger equation

$$h_k\,\varphi_k(\boldsymbol{r};\boldsymbol{R}) = \varepsilon_k(R)\,\varphi_k(\boldsymbol{r};\boldsymbol{R})\,, \tag{6.70}$$

$$h_k = -\nabla_k^2/2 + v_k(\boldsymbol{r}_1, \boldsymbol{r}_2; \boldsymbol{R}) \tag{6.71}$$

where $v_k(\boldsymbol{r}_1, \boldsymbol{r}_2; \boldsymbol{R})$ is the effective two-centre potential acting on the electron.

The best single-particle two-centre potential v_k is generated by the Hartree-Fock method. In this case v_k has a non-local character. The corresponding many-electron wave function is represented by a Slater-type determinant, in which the MO's φ_k are varied so as to minimize the total molecular energy. The resulting Hartree-Fock MO's are adiabatic in character, i.e. the corresponding MO energies $\varepsilon_k(R)$ obey the von Neumann-Wigner non-crossing rule.

Since in many-electron diatomic systems the separation constant λ is no longer conserved, the non-crossing rule is formulated with respect to the geometrical (and, in the case of homonuclear systems, the inversion) symmetry only. Therefore, the number of pseudocrossings in many-electron diatomic systems is much larger than in the (Z_1, e, Z_2) system.

Another approach to the determination of the effective two-centre potential $v_k(\boldsymbol{r}_1, \boldsymbol{r}_2; \boldsymbol{R})$ is to generalize the Thomas-Fermi model to diatomic systems [6.26]. In that case, $v_k(\boldsymbol{r}_1, \boldsymbol{r}_2; \boldsymbol{R})$ can be represented in the form

$$v_k = \tilde{v}_{k_1}(r_1; R) + \tilde{v}_{k_2}(r_2; R) \qquad (6.72)$$

where

$$\tilde{v}_{k_j}(r_j; R) = -(Z_j/r_j)\, \chi_j[r_j; \alpha_j(r_j; R)] \qquad (6.73)$$

where $\chi_j(r_j; \alpha_j)$ is the generalized Thomas-Fermi screening function, which in the united atom (UA) and separated atom (SA) limits goes over into the corresponding atomic Thomas-Fermi screening functions. The interpolation between the UA and SA limits is provided by a suitable choice of a set of variable screening parameters α_j. *Eichler* and *Wille* [6.27] have proposed the following form of $\alpha_j(r_j, R)$:

$$\alpha_j = [a^2 \alpha_j^{SA} + (r_j/R)^2 \alpha_j^{UA}]/[a^2 + (r_j/R)^2], \quad j = 1, 2 \qquad (6.74)$$

where a^2 is roughly equal to unity and α_j^{SA}, α_j^{UA} are parameters which provide correct asymptotic (at $r_j \to \infty$) behaviour of the atomic Thomas-Fermi screening function [6.28]. The above variable-screening model for $v_k(\boldsymbol{r}_1, \boldsymbol{r}_2; \boldsymbol{R})$ is able to provide adiabatic MO energies in close agreement with the Hartree-Fock method [6.27]. For diatomic systems with significantly different nuclear charges, the variable-screening model possesses advantages with respect to the Hartree-Fock method, since it is able to furnish MO's with the asymptotically correct behaviour. This aspect is very important from the point of view of collision processes involving highly charged ions.

In connection with the identification of the transition mechanisms and the asymptotic states in the colliding many-electron diatomic systems, the construction of correct correlation diagrams for these systems is an important issue. In a collision process, the molecular orbital tends to conserve its character [i.e. model structure, smoothness of $\varepsilon_k(R)$, etc.], which depends on the ratio of the variation of the adiabatic MO energy and the collision velocity. The adiabatic interactions (giving rise to pseudo-crossings), usually arise from small terms in the electronic Hamiltonian and, for considerable collision energies, can be neglected. The pseudocrossings then become true crossings and the corresponding MO's are called *diabatic.* The molecular states, built up from diabatic MO's are slowly varying functions of R and conserve their character throughout when R varies. For many collisional problems these states provide a much more adequate description of the physical situation than the adiabatic states which diagonalize the electronic Hamiltonian. Although the concept of diabatic states was introduced years ago [6.29, 30], the diabatic MO's, arising from the two-centre Coulomb problem, became a powerful means of analysis and description of ion-atom collision processes only after the work of *Lichten* [6.31].

An important issue in this context is the construction of correct diabatic correlation diagrams for the colliding system. For many-electron homonuclear systems, *Fano* and *Lichten* [6.32] have proposed rules for constructing diabatic correlation diagrams in analogy with the correlation rules for the H_2^+ system, i.e. allowing MO energies $\varepsilon_k(r)$, belonging to states with the same geometrical symmetry, to cross. *Barat* and *Lichten* [6.33] have extended the correlation rules for non-symmetrical (Z_1, e, Z_2) systems to many-electron heteronuclear diatomic systems. However, since in the separated atom limit the single-electron state is not described by parabolic but rather by spherical quantum numbers, the correlation rules are reduced to the requirement that only the nodal structure of the radial part of the electronic wave function be conserved. Thus, the diabatic correlation of the levels in the united and separated atom limits is subject to the rule

$$n_r^{\mathrm{UA}} = n_r^{\mathrm{SA}}, \tag{6.75}$$

where n_r is the radial quantum number $(n_r^{\mathrm{UA}} = N - l - 1,\ n_r^{\mathrm{SA}} = n_{at} - l_{at} - 1)$.

Systematic investigations of the adiabatic potential energy curves $\varepsilon_j(R)$, generated both with the Hartree-Fock and with the variable-screening models for different diatomic systems [6.34], have shown that the correlation rule (6.75) is a correct prescription only for slightly charge-asymmetric systems, and when the subshell splitting of the SA levels due to screening effects is large compared to their Stark splitting. For strongly asymmetric systems, when the Stark splitting is larger than the splitting due to the

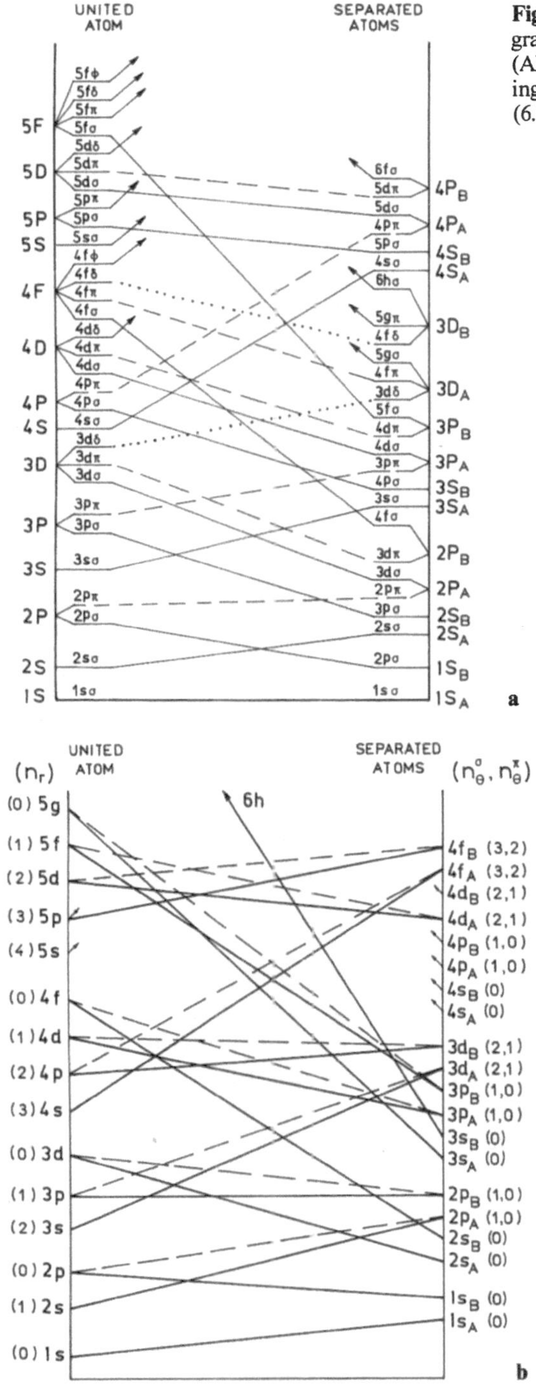

Fig. 6.3 a, b. Diabatic correlation diagrams for a slightly asymmetric $(AB)^{Z+}$ molecular system: (**a**) according to (6.75) [6.33], (**b**) according to (6.76) [6.34]

screening effects, a better diabatic correlation is provided if the following rule is postulated [6.34]:

$$n_r^{\mathrm{UA}} = n_\theta^{\mathrm{SA}} , \tag{6.76}$$

where $n_\theta^{\mathrm{SA}} = (l-m)_{\mathrm{SA}}$ is the number of angular nodes of the wave function corresponding to the polar angle θ. For a given SA principal shell, the rule (6.75) leads to a minimal promotion of the MO's belonging to different subshells, whereas the diabatic correlation rule (6.76) leads to their maximal promotion. In Figs. 6.3a, b, the diabatic correlation diagrams for a slightly asymmetric system, constructed respectively according to the rules (6.75, 76) are shown.

The correlation rules (6.75, 76) have not been rigorously proved. Their validity depends on the charge asymmetry, collision velocity, strength of the spin-orbit and other relativistic perturbations in heavy atomic systems. It is especially difficult to establish the diabatic correlation in the case of a Rosenthal pseudocrossing [6.35], when one cannot decide whether the pseudocrossing is diabatic or adiabatic. For the outer electronic shells of a many-electron diatomic system, the construction of diabatic states and the corresponding correlation diagrams become even more difficult and uncertain. These problems are discussed in more detail elsewhere [6.36–38].

6.2 Close-Coupling Methods

Among all the general theoretical methods of atomic collision theory, the close-coupling methods provide, at least in principle, the most accurate and physically the most transparent description of collision dynamics [6.39–41]. The close-coupling methods are based on an expansion of the total electron wave function in terms of some complete set of functions, and transformation of the Schrödinger equation into an infinite system of coupled integro-differential or first-order differential equations for the expansion coefficients. Different physical arguments are used in the choice of an adequate basis of finite size, which reduces the number of coupled equations and makes the problem numerically tractable. The close-coupling methods among themselves differ mainly in the type of restricted basis set used in the calculations.

The collision problem, in terms of a system of coupled equations, can be formulated either in a full quantum-mechanical way (i.e. describing both the electronic and nuclear motion quantum-mechanically), or by using the semi-classical approximation (i.e. using classical mechanics to describe the relative nuclear motion). The semi-classical description is justified only when the wave number of the nuclear motion is much greater than unity which is fulfilled when the relative velocity of the nuclei is greater than $\sim 10^{-2} v_0$, v_0 being the classical velocity of the electron in its initial state. For $v_0 = 1$ a.u. (ground-state hydrogen target atom) this condition is satisfied for collision energies above ~ 10 eV/amu (amu = atomic mass unit). Since most of the theoretical and experimental investigations of the collision processes involving multiply charged ions have been performed in this energy region, our presentation of the close-coupling methods will be confined to the semi-classical approximation. In this approximation, the relative nuclear motion is described by the classical trajectory $\boldsymbol{R} = \boldsymbol{R}(t)$, where $\boldsymbol{R}$ is the vector of the internuclear distance.

6.2.1 General Formulation of the Close-Coupling Methods

Within the semi-classical approximation, the collision problem in the system $A + B^{Z+}$ is described by the time-dependent Schrödinger equation

$$(H - \mathrm{i}\,\partial/\partial t)\,\Psi = 0\,, \tag{6.77}$$

where H and Ψ are the electronic Hamiltonian and wave function. Let us expand Ψ in terms of a complete basis set $\{\chi_k\}$:

$$\Psi = \sum_k a_k(t)\,\chi_k \tag{6.78}$$

where χ_k are linearly independent functions (normalized to unity) which describe the electronic motion. By projecting (6.77) with the functions χ_j, one readily obtains coupled equations for the amplitudes $a_k(t)$. Using a matrix notation, the coupled equation can be written as

$$\mathrm{i}\,\underline{S} \cdot \dot{\underline{a}} = \underline{M} \cdot \underline{a} \tag{6.79}$$

where $\underline{a}$ is a column vector which consists of the coefficients $a_k(t)$, $\underline{S}$ is the overlap matrix with elements

$$S_{kj} = \langle \chi_j \mid \chi_k \rangle\,, \tag{6.80}$$

and $\underline{M}$ is the coupling matrix with elements

$$M_{kj} = \langle \chi_j \mid H - \mathrm{i}\,\partial/\partial t \mid \chi_k \rangle\,. \tag{6.81}$$

For an arbitrary choice of the basis functions χ_k, the coupling matrix contains two terms: a "potential" coupling $\langle \chi_j | \, H \, | \chi_k \rangle$, and a "dynamic" coupling $\langle \chi_j | -\mathrm{i}\,\partial/\partial t\,| \chi_k \rangle$. These two couplings induce transitions between the basis states.

In general, the overlap matrix $\underline{S}$ is not diagonal and, consequently, the coupling matrix $\underline{M}$ is not a Hermitian matrix. In such a situation, the condition for conservation of probability (or the unitarity condition) is expressed by [6.42]

$$(d/dt)\,\langle \Psi | \Psi \rangle = 0\,, \tag{6.82}$$

which reduces to

$$\underline{M} - \underline{M}^{\dagger} + \mathrm{i}\,(d/dt)\,\underline{S} = 0\,. \tag{6.83}$$

Thus, $\underline{M}$ is Hermitian ($\underline{M} = \underline{M}^{\dagger}$) if and only if the S_{kj} do not depend on time. If the basis functions χ_k are chosen in such a way that Ψ satisfies correct asymptotic (i.e. for $|t| \to \infty$) boundary conditions, and if the initial condition for the problem is

$$a_k(-\infty) = \delta_{k0}\,, \tag{6.84}$$

then the probability for the $0 \to n$ electronic transition is given by

$$P_n(\varrho) = |\,a_n(+\infty)\,|^2 \tag{6.85}$$

where ϱ is the classical impact parameter. The cross section for the process is then

$$\sigma_n = \int P_n\, d^2\varrho = 2\pi \int_0^{\infty} P_n(\varrho)\,\varrho\, d\varrho\,. \tag{6.86}$$

Integration of the system of coupled equations (6.79) requires that the trajectory $\boldsymbol{R} = \boldsymbol{R}(\varrho, v, t)$ of the relative nuclear motion be known for each of the reaction channels. If the collision energy is sufficiently higher than the internal energy changes in the system during the collision, one can use a unique trajectory for all the channels, determined by some average potential. It is often sufficient to employ a screened Coulomb potential to determine the common classical trajectory. At a still higher collision energy (e.g. above ~ 500 eV/amu), one can use a straight-line trajectory ($R^2 = \varrho^2 + v^2 t^2$) to describe the nuclear motion. In this context, it should be noted that the internuclear potential, apart from its role in determining the nuclear trajectory at low energies, does not affect the calculation of the total cross section (6.86), since it can be removed from (6.77) by a phase transformation of the total wave function. It is, however, important in the calculations of the differential cross sections, since these also depend on the phase of the amplitudes $a_k(t)$ [6.41].

For a complete basis $\{\chi_k\}$, the system of coupled equations (6.79) is equivalent to the Schrödinger equation (6.77). For the atom-multicharged ion system, $\{\chi_k\}$ contains an infinite number of terms (including the discrete and continuum spectrum). In order to make the problem tractable, i.e. to reduce the number of coupled equations (6.79) to a finite size, one has to include in the basis set $\{\chi_k\}$ only those states which are important for the particular process. The truncation of the basis makes the choice of the basis functions χ_k critical. Two general criteria should be satisfied when choosing the basis functions. First, χ_k should preserve (I) the *character* of the electronic motion during the collision, and (II) the Galilean invariance of the system of coupled equations. Condition (I) leads to minimization of the coupling between the basis states (which is important in the practical integration of coupled equations), whereas condition (II) provides fulfillment of the proper boundary conditions for the problem (and related to this, the independence of the results from the choice of reference frame). These two conditions are mutually independent, which allows χ_k to be represented in the form

$$\chi_k = F_k \, \psi_k \, , \tag{6.37}$$

where ψ_k is the electron function in the proper sense (i.e. eigenfunction of some stationary Hamiltonian), and F_k is a function which provides fulfillment of the Galilean invariance. In physical terms, the function F_k has to describe the translational motion of the bound electron when the internuclear distance varies. Therefore, in general, F_k has to be a phase function, $F_k = \exp(\mathrm{i}\,\alpha_k)$, called the *electron translation factor* (ETF).

The character of the functions ψ_k is determined by the ratio v/v_0, where v is the collision velocity and v_0 is the classical velocity of the bound electron. If the collision is slow ($v \ll v_0$), an appropriate choice for the ψ_k is the one-electron molecular orbitals. If the collision is fast ($v \gtrsim v_0$), the electron is bound around one of the nuclei for most of the duration of the collision and the atomic states provide an adequate representation for ψ_k. These two cases will be separately considered below.

6.2.2 Molecular-Orbital Close-Coupling (MO-CC) Methods

In the context of slow ion-atom collisions, two basic representations may be used for the electronic states of the quasi-molecular complex: adiabatic and diabatic. The electronic wave functions in the adiabatic representation diagonalize the electronic Hamiltonian H, while in the diabatic representation they are not eigenfunctions of H. In describing the collision dynamics, the adiabatic representation is more adequate at very low collision velocities ($v \ll v_0$), while at higher velocities ($v \lesssim v_0$) the diabatic representation is more appropriate. The structure of the coupled equations obtained from the expansion of the electron wave function in terms of adiabatic or diabatic basis states is different in these two cases. The couplings which induce transitions between the electronic states are also different.

a) Adiabatic Molecular Basis

The molecular orbitals in this basis are given by

$$\psi_k^{\mathrm{MO}}(\boldsymbol{r};\boldsymbol{R}) = \Phi_k(\boldsymbol{r};\boldsymbol{R}) \exp\left[-\mathrm{i}\int_{-\infty}^{t} E_k(R)\,dt'\right] \tag{6.88}$$

where $\Phi_k(\boldsymbol{r};\boldsymbol{R})$ and $E_k(R)$ are, respectively, the eigenfunctions and the eigenenergies of the electronic Hamiltonian H in the Born-Oppenheimer approximation,

$$H\,\Phi_k(\boldsymbol{r};\boldsymbol{R}) = E_k(R)\,\Phi_k(\boldsymbol{r};\boldsymbol{R}) \tag{6.89}$$

and $\boldsymbol{r}$ is the position vector with respect to the origin of a space-fixed reference frame placed on the internuclear line. The adiabatic functions Φ_k form an orthonormal set. Let us introduce two classes of orbitals ψ_k^{MO}, according to their asymptotic behaviour:

$$\psi_{k,\mathrm{A}}^{\mathrm{MO}}(\boldsymbol{r};\boldsymbol{R}) \xrightarrow[R\to\infty]{} \phi_k^{\mathrm{A}}(\boldsymbol{r}_{\mathrm{A}}) \exp(-\mathrm{i}\,\varepsilon_k^{\mathrm{A}}\,t)\,, \tag{6.90 a}$$

$$\psi_{k,\mathrm{B}}^{\mathrm{MO}}(\boldsymbol{r};\boldsymbol{R}) \xrightarrow[R\to\infty]{} \phi_k^{\mathrm{B}}(\boldsymbol{r}_{\mathrm{B}}) \exp(-\mathrm{i}\,\varepsilon_k^{\mathrm{B}}\,t)\,, \tag{6.90 b}$$

where $\phi_k^{\lambda}(\boldsymbol{r}_{\lambda})$ ($\lambda = \mathrm{A}, \mathrm{B}$) is the atomic orbital centred on the nucleus λ, and ε_k^{λ} is the corresponding energy. In the case of a homonuclear system ($\varepsilon_k^{\mathrm{A}} = \varepsilon_k^{\mathrm{B}}$), (6.90 a, b) should be replaced by

$$2^{-1/2}\left(\psi_{k,\mathrm{g}}^{\mathrm{MO}} \pm \psi_{k,\mathrm{u}}^{\mathrm{MO}}\right) \xrightarrow[R\to\infty]{} \phi_k^{\lambda}(\boldsymbol{r}_{\lambda}) \exp(-\mathrm{i}\,\varepsilon_k^{\lambda}\,t) \tag{6.90 c}$$

where g and u designate the MO's with an even (*gerade*) and odd (*ungerade*) symmetry, respectively, and λ is either A or B. With the adiabatic functions $\psi_{k,\mathrm{A}}^{\mathrm{MO}}$ and $\psi_{k,\mathrm{B}}^{\mathrm{MO}}$, the following electron translation factors are associated:

$$F_k^{\mathrm{A}}(\boldsymbol{r},\boldsymbol{R}) = \exp\left[\mathrm{i}\,\alpha_k(\boldsymbol{r},\boldsymbol{R})\right]\,, \tag{6.91 a}$$

$$F_k^{\mathrm{B}}(\boldsymbol{r},\boldsymbol{R}) = \exp\left[\mathrm{i}\,\beta_k(\boldsymbol{r},\boldsymbol{R})\right]\,. \tag{6.91 b}$$

In order to determine the form of the functions $\alpha_k(\boldsymbol{r},\boldsymbol{R})$ and $\beta_k(\boldsymbol{r},\boldsymbol{R})$, let us assume that the origin of the space-fixed coordinate system (x, y, z) divides the internuclear axis in the ratio $p:q$. We orient the z axis along the

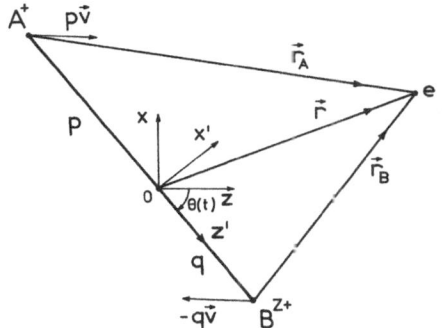

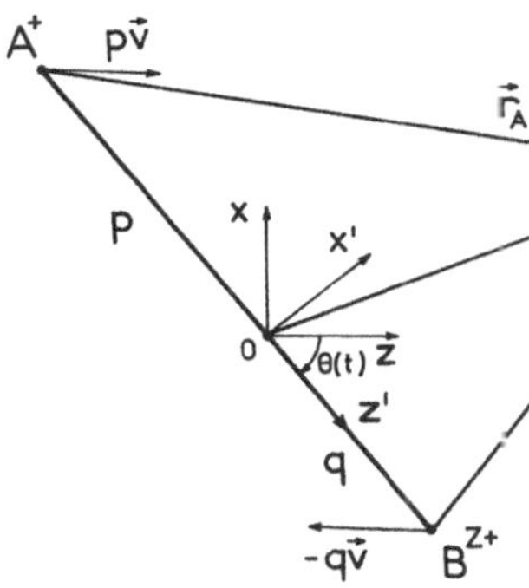

Fig. 6.4. Space-fixed (x, y, z) and rotating (x', y', z') coordinate systems for analysis of the $(A^+, e) + B^{Z+}$ collision

vector $\boldsymbol{v}$ of the relative collision velocity and define the collision plane as the (x, z) plane (see Fig. 6.4). We shall also assume that the trajectory is rectilinear: $\boldsymbol{R} = \boldsymbol{\varrho} + \boldsymbol{v}\, t$. If $p\, \boldsymbol{v}$ and $-q\, \boldsymbol{v}$ are the velocities of the nuclei of A and B^{Z+} at $R \to \infty$, then the condition that the functions $\chi_{k,A}(\boldsymbol{r}; \boldsymbol{R})$ and $\chi_{k,B}(\boldsymbol{r}; \boldsymbol{R})$ be translationally invariant identifies the functions F_k^A and F_k^B as the ordinary Galilean transforms

$$F_k^A(\boldsymbol{r}, \boldsymbol{R}) = \exp(\mathrm{i}\, p\, \boldsymbol{v} \cdot \boldsymbol{r} - \mathrm{i}\, p^2 v^2 t/2)\,, \tag{6.92a}$$

$$F_k^B(\boldsymbol{r}, \boldsymbol{R}) = \exp(-\mathrm{i}\, q\, \boldsymbol{v} \cdot \boldsymbol{r} - \mathrm{i}\, q^2 v^2 t^2/2)\,. \tag{6.92b}$$

The plane-wave translational factors (6.92) have been introduced by *Bates* and *McCarroll* [6.43]. These authors were the first to point out that the MO-CC equations do not satisfy translational invariance if constructed on the basis $\{\psi_k\}$ only. In calculating the matrix elements M_{kj} with the functions $\chi_k = F_k\, \psi_k^{MO}$, one has to take into account the orthogonality of Φ_k and the relations

$$\begin{aligned}(-1/2)\nabla_r^2\, \phi_k^A \exp(\mathrm{i}\, p\, \boldsymbol{v} \cdot \boldsymbol{r})\\ = \exp(\mathrm{i}\, p\, \boldsymbol{v} \cdot \boldsymbol{r})\, (-\nabla_r^2/2 - \mathrm{i}\, p\, \boldsymbol{v} \cdot \nabla_r + p^2 v^2/2)\, \phi_k^A\,,\end{aligned} \tag{6.93a}$$

$$\partial/\partial t_r = \partial/\partial t_{r_A} + p\, \boldsymbol{v} \cdot \nabla_r|_R\,, \quad \partial/\partial t_r = \partial/\partial t_{r_B} - q\, \boldsymbol{v} \cdot \nabla_r|_R \tag{6.93b}$$

where the subscript in the differentiation operators indicates that the corresponding radius vector is held fixed. The overlap and the coupling matrix elements for the states $\chi_{kA} = F_k^A\, \psi_{k,A}^{MO}$, separating at $R \to \infty$ to the atom A, are given by

$$S_{kj}^{AA} = \langle \chi_{j,A} \,|\, \chi_{k,A} \rangle = \delta_{kj}\,, \tag{6.94}$$

$$\begin{aligned}M_{kj}^{AA} &= \langle \Phi_j^A |\, (-\mathrm{i}\, \partial/\partial t_r - \mathrm{i}\, p\, \boldsymbol{v} \cdot \nabla_r)\, | \Phi_k^A \rangle \exp\left[-\mathrm{i} \int_{-\infty}^{t} (E_k^A - E_j^A)\, dt'\right]\\ &= \langle \Phi_j^A |\, (-\mathrm{i}\, \partial/\partial t_{r_A})\, | \Phi_k^A \rangle \exp\left[-\mathrm{i} \int_{-\infty}^{t} (E_k^A - E_j^A)\, dt'\right].\end{aligned} \tag{6.95}$$

For the states $\chi_{k,\mathrm{B}}$, separating at $R \to \infty$ to the ion B^{Z+}, the overlap integral S_{kj}^{BB} and the "direct" matrix elements M_{kj}^{BB} are obtained from S_{kj}^{AA} and M_{kj}^{AA} by the replacement $\mathrm{A} \to \mathrm{B}$, $p \to -q$. From the expression (6.95) for M_{kj}^{AA} it can be seen that the ETF has the effect of translating the coordinate origin to the nucleus A, so that the boundary condition at $R \to \infty$ is trivially satisfied. Analogous conclusions are valid for the matrix elements M_{kj}^{BB}.

The overlap integral and the "exchange" coupling matrix elements for the states separating to different nuclei are given by

$$S_{kj}^{\mathrm{AB}} = \langle \Phi_j^{\mathrm{B}} | \exp(\mathrm{i}\, \boldsymbol{v} \cdot \boldsymbol{r}) | \Phi_k^{\mathrm{A}} \rangle \times \exp\left[-\mathrm{i}\,(p^2 - q^2)\, v^2\, t/2 - \mathrm{i} \int_{-\infty}^{t} (E_k^{\mathrm{A}} - E_j^{\mathrm{B}})\, dt' \right], \tag{6.96}$$

$$M_{kj}^{\mathrm{AB}} = \langle \Phi_j^{\mathrm{B}} | \exp(\mathrm{i}\, \boldsymbol{v} \cdot \boldsymbol{r})\, (-\mathrm{i}\, \partial/\partial t_{r_{\mathrm{A}}}) | \Phi_k^{\mathrm{A}} \rangle \times \exp\left[-\mathrm{i}\,(p^2 - q^2)\, v^2\, t/2 - \mathrm{i} \int_{-\infty}^{t} (E_k^{\mathrm{A}} - E_j^{\mathrm{B}})\, dt' \right], \tag{6.97a}$$

$$M_{kj}^{\mathrm{BA}} = \langle \Phi_j^{\mathrm{A}} | \exp(-\mathrm{i}\, \boldsymbol{v} \cdot \boldsymbol{r})\, (-\mathrm{i}\, \partial/\partial t_{r_{\mathrm{B}}}) | \Phi_k^{\mathrm{B}} \rangle \times \exp\left[\mathrm{i}\,(p^2 - q^2)\, v^2\, t/2 - \mathrm{i} \int_{-\infty}^{t} (E_k^{\mathrm{B}} - E_j^{\mathrm{A}})\, dt' \right]. \tag{6.97b}$$

For the direct transitions, S_{kj}^{AA} is diagonal and, consequently, the matrix M_{kj}^{AA} is Hermitian. For the exchange transitions, M_{kj}^{AB} (or M_{kj}^{BA}) is not Hermitian [see (6.96) for S_{kj}^{AB}]; however, the unitarity condition $M_{kj} - M_{jk}^{*} = -\mathrm{i}\,(d/dt)\, S_{kj}$ is still fulfilled.

The attachment of the plane-wave translational factors (6.92) to the adiabatic molecular orbitals Φ_k for all values of R is inconsistent. While the Φ_k are determined by the fields of both nuclei, the plane-wave ETF's constrain the electron to move with one or other of the nuclei. To resolve this inconsistency, *Schneiderman* and *Russek* [6.44] have proposed that the following ETF be associated with all the Φ_k's:

$$F(\boldsymbol{r}; \boldsymbol{R}) = \exp[\mathrm{i}\, \boldsymbol{v} \cdot \boldsymbol{r}\, f(\boldsymbol{r}, \boldsymbol{R})]. \tag{6.98}$$

In order to preserve the translational invariance of the coupled equations, $f(\boldsymbol{r}, \boldsymbol{R})$ should have the following switching properties: $f \to p$ when $R \to \infty$ and $r_{\mathrm{A}} \ll r_{\mathrm{B}}$, $f \to -q$ when $R \to \infty$ and $r_{\mathrm{B}} \ll r_{\mathrm{A}}$, and $f \to 0$ when $(R/r) \to 0$. Elsewhere, the form of the switching function $f(\boldsymbol{r}, \boldsymbol{R})$, which usually contains adjustable parameters, can be determined on the basis of physical arguments [6.44–47]. The original switching function suggested by *Schneiderman* and *Russek* [6.44] is

$$f(\boldsymbol{r}, \boldsymbol{R}) = \cos\theta\, [R^2/(a^2 + R^2)], \tag{6.99}$$

where θ is the angle between $\boldsymbol{r}$ and $\boldsymbol{R}$, and a is a cutt-off parameter. An important property of the common ETF (6.98) is that it makes the basis $\{F_k \psi_k^{\mathrm{MO}}\}$ orthogonal and the coupling matrix $\underline{M}$ Hermitian. The matrix elements are now given by

$$M_{kj} = -\langle \Phi_j | \, \partial/\partial t_r + \nabla_r (f \, \boldsymbol{v} \cdot \boldsymbol{r}) \nabla_r + \nabla_r^2 (f \, \boldsymbol{v} \cdot \boldsymbol{r})/2$$
$$+ \mathrm{i} \, [\nabla_r (f \, \boldsymbol{v} \cdot \boldsymbol{r})]^2/2 + \mathrm{i} \, (\partial/\partial t_r) \, (f \, \boldsymbol{v} \cdot \boldsymbol{r}) \, | \Phi_k \rangle$$
$$\times \exp \left[-\mathrm{i} \int_{-\infty}^{t} (E_k - E_j) \, dt' \right] . \tag{6.100}$$

In principle, an optimum choice of the translational factors $F_k(\boldsymbol{r}, \boldsymbol{R})$ for each coupled channel can be made by using the variational method [6.48]. The appropriate Euler-Lagrange equations for $F_k(\boldsymbol{r}, \boldsymbol{R})$ have to be solved simultaneously with the coupled equations for the amplitudes. The variational procedure can be significantly simplified if the $F_k(\boldsymbol{r}, \boldsymbol{R})$ are taken in the form [6.49, 50]

$$F_k(\boldsymbol{r}, \boldsymbol{R}) = \exp [\mathrm{i} \, \boldsymbol{v} \cdot \boldsymbol{R} \, f_k(R)] , \tag{6.101}$$

i.e. independent of $\boldsymbol{r}$. Variational estimates for $f_k(R)$ can now be obtained before solving the coupled equations. It seems, however, that the electron translation factor is a property of the electron configuration space only, which implies that the use of a single ETF for all channels is preferable [6.51].

For the case of the one-electron two-Coulomb centre system ($Z_{\mathrm{A}}, e, Z_{\mathrm{B}}$), an analytic, parameter-free common ETF has been proposed in the form [6.52]

$$f(\boldsymbol{r}, \boldsymbol{R}) = (1/2) \, [(Z_{\mathrm{B}} r_{\mathrm{A}}^3 - Z_{\mathrm{A}} r_{\mathrm{B}}^3)/(Z_{\mathrm{B}} r_{\mathrm{A}}^3 + Z_{\mathrm{A}} r_{\mathrm{B}}^3)$$
$$+ (Z_{\mathrm{A}} - Z_{\mathrm{B}})/(Z_{\mathrm{A}} - Z_{\mathrm{B}})] . \tag{6.102}$$

This switching function ensures that the electron shares the translational motion of the nuclei Z_{A}, Z_{B} and the centre of charge when $r_{\mathrm{A}} \ll r_{\mathrm{B}}$, $r_{\mathrm{B}} \ll r_{\mathrm{A}}$ and in the united atom limit, respectively.

b) Perturbed Stationary State (PSS) Method

If, in the basis $\{F_k \psi_k^{\mathrm{MO}}\}$, one sets $F_k = 1$ and takes $\psi_k^{\mathrm{MO}} = \Phi_k$, the system of coupled equations (6.79) becomes

$$(d/dt) \, a_j = -\sum_{k \neq j} a_k(t) \, \Gamma_{kj} \exp \left[-\mathrm{i} \int_{-\infty}^{t} (E_k - E_j) \, dt' \right] \tag{6.103}$$

with

$$\Gamma_{kj} = \langle \Phi_j | \partial/\partial t_r | \Phi_k \rangle . \tag{6.104}$$

In this form, the MO-CC method was first introduced by *Mott* [6.53] and, since $\Phi_k(\boldsymbol{r}; \boldsymbol{R})$ are the stationary Born-Oppenheimer electronic states, it is called the *perturbed stationary state* (PSS) method.

As has been discussed before, the basis $\{\Phi_k\}$ is not translation-invariant, i.e. the matrix elements Γ_{kj} depend on the choice of the coordinate origin on the internuclear line. However, if the reference frame is fixed on the nucleus of atom A ($p = 0$) then the molecular functions Φ_k^{A}, going over into the atomic orbitals ϕ_k^{A} as $R \to \infty$, trivially satisfy the asymptotic boundary condition. The matrix elements $\Gamma_{kj}^{\mathrm{AA}}$ for the direct (excitation) transition satisfy the relation

$$\Gamma_{kj}^{\mathrm{AA}} = \langle \Phi_j^{\mathrm{A}} | \partial/\partial t_{r_{\mathrm{A}}} | \Phi_k^{\mathrm{A}} \rangle \xrightarrow[R \to \infty]{} 0 , \tag{6.105}$$

i.e. they have the correct asymptotic behaviour. In this reference frame, however, the matrix elements $\Gamma_{kj}^{\mathrm{BB}}$ and $\Gamma_{kj}^{\mathrm{AB}}$ may have an incorrect asymptotic behaviour. Thus, for instance, $\Gamma_{kj}^{\mathrm{BB}}$ is

$$\Gamma_{kj}^{\mathrm{BB}} = \langle \Phi_j^{\mathrm{B}} | \partial/\partial t_{r_{\mathrm{A}}} | \Phi_k^{\mathrm{B}} \rangle \xrightarrow[R \to \infty]{} \langle \phi_j^{\mathrm{B}} | \boldsymbol{v} \cdot \nabla_{r_{\mathrm{B}}} | \phi_k^{\mathrm{B}} \rangle , \tag{6.106}$$

where $\phi_k^{\mathrm{B}}(\boldsymbol{r}_{\mathrm{B}})$ is the atomic orbital centred on B^{Z+}. The dipole matrix element on the right-hand side of (6.106) gives a constant value for states between which dipole transitions are allowed. Similar non-vanishing asymptotic values may have some of the coupling matrix elements $\Gamma_{kj}^{\mathrm{AB}}$ (and $\Gamma_{kj}^{\mathrm{BA}}$), corresponding to the electron rearrangement channels. These difficulties arise in the PSS method due to the fact that the molecular states Φ_k^{B}, which at $R \to \infty$ do not satisfy the proper boundary conditions, are not proper physical states. These "unphysical" states may always be expressed as a linear combination of the proper physical states if the basis set $\{\Phi_k\}$ is complete. In practical calculations, however, the basis is never complete, and the above difficulties in the PPS method can only be eliminated by introducing appropriate electron translation factors $F_k(\boldsymbol{r}, \boldsymbol{R})$.

The above discussion shows that the use of the PSS method, with an adiabatic basis centred on the target atom, enables one to calculate the probabilities P_n^{ex} for excitation of the individual target levels correctly, provided the basis is sufficiently large. Moreover, as pointed out by *Piacentini* and *Salin* [6.54], by using this method and the assumption of "near-completeness" of the basis, one can also calculate the total probability for electron capture by using the relation

$$P_{\mathrm{tot}}^{\mathrm{cx}} = 1 - \sum_n P_n^{\mathrm{ex}} . \tag{6.107}$$

An alternative choice for the position of the coordinate origin in the PSS method is to place it on the nucleus of the B^{Z+} ion ($q = 0$, $\boldsymbol{r} = \boldsymbol{r}_B$). In that case the states

$$\Phi_k^B \xrightarrow[R\to\infty]{} \phi_k^B(\boldsymbol{r}_B)$$

are well defined (i.e. satisfy exactly the asymptotic boundary conditions for the problem), but the specification of the states Φ_k^A, which at $R \to \infty$ separate into the atomic states ϕ_k^A, encounters difficulties.

Let us consider now in more detail the structure of the coupling matrix element Γ_{kj}. At low collision energies the adiabatic electron functions "follow" the variations of the internuclear distance vector $\boldsymbol{R}(t)$, and the internuclear axis is a natural quantization axis.

If (x, z) is the collision plane in the space-fixed reference frame and (x', y', z') is a rotating reference frame with the z' axis along the internuclear line, then the electron position vector $\boldsymbol{r}'$ in the rotating frame has the following coordinates (see Fig. 6.4):

$$\begin{aligned} x' &= x\cos\Theta(t) + z\sin\Theta(t) \\ y' &= y \\ z' &= -x\sin\Theta(t) + z\cos\Theta(t) \end{aligned} \tag{6.108}$$

where $\Theta(t)$ is the angle of rotation. The time derivative $\partial/\partial t_r$ in rotating coordinates has the form

$$\partial/\partial t_r = v_R(\partial/\partial R_{r'}) + \dot{\Theta}(\partial/\partial\Theta_{r'}) \quad \text{with} \tag{6.109}$$

$$v_R = dR/dt\,, \quad \dot{\Theta} = -\varrho v/R^2\,, \tag{6.110}$$

where ϱ and v are the impact parameter and collision velocity, respectively. Keeping in mind the relation $\partial/\partial\Theta_{r'} = -\mathrm{i}\,L_{y'}$, where $L_{y'}$ is the component of the electron angular momentum along the y' axis, the matrix element Γ_{kj} can be written as

$$\Gamma_{kj} = v_R\,\Gamma_{kj}^R + (\varrho v/R^2)\,\Gamma_{kj}^L\,, \tag{6.111}$$

$$\Gamma_{kj}^R = \langle\Phi_j\,|\,\partial/\partial R_{r'}|\,\Phi_k\rangle, \quad \Gamma_{kj}^L = \langle\Phi_j\,|\,\mathrm{i}\,L_{y'}|\,\Phi_k\rangle, \tag{6.112}$$

where $\Phi_k = \Phi_k(\boldsymbol{r}'; \boldsymbol{R})$. The quantities $\partial/\partial R_{r'}$ and $\mathrm{i}\,L_{y'}$ are called operators for *radial* and *angular* (or *rotational*) *coupling* of the molecular states, respectively. The radial coupling operator connects states having the same angular symmetry ($\sigma \leftrightarrow \sigma$, $\pi \leftrightarrow \pi$, etc.), while the rotational coupling operator connects states for which the projection of the angular momentum on the internuclear axis differs by one ($\sigma \leftrightarrow \pi$, $\pi \leftrightarrow \delta$, etc.). Additionally, both $\partial/\partial R_{r'}$ and i $L_{y'}$ couple states which have the same parity. Note that the transformation (6.108) from a space-fixed to a rotating reference frame holds, independent of the position of the coordinate origin; in particular it can be placed on the target nucleus ($\boldsymbol{r} = \boldsymbol{r}_A$). Thus, the relations (6.109–112) also remain valid for the operator $\partial/\partial t_{r_A}$.

For the (Z_1, e, Z_2) system, the adiabatic functions $\Phi_k(\boldsymbol{r}'; \boldsymbol{R})$ are usually determined in prolate spheroidal coordinates (ξ, η, ϕ). Setting the coordinate origin in the mid-point of the internuclear line, the operators $\partial/\partial R_{r'}$ and i $L_{y'}$ have the following forms in the (ξ, η, ϕ) coordinates [6.54]:

$$\frac{\partial}{\partial R_{r'}} = -\frac{2[(Z_1+Z_2)\,\xi + (Z_2 - Z_1)\,\eta]}{(E_j - E_k)\,R^2\,(\xi^2 - \eta^2)} + \frac{\xi(1-\xi^2)(\partial/\partial\xi) + \eta(\eta^2-1)(\partial/\partial\eta)}{R(\xi^2-\eta^2)} \qquad (6.113\text{a})$$

$$\mathrm{i}\,L_{y'} = \frac{(\xi^2-1)^{1/2}(1-\eta^2)^{1/2}}{\xi^2-\eta^2}\cos\phi\,(\eta\,\partial/\partial\xi - \xi\,\partial/\partial\eta) - \frac{\xi\,\eta}{(\xi^2-1)^{1/2}(\eta^2-1)^{1/2}}\sin\phi(\partial/\partial\phi)\,. \qquad (6.113\text{b})$$

These expressions can easily be generalized for the case of an arbitrary position of the coordinate origin on the internuclear line.

The properties of the matrix elements Γ_{kj}^R and Γ_{kj}^L will be discussed in more detail in the next chapter (see also [6.1, 36, 55]. We would like here only to point out that in the region of a pseudocrossing $\Gamma_{kj}^R(R)$ is strongly peaked, while $\Gamma_{kj}^L(R)$ changes its behaviour drastically. The behaviour of Γ_{kj}^R and Γ_{kj}^L as functions of R creates serious difficulties in the numerical integration of the PSS coupled equations. These practical difficulties can be resolved by choosing another representation of the electronic basis states.

c) Diabatic Molecular Basis

By using an appropriate unitary transformation, the adiabatic basis $\{\Phi_k\}$ can be transformed into a new basis $\{\varphi_k\}$ such that either the radial or the angular coupling vanishes exactly. In the new representation, called *diabatic* [6.56], the basis does not diagonalize the electronic Hamiltonian and its diagonal matrix elements (diabatic energies) do not obey the von Neumann-Wigner "non-crossing" theorem. Such an approach to constructing the diabatic basis has often undesirable properties. If, for example, one diagonalizes the operator ($\partial/\partial R_r$), the corresponding diabatic functions φ_k do not depend on R (except through some phase factors) [6.57], and for an atom-fully stripped ion system such a basis is equivalent to an atomic basis.

The construction of an appropriate diabatic molecular basis depends on the particular problem and has been discussed elsewhere [6.10, 36, 55, 58]. The following criteria are imposed on the choice of diabatic basis: (I) $\varphi_k(\boldsymbol{r};\boldsymbol{R})$ should minimize the dynamic coupling $(\partial/\partial \dot{t}_r)$ between the states, and (II) $\varphi_k(\boldsymbol{r};\boldsymbol{R})$ should go over into the atomic orbitals $\phi_k^{\mathrm{A}}(\boldsymbol{r}_A)$ or $\phi_k^{\mathrm{B}}(\boldsymbol{r}_B)$ (or into linear combinations of them) when $R \to \infty$.

In the formulation of collision problems in the diabatic basis, the problem of electron translation factors $F_k(\boldsymbol{r},\boldsymbol{R})$, associated with the diabatic functions $\varphi_k(\boldsymbol{r};\boldsymbol{R})$, remains the same as when an adiabatic basis is used. If the function Ψ is expanded in terms of the basis set $\{F_k\,\varphi_k\}$, the expansion coefficients $c_k(t)$ satisfy the following coupled equations (in matrix notation):

$$\mathrm{i}\,\underline{\dot{S}} = \tilde{M}\,\underline{\dot{c}}\,,\quad \text{with} \tag{6.114}$$

$$S_{kj} = \langle \varphi_j F_j |\, F_k\,\varphi_k \rangle \exp\left[-\mathrm{i}\int_{-\infty}^{t} (\tilde{V}_{kk}-\tilde{V}_{jj})\,dt'\right], \tag{6.115}$$

$$\tilde{M}_{kj} = \tilde{V}_{kj}\exp\left[-\mathrm{i}\int_{-\infty}^{t} (\tilde{V}_{kk}-\tilde{V}_{jj})\,dt'\right], \tag{6.116}$$

$$\tilde{V}_{kj} = \langle \varphi_j F_j\,|H|\,F_k\,\varphi_k\rangle. \tag{6.117}$$

If $\{F_k\,\varphi_k\}$ is constructed to form a complete orthonormal set, then $\tilde{V}_{kj}$ is a Hermitian matrix and (6.114) becomes

$$\mathrm{i}\,\dot{c}_j = \sum_{k\neq j} c_k(t)\,\tilde{V}_{kj}\exp\left[-\mathrm{i}\int_{-\infty}^{t} (\tilde{V}_{kk}-\tilde{V}_{jj})\,dt'\right] \tag{6.118}$$

with

$$\sum_k |c_k(t)|^2 = 1. \tag{6.119}$$

The system of coupled equations (6.118) holds when $F_k = 1$ and $\{\varphi_k\}$ is complete and orthonormal. However, the same form of the diabatic MO-CC equations can be obtained if plane-wave translational factors are used and an orthogonalization procedure is applied to the functions φ_k, which separate to different centres when $R \to \infty$ [6.59, 60].

6.2.3 Atomic-Oribtal Close-Coupling (AO-CC) and Related Methods

a) The AO-CC Method

At intermediate and high collison velocities the one-electron atomic orbitals ϕ_k, weighted by plane-wave translational factors, constitute a physically appropriate basis for the expansion of electron wave functions. If the collision geometry is as shown in Fig. 6.4, the expansion has the form

$$\begin{aligned}\Psi = &\sum_n a_n(t)\,\phi_n^A(\boldsymbol{r}_A)\exp[\mathrm{i}\,p\,\boldsymbol{v}\cdot\boldsymbol{r} - \mathrm{i}(\varepsilon_n^A + p^2 v^2/2)\,t]\\ &+ \sum_m b_m(t)\,\phi_m^B(\boldsymbol{r}_B)\exp[-\mathrm{i}\,q\,\boldsymbol{v}\cdot\boldsymbol{r} - \mathrm{i}(\varepsilon_m^B + q^2 v^2/2)\,t]\,.\end{aligned} \tag{6.120}$$

In (6.120), $\phi_n^A(\boldsymbol{r}_A)$ and ε_n^A (or $\phi_m^B(\boldsymbol{r}_B)$ and ε_m^B) are, respectively, the eigenfunction and eigenenergies of the effective one-electron "atomic" Hamiltonian H_0^A (or H_0^B). H_0^A and H_0^B correspond to the decomposition of H

$$H = H_0^A + V^B = H_0^B + V^A\,. \tag{6.121}$$

The use of the two-centre expansion (6.120) in (6.77) leads to coupled equations for the amplitudes $a_n(t)$ and $b_m(t)$ in the form of (6.79). However, making use of the separation (6.121) for H, as well as of a variational principle of the Hulthén-Kohn type, the coupled equations can also be obtained in the form [6.61] (for convenience we take $p = q = 1/2$)

$$\begin{aligned}\mathrm{i}(\dot{\underline{a}} + \underline{s}\,\dot{\underline{b}}) &= \underline{H}\,\underline{a} + \underline{W}\,\underline{b}\,,\\ \mathrm{i}(\underline{s}^\dagger\,\dot{\underline{a}} + \dot{\underline{b}}) &= \underline{\tilde{H}}\,\underline{a} + \underline{\tilde{W}}\,\underline{b}\,,\end{aligned} \tag{6.122}$$

where $\underline{a}$ and $\underline{b}$ are the column vectors of the coefficients $a_n(t)$ and $b_m(t)$, and the elements of the matrices $\underline{s}$, $\underline{H}$, $\underline{W}$, $\underline{\tilde{H}}$ and $\underline{\tilde{W}}$ are given by

$$s_{nm} = \langle\phi_m^B|\exp(\mathrm{i}\,\boldsymbol{v}\cdot\boldsymbol{r})\,|\phi_n^A\rangle\exp[-\mathrm{i}(\varepsilon_n^A - \varepsilon_m^B)\,t]\,, \tag{6.123}$$

$$H_{nn'} = \langle\phi_{n'}^A|\,V^B\,|\phi_n^A\rangle\exp[-\mathrm{i}(\varepsilon_n^A - \varepsilon_{n'}^A)\,t]\,, \tag{6.124a}$$

$$\tilde{H}_{mm'} = \langle\phi_{m'}^B|\,V^A\,|\phi_m^B\rangle\exp[-\mathrm{i}(\varepsilon_m^B - \varepsilon_{m'}^B)\,t]\,, \tag{6.124b}$$

$$W_{mn} = \langle\phi_n^A|\exp(-\mathrm{i}\,\boldsymbol{v}\cdot\boldsymbol{r})\,V^B\,|\phi_m^B\rangle\exp[-\mathrm{i}(\varepsilon_m^B - \varepsilon_n^A)\,t]\,, \tag{6.124c}$$

$$\tilde{W}_{nm} = \langle\phi_m^B|\exp(\mathrm{i}\,\boldsymbol{v}\cdot\boldsymbol{r})\,V^A\,|\phi_n^A\rangle\exp[-\mathrm{i}(\varepsilon_n^A - \varepsilon_m^B)\,t]\,. \tag{6.124d}$$

The condition for the conservation of probability is given by the relation $\underline{s} = \mathrm{i}(\underline{W} - \underline{\tilde{W}}^\dagger)$.

The AO-CC equations (6.122) have to be solved subject to the boundary conditions

$$a_n(-\infty) = \delta_{no}\,,\quad b_m(-\infty) = 0\,, \tag{6.125}$$

if the system is initially in the state $\phi_0^A(\boldsymbol{r}_A)$. The probabilities for the $0 \to n$ excitation and for the $n = 0 \to m$ electron-transfer process are given by

$$P_n^{\mathrm{ex}} = |a_n(+\infty)|^2\,, \tag{6.126a}$$

$$P_m^{\mathrm{cx}} = |b_m(+\infty)|^2\,. \tag{6.126b}$$

The matrix elements (6.124) are written in a coordinate system fixed at the mid-point of the internuclear line. Alternatively, one can take a rotating coordinate system (x', y', z') with the z'-axis oriented along the vector $\boldsymbol{R}$. The matrix elements $H_{nn'}$ and W_{mn} in this reference frame have the form

$$H_{nn'} = \langle \phi_{n'}^{A} | \, [V^{B}(\boldsymbol{r}', \boldsymbol{R}) + (\varrho\, v/R^2)(\mathrm{i}\, L_{y'})] \, | \phi_{n}^{A} \rangle \exp[-\mathrm{i}\,(\varepsilon_{n}^{A} - \varepsilon_{n'}^{A})\, t] , \quad (6.127\,\mathrm{a})$$

$$W_{mn} = \langle \phi_{n}^{A} | \exp(-\mathrm{i}\, \boldsymbol{v} \cdot \boldsymbol{r}) \, [V^{B} + (\varrho\, v/R^2)(\mathrm{i}\, L_{y'})] \, | \phi_{m}^{B} \rangle \exp[-\mathrm{i}\,(\varepsilon_{m}^{B} - \varepsilon_{n}^{A})\, t] , \quad (6.127\,\mathrm{b})$$

and similarly for $\tilde{H}_{mm}$ and $\tilde{W}_{nm}$. The use of a rotating reference frame may have advantages with respect to a space-fixed reference frame at relatively low velocities when the wave function Ψ can follow the slow rotation of the internuclear axis. With increasing velocity, the angular coupling terms in the H_{nn}, W_{mn} matrix elements also increase and tend to decouple the electronic wave function from the rotating axis. A space-fixed reference frame for the basis is then more natural.

The two-centre atomic basis expansion method encounters some difficulties in its practical applications. With increasing collision velocity, the ionization channel becomes increasingly more important, and continuum states have to be included in the basis. This implies serious difficulties in the integration of coupled equations. If the discrete states only are included in the basis, the convergence of the method is poor for small values of R, since the maximum of the electron distribution of states with high n and m is far from the nuclei. To resolve these difficulties in the two-centre AO-CC method, orther expansions of Ψ have been proposed, some of which we discuss below. We would like, however, to note first that instead of using a two-centre expansion of the type given by (6.120), a single-centre expansion may be used alternatively provided the basis is sufficiently large. In using such an expansion, the electron rearrangement channels cannot be well described unless the basis contains in infinite number of terms. If, however, one constructs a complete single-centre basis on the projectile (i.e. including the projectile continuum states), to which an orthogonalized target state is added (but centred again on the projectile), the electron-capture channels are adequately described, including the "capture into continuum" [6.59, 60]. Evidently such a basis is particularly convenient for the charge-exchange problem involving highly charged projectile ions.

b) Sturmian Basis

To take into account the effects of the continuum and high-lying discrete states in an "atomic" type coupled-channel formulation of the collision problem, *Gallaher* and *Wilets* [6.62] suggested the use of the Sturmian functions as an expansion basis. These functions form an infinite discrete set which, in its entirety, is complete. The Sturmian functions are scaled hydrogenic functions

$$\begin{aligned} S_{nl}(r;\varepsilon) &= \alpha_{nl}^{1/2} R_{nl}(\alpha_{nl} r)\,, \\ \alpha_{nl} &= n(-2\,\varepsilon)^{1/2}\,, \end{aligned} \tag{6.128}$$

where ε is an arbitrary parameter. S_{nl} are solutions of the radial equation for the hydrogen-like system with a "charge" α_{nl} and "energy" ε. The R_{nl} in (6.128) are the radial hydrogenic functions. The parameter ε can be chosen to give some useful scalings. However, since the functions $S_{nl}(r;\varepsilon)$ are not eigenfunctions of the "atomic" Hamiltonian $H_0^{\rm A}$ (or $H_0^{\rm B}$), they do not satisfy the boundary conditions for the problem. As a consequence, the expansion coefficients have an incorrect (oscillatory) behaviour as $|t| \to \infty$, which is reflected equally in the transition amplitudes. This is a result of the fact that Sturmian functions do not describe "physical" states and, if a finite basis is used, the physical states (the proper atomic orbitals) cannot be reconstructed from the basis functions. The oscillatory terms in the transition amplitudes can, however, be extracted and, in general, can be neglected [6.62].

c) Expansions in Pseudostates

Another possible way of simulating the effects of the neglected part of the spectrum is to add to a finite atomic basis set some functions ("pseudo-states"), which are conveniently selected. The pseudostate functions should be chosen in such a way as to overlap as much as possible with the states whose effect on the reaction channels is expected to be large. For example, if one wants to improve the convergence of the two-centre AO-CC method, limited by the poor representation of the electronic motion at small internuclear distances, the pseudostates have to be chosen so that they overlap as much as possible with the low-lying united atom orbitals [6.63], or are even identical with them [6.64]. On the other hand, if one wants to include the effects of the continuum states on the reaction channels represented by AO's in the basis, then the discrete pseudostate functions have to be chosen to overlap with the continuum functions as much as possible over the relevant energy [6.65, 66]. It is important to note that in the pseudostate expansion method, the transition amplitudes for the atomic states represented in the basis have the correct asymptotic behaviour.

A specific application of the pseudostate expansion method is the so-called "one and a half centred expansion" (OHCE) [6.67]. The wave function is expanded in the form [compare with (6.120)]

$$\Psi = \sum_{n}^{N} a_n(t)\,\phi_n^{\mathrm{A}}(\boldsymbol{r}_{\mathrm{A}})\,F_n^{\mathrm{A}}(\boldsymbol{r},t) + \sum_{m}^{M} b_m(t_L)\,\beta_m(t_L)\,\phi_m^{\mathrm{B}}(\boldsymbol{r}_{\mathrm{B}})\,F_m(\boldsymbol{r},t)\,, \tag{6.129}$$

where the basis set of functions ϕ_n^{A} is large enough to represent the wave function in the interaction region $R(t) \leqq R(t_L)$, and t_L is the time beyond which the electron exchange effects can be neglected. The functions $\beta_m(t)$ are arbitrary, subject to the conditions $\beta_m(t) = 0$ for $t \leqq -t_L$, and $\beta_m(t) = 1$ for $t \geqq t_L$. The constants $b_m(t_L)$ are the electron transfer amplitudes and F_n, F_m are the plane-wave ETF's. By projecting the Schrödinger equation with the functions ϕ_n^{A} and ϕ_m^{B}, coupled differential equations are obtained for the amplitudes $a_n(t)$ and algebraic equations for $b_m(t_L)$. In contrast to the two-centre expansion methods, only single-centre integrals of the potential matrix are required in the OHCE method. As compared with the single-centred expansion methods, the OHCE method allows for the rearrangement channels explicitly, despite the finiteness of the basis.

6.2.4 Two-State Close-Coupling Models and Approximations

The potential couplings $\tilde{V}_{kj}$ in the diabatic close-coupled equations and the dynamic couplings Γ_{kj}^{R} and Γ_{kj}^{L} in the adiabatic MO-CC method are usually localized in narrow regions of the internuclear coordinate R. For example, the radial coupling Γ_{kj}^{R} is effective only in the region of a pseudocrossing, while Γ_{kj}^{L} is mostly pronounced at very small R, where groups of states with different projections of the angular momentum on the internuclear axis become degenerate (or almost degenerate). These circumstances may make it possible to decouple approximately the multistate close-coupling problem of the low-energy ion-atom collisions and to represent it by a set of mutually separated (in time) two-state close-coupling problems with an adiabatic (or diabatic) evolution of the system in the remaining part of the collision. The semi-classical S matrix can be thus represented as a product of the transition matrices $T_{kj}(\delta R_{kj})$ for the separated two-state close-coupling problems in the strong coupling regions δR_{kj}, and the adiabatic (or diabatic) evolution matrices $U_k(R_{kj}, R_{k'j'})$ [6.10]. Such an approximate decomposition of the multichannel close-coupling problem facilitates the practical calculations a great deal and often provides an adequate description of the collision dynamics. The two-state close-coupling problem is easy to solve numerically. Moreover, in certain typical physical situations, the coupling matrix elements and the potential energies in the strong coupling region δR_{kj} can be approximated by analytical functions which permit an exact solution of the coupled equations. Such models are particularly useful since they provide analytical information about the dependence of the transition probability (or the cross section) on the parameters of the collision problem.

The two-state models which are most frequently met in different physical problems are the *Landau-Zener-Stueckelberg* model [6.68–70], the *Rosen-Zener-Demkov* model [6.71, 72] and the *Nikitin* model [6.73]. These models have been widely discussed in the standard monographs on the theory of atomic collisions [6.10, 39–41], and we shall give here only the expressions for the transition probability which they provide.

If the two interacting states are considered to form a complete basis, then the adiabatic and diabatic states are connected by a unitary transformation and the diabatic and adiabatic formulations of the collision problem are equivalent. The close-coupled equations in the diabatic formulation are [see (6.118)]

$$\mathrm{i}\,\dot{c}_1 = c_2\,\tilde{V}_{12}\exp\left[\mathrm{i}\int_{-\infty}^{t}\omega_{12}\,dt'\right], \tag{6.130a}$$

$$\mathrm{i}\,\dot{c}_2 = c_1\,\tilde{V}_{12}\exp\left[-\mathrm{i}\int_{-\infty}^{t}\omega_{12}\,dt'\right], \tag{6.130b}$$

$$\omega_{12} = \tilde{V}_{11} - \tilde{V}_{22}, \quad \tilde{V}_{21}^{*} = V_{21} \quad \text{with} \tag{6.130c}$$

$$|c_1(t)|^2 + |c_2(t)|^2 = 1\,, \tag{6.131}$$

where Hermiticity and unitarity are assumed. The matrix elements $\tilde{V}_{kj}$ $(k, j = 1, 2)$ are, in general, calculated with basis functions containing electron translational factors.

The solutions $c_1(t)$ and $c_2(t)$ depend only on two functions, $\omega_{12}(t)$ and $\tilde{V}_{12}(t)$. If the initial conditions for the system (6.130) are

$$c_1(-\infty) = 1\,, \quad c_2(-\infty) = 0\,, \tag{6.132}$$

then the transition probability is given by

$$P_{12} = |c_2(+\infty)|^2\,. \tag{6.133}$$

Let us take the functions $\omega_{12}(t)$ and $\tilde{V}_{12}(t)$ in the form

$$\omega_{12}(t) = \Delta\varepsilon - A\cos\chi\,\mathrm{e}^{-\gamma t}\,, \tag{6.134a}$$

$$2\,\tilde{V}_{12}(t) = A\sin\chi\,\mathrm{e}^{-\gamma t} \tag{6.134b}$$

where A, $\Delta\varepsilon$, χ and γ are constants. (The constant A is connected with the zero of the time-scale and will not appear in the results.) For $t = 0$ the coupling is the strongest. With the above functions, the coupled equations (6.130), subject to the initial conditions (6.132), can be solved exactly. The transition probability is obtained in the form [6.73] (*Nikitin*'s formula)

$$P_N = 4[\sinh(2\lambda)]^{-2}\exp(2\lambda\cos\chi)\sinh[\lambda(1-\cos\chi)] \times \sinh[\lambda(1+\cos\chi)]\sin^2\eta \tag{6.135}$$

where

$$\lambda = \frac{\pi\,\Delta\varepsilon}{2\gamma}, \quad \eta = \int (\omega_{12}^2 + 4\,\tilde{V}_{12}^2)^{1/2}\,dt. \tag{6.136}$$

The Landau-Zener and Demkov formulae can be obtained as limitting cases of (6.135). If $\chi \ll 1$, $\lambda \gg 1$ and $A = \Delta\varepsilon$, then in the transition zone around $t = t_0\,(=0)$ one has

$$\omega_{12}(t) = \gamma\,\Delta\varepsilon\, t = (v_{R_0}\,\Delta F)|_{R_0}\cdot t, \tag{5.137a}$$

$$2\,\tilde{V}_{12} = \Delta\varepsilon\,\chi = \text{const}, \tag{6.137b}$$

where $V_{R_0} = (dR/dt)|_{t=t_0}$, $\Delta F_0 = (d\omega_{12}/dR)|_{R=R_0}$ and $R_0 = R(t_0)$. The relations (6.137) constitute the physical conditions of the Landau-Zener model. In this case the expression (6.135) for P_{12} reduces to

$$P_{LZS} = 4\,p(1-p)\sin^2\eta, \tag{6.138}$$

$$p = \exp[-\pi\,\Delta^2/(2\,v_R\,\Delta F)]|_{R=R_0}, \quad \Delta = 2\,\tilde{V}_{12}. \tag{6.139}$$

Averaging (6.138) over the phase shift η (if η is large), one obtains in the Landau-Zener formula. By setting $\chi = \pi/2$ in (6.134), one has

$$\omega_{12} = \Delta\varepsilon = \text{const}, \quad 2\,\tilde{V}_{12} = A\exp(-\gamma t), \tag{6.140}$$

i.e. the conditions for the Demkov coupling problem are obtained. The expression (6.135) now reduces to

$$P_D = \operatorname{sech}^2(\pi\,\Delta\varepsilon/2\gamma)\sin^2\eta_D \quad \text{with} \tag{6.141}$$

$$\eta_D = \int_{-\infty}^{+\infty} \tilde{V}_{12}\,dt. \tag{6.142}$$

The Rosen-Zener formula is obtained from (6.141) by averaging over η_D.

The system of coupled equations (6.130) can be approximately solved under more general assumptions about the form of the functions $\omega_{12}(t)$ and $\tilde{V}_{12}(t)$. For example, if $|\tilde{V}_{12}| \ll |\omega_{12}|$, a perturbational approach (with respect to $\tilde{V}_{12}$) can be used for its solution, and when $|\tilde{V}_{12}| \gg |\omega_{12}|$, adiabatic perturbation theory can be applied. The system (6.130) of first-order differential equations can also be transformed into one second-order differential equation for the coefficient $c_2(t)$ [or $c_1(t)$]. This can be treated by quasi-classical methods under very general assumptions about the analytical properties of ω_{12} and $\tilde{V}_{12}$, if the inverse collision velocity can be considered a large parameter. The perturbational treatments of the ion-atom collision problem will be briefly considered in the next section, while for the asymptotic (quasi-classical) approach we refer to the standard textbooks [6.10, 74].

An approximate non-perturbative approach for solving the system of coupled equations (6.130) within a wide region of validity has been proposed by *Vainshtein* et al. [6.75]. Introducing a new function $K(t)$ by the relation

$$K(t) = c_2(t)/c_1(t), \tag{6.143}$$

the system (6.130) is transformed into a non-linear Riccati equation,

$$dK/dt = -\mathrm{i}\left[\tilde{V}_{12}\exp\left(\mathrm{i}\int\limits_{-\infty}^{t}\omega_{12}\,dt'\right) - \tilde{V}_{12}\exp\left(-\mathrm{i}\int\limits_{-\infty}^{t}\omega_{12}\,dt'\right)K^2(t)\right]. \tag{6.144}$$

Using (6.143, 131), one obtains

$$|c_1(t)|^2 = \frac{1}{1+|K(t)|^2}, \quad |c_2(t)|^2 = \frac{|K(t)|^2}{1+|K(t)|^2}, \tag{6.145}$$

i.e. unitarity is preserved independent of the accuracy to which $K(t)$ is calculated. An approximate solution of the Riccati equation (6.144) leads to the following transition probability $|c_2(+\infty)|^2$:

$$P_{\mathrm{VPS}} = \left|\int\limits_{-\infty}^{\infty} dt\,\tilde{V}_{12}\exp\left[i\int\limits^{t}(\omega_{12}^2 + 4\,\tilde{V}_{12}^2)^{1/2}\,dt'\right]\right|^2. \tag{6.146}$$

The probability P_{VPS} depends on the analytic properties of the "frequency" function

$$\Omega(t) = (\omega_{12}^2 + 4\,\tilde{V}_{12}^2)^{1/2}, \tag{6.147}$$

and its value is dominantly determined by those zeroes of $\Omega(t)$ in the complex t-plane which lie close to the real axis. The interaction zone is thus determined by the equation

$$\Omega(t_c) = 0. \tag{6.148}$$

Assuming $\omega_{12}(t)$ to be a slowly varying function with respect to $\tilde{V}_{12}$ and expanding $\Omega(t)$ as

$$\Omega(t) \approx \omega_{12}(t) + 2\,\tilde{V}_{12}, \tag{6.149}$$

one can obtain

$$P_{\mathrm{VPS}} = \mathrm{e}^{-2\zeta} \sin^2\left(\int_{-\infty}^{+\infty} \tilde{V}_{12}\, dt\right), \quad \text{where} \tag{6.150}$$

$$\zeta = \mathrm{Im}\left\{\int^{t_c} \omega_{12}\, dt\right\} \tag{6.151}$$

and t_c is the zero of $\Omega(t)$, (6.147), closest to the real axis.

Let us note that in the adiabatic region a general formula for the probability of the quasi-resonant transitions in a two-state system has been derived by *Crothers* [6.76]:

$$P_C = \mathrm{sech}^2(\mathrm{Im}\{S\}) \sin^2(\mathrm{Re}\{S\}) \tag{6.152}$$

where

$$S = \int_0^{t_c} \Omega(t)\, dt. \tag{6.153}$$

6.3 Perturbation Methods

6.3.1 Formal Theory of Atomic Reactions

The formal theory of atomic reactions [6.40, 41] provides an elegant framework for presenting some consistent approximate descriptions of direct and rearrangement channels in an atom-ion collision. The Hamiltonian H of the interacting system $A + B^{Z+}$ can be represented in the two forms

$$\begin{aligned} H &= H_d + V_d \\ &= H_r + V_r, \end{aligned} \tag{6.154}$$

where H_d is the total Hamiltonian of the isolated subsystems A and B^{Z+}, including their relative motion, and V_d is their interaction. H_r and V_r have analogous meanings for the electron-rearranged system $A^+ + B^{(Z-1)+}$. The Hamiltonians H_d and H_r generate sets of eigenfunctions $\{\psi_n^d\}$, $\{\psi_m^r\}$ and corresponding energies E_n^d, E_m^r. In general, the functions ψ_n^d and ψ_m^r are not orthogonal. The scattering problem is described by the Schrödinger equation

$$(E-H)\,\Psi=0\,, \tag{6.155}$$

where Ψ and E are the total wave function and total energy of the system. Denoting by Ψ^+ the solution of (6.155) which satisfies the initial condition

$$\Psi^+ \xrightarrow[t\to-\infty]{} \psi_\mathrm{i}^\mathrm{d} \tag{6.156}$$

and an "outgoing-wave boundary condition" after the collision, and introducing the Green's operators

$$G_\mathrm{d}^+=(E-H_\mathrm{d}+\mathrm{i}\,\eta)^{-1}\,,\quad G_\mathrm{r}^+=(E-H_\mathrm{r}+\mathrm{i}\,\eta)^{-1} \tag{6.157}$$

with η being a small positive quantity, the scattering problem can be written in the form of a Lippman-Schwinger equation [6.40]

$$\Psi^+=\psi_\mathrm{i}^\mathrm{d}+G_\mathrm{d}^+\,V_\mathrm{d}\,\Psi^+\,, \tag{6.158}$$

and for the rearrangement channel only

$$\Psi^+=G_\mathrm{r}^+\,(\mathrm{i}\,\eta\,\psi_\mathrm{i}^\mathrm{d}+V_\mathrm{r}\,\Psi^+)\,. \tag{6.159}$$

Equation (6.158) has a formal solution in the form

$$\Psi^+=(1+\mathcal{G}^+\,V_\mathrm{d})\,\psi_\mathrm{i}^\mathrm{d} \quad \text{where} \tag{6.160}$$

$$\mathcal{G}^+=(E-H+\mathrm{i}\,\eta)^{-1} \tag{6.161}$$

is the complete Green's operator.

Equations analogous to those represented by (6.158, 160) can be obtained for the solution Ψ^- of (6.155) which, at $t\to\infty$, goes over into a state $\psi_\mathrm{f}^\mathrm{r}$, and at $t\to-\infty$ satisfies an "incoming-wave boundary condition". To this end one should introduce the Green's operators G_r^- and $\mathcal{G}^-$ by changing the sign of $\mathrm{i}\,\eta$ in (6.157, 161), respectively.

For a transition from the initial state $\psi_\mathrm{i}^\mathrm{d}$ into a final state $\psi_\mathrm{f}^\mathrm{d}$ or $\psi_\mathrm{f}^\mathrm{r}$, the transition matrix $\underline{T}$ on the energy shell ($E_\mathrm{i}^\mathrm{d}=E_\mathrm{f}^\mathrm{d}=E$, $E_\mathrm{i}^\mathrm{r}=E_\mathrm{f}^\mathrm{r}=E$) has the form [6.40, 41]

$$T_\mathrm{if}^\mathrm{d}=\langle\psi_\mathrm{f}^\mathrm{d}|\,V_\mathrm{d}\,|\Psi^+\rangle\,, \tag{6.162a}$$

$$T_\mathrm{if}^\mathrm{r}=\langle\psi_\mathrm{f}^\mathrm{r}|\,V_\mathrm{r}\,|\Psi^+\rangle \tag{6.162b}$$

$$=\langle\Psi^-|\,V_\mathrm{d}\,|\psi_\mathrm{i}^\mathrm{d}\rangle\,. \tag{6.162c}$$

The scattering amplitudes f_{if}^{d} and f_{if}^{r} for the direct and the rearrangement channels are expressed in terms of T_{if}^{d} and T_{if}^{r} as

$$f_{if}^{d}(\hat{k}_i, \hat{k}_f) = -(M_d/2\pi)\, T_{if}^{d}, \tag{6.163a}$$

$$f_{if}^{r}(\hat{k}_i, \hat{k}_f) = -(M_r/2\pi)\, T_{if}^{r}, \tag{6.163b}$$

where M_d, M_r are the reduced masses of the system in the corresponding channels, and k_i, k_f are the wave vectors of the relative nuclear motion in the initial and final states. If ϑ is the angle between the vectors k_i and k_f, then the differential cross sections for the direct and rearrangement process are

$$\sigma_{if}^{d}(\vartheta) = (k_f/k_i)\, |f_{if}^{d}(\hat{k}_i, \hat{k}_f)|^2, \tag{6.164a}$$

$$\sigma_{if}^{r}(\vartheta) = (M_d k_f / M_r k_i)\, |f_{if}^{r}(\hat{k}_i, \hat{k}_f)|^2. \tag{6.164b}$$

6.3.2 Born Series

Let us note that the operators H_d and H_r in (6.154) contain both the operators of the relative motion as well as the unperturbed "atomic" Hamiltonians

$$H_d = T_R^d + H_0^d, \quad H_r = T_R^r + H_0^r \quad \text{with} \tag{6.165}$$

$$T_R^d = (2M_d)^{-1} \nabla_R^2, \quad T_R^r = -(2M_r)^{-1} \nabla_R^2. \tag{6.166}$$

If ϕ_n^d and ϕ_m^r are eigenfunctions of the Hamiltonians H_0^d and H_0^r, and ε_n^d, ε_m^r are the corresponding eigenenergies, then the decompositions (6.155) imply

$$\psi_n^d = \xi_n(k_n, R_d)\, \phi_n^d(r_d), \quad \psi_m^r = \xi_m(k_m, R_r)\, \phi_m^r(r_r), \tag{6.167a}$$

$$E_n^d = (k_n^2/2M_d) + \varepsilon_n^d, \quad E_m^r = (k_m^2/2M_r) + \varepsilon_m^r, \tag{6.167b}$$

where ξ_n, ξ_m describe the relative motion of the heavy particles, k_n, k_m are the respective wave vectors of the relative motion for the nth direct and the mth rearrangement channel, and R_d and R_r are the position vectors of the projectile's centre of mass with respect to the centre of mass of the target in the direct and rearrangement channels, respectively.

If the potential V_d is a small perturbation, (6.158) can be solved by iteration:

$$\Psi^+ = \psi_i^d + G_d^+ V_d \psi_i^d + G_d^- V_d G_d^+ V_d \psi_i^d + \dots. \tag{6.168}$$

Inserting this expansion into (6.162), one obtains

$$T^{\rm d}_{\rm if} = \langle \psi^{\rm d}_{\rm f} | V_{\rm d} | \psi^{\rm d}_{\rm i} \rangle + \langle \psi^{\rm d}_{\rm f} | V_{\rm d} G^{+}_{\rm d} V_{\rm d} | \psi^{\rm d}_{\rm i} \rangle + \dots , \tag{6.169a}$$

$$T^{\rm r}_{\rm if} = \langle \psi^{\rm r}_{\rm f} | V_{\rm r} | \psi^{\rm d}_{\rm i} \rangle + \langle \psi^{\rm r}_{\rm f} | V_{\rm r} G^{+}_{\rm d} V_{\rm d} | \psi^{\rm d}_{\rm i} \rangle + \dots . \tag{6.169b}$$

Equations (6.169a, b) are the Born series for the T matrix of the direct and rearrangement reaction channels. The nth Born approximation for the corresponding process is obtained by retaining n terms in the above series.

The expansion (6.168) is valid when the effective magnitude of the potential is small compared to the kinetic energy of relative motion. This condition is frequently expressed as $v \gg v_0$, where v_0 is the classical velocity of the bound electron in its initial state. It has been proved that, when this condition holds, the Born series for $T^{\rm d}_{\rm if}$ is convergent. However, the convergence of the Born series for the rearrangement channel $T^{\rm r}_{\rm if}$ is questionable [6.77]. For potentials decreasing at infinity faster than the Coulomb potential, the full Born series converges toward the second Born approximation for asymptotically high collision velocities [6.78]. There are some indications that this is also true for the real ion-atom rearrangement collisions [6.79].

An important feature of the first Born approximation is that the expressions

$$T^{\rm r\,B1}_{\rm if} = \langle \psi^{\rm r}_{\rm f} | V_{\rm r} | \psi^{\rm d}_{\rm i} \rangle \tag{6.170a}$$

$$= \langle \psi^{\rm r}_{\rm f} | V_{\rm d} | \psi^{\rm d}_{\rm i} \rangle \tag{6.170b}$$

are mutually equal provided exact eigenfunctions $\psi^{\rm d}_{\rm i}$ and $\psi^{\rm r}_{\rm f}$ are used in their calculations [6.42]. The expressions (6.170a, b) are respectively called "post-" and "prior-interaction" representations of $T^{\rm r\,B1}_{\rm if}$. Let us note that in the first Born approximation the functions $\xi_n(\boldsymbol{k}_n, \boldsymbol{R}_d)$ and $\xi_m(\boldsymbol{k}_m, \boldsymbol{R}_{\rm r})$ for the relative motion of the heavy particles are in the form of plane waves:

$$\xi_j(\boldsymbol{k}_j, \boldsymbol{R}_\lambda) = \exp(-i\,\boldsymbol{k}_j \cdot \boldsymbol{R}_\lambda), \quad j = 1, 2;\ \lambda = \rm d, r. \tag{6.171}$$

6.3.3 Distorted-Wave Method

The basic idea of the distorted-wave method is to include some parts of the potentials $V_{\rm d}$ and $V_{\rm r}$ into the zero-order Hamiltonian and to treat the rest of the interaction by perturbation theory. Introducing distorting potentials $U_{\rm d}$ and $U_{\rm r}$ in the direct and rearrangement channels by the identity transformations

$$V_{\rm d} = U_{\rm d} + (V_{\rm d} - U_{\rm d}), \quad V_{\rm r} = U_{\rm r} + (V_{\rm r} - U_{\rm r}), \tag{6.172}$$

and Green's operators g_d^+, g_r^- by

$$g_d^+ = (E - H_d - U_d + i\,\eta)^{-1}, \quad g_r^- = (E - H_r - U_r - i\,\eta)^{-1}, \tag{6.173}$$

one can define distorted channel states $\chi_{i,f}^+$ and χ_f^- as

$$\chi_{i,f}^+ = (1 + g_d^+\, U_d)\, \psi_{i,f}^d, \tag{6.174a}$$

$$\chi_f^- = (1 + g_r^-\, U_r)\, \psi_f^r. \tag{6.174b}$$

χ^+ and χ^- satisfy "outgoing-wave" and "incoming-wave" boundary conditions, respectively. Equations (6.174a, b) can be obtained similarly to (6.160). If U_d is chosen in such a way as to have no component from the rearrangement channels, the transition matrix for the direct channels is

$$T_{if}^{d,DW} = \langle \chi_f^+ |\, V_d - U_d \,| \Psi^+ \rangle \tag{6.175}$$

where, similarly to (6.158), Ψ^+ satisfies the equation

$$\Psi^+ = \chi_i^+ + g_d^+ (V_d - U_d)\, \Psi^+. \tag{6.176}$$

In an analogous manner, the T matrix for the rearrangement channels is obtained in the form

$$T_{if}^{r,DW} = \langle \chi_f^- |\, V_r - U_r \,| \Psi^+ \rangle. \tag{6.177}$$

Assuming $(V_d - U_d)$ to be small, and solving (6.176) by iteration, one obtains perturbational series for $T_{if}^{d,DW}$ and $T_{if}^{r,DW}$.

The first terms in these series give the distorted-wave Born approximation (DWBA)

$$T_{if}^{d,DWBA} = \langle \chi_f^+ |\, V_d - U_d \,| \chi_i^+ \rangle, \tag{6.178a}$$

$$T_{if}^{r,DWBA} = \langle \chi_f^- |\, V_r - U_r \,| \chi_i^+ \rangle. \tag{6.178b}$$

Similarly to the Born series, the perturbation series for $T_{if}^{d,DW}$ converges toward $T_{if}^{d,DWBA}$, whereas the convergence of the series for $T_{if}^{r,DW}$ is uncertain [6.80].

There is some arbitrariness in the choice of the distorting potentials U_d and U_r. Possible options are the averaged static ion-atom interaction (for U_d), the internuclear interaction (for U_r), etc. The actual choice is guided by physical intuition and depends on the particular collision problem.

6.3.4 Perturbation Treatment of Close-Coupled Equations

In the semi-classical approximation, the Born and distorted-wave Born approximations can be derived from the system of coupled equations given either in a molecular or atomic basis. A perturbational approach to the close-coupled equations presumes a weak coupling between the channels for the duration of the collision. Starting with the diabatic close-coupled equations (6.118) subject to the initial condition $c_k(-\infty) = \delta_{k0}$, the first-order perturbation result for the amplitude of a $0 \to n$ transition at $t \to \infty$ is

$$c_n(+\infty) = -\mathrm{i} \int_{-\infty}^{+\infty} dt\, \tilde{V}_{0n} \exp\left[-\mathrm{i} \int_{-\infty}^{t} (\tilde{V}_{00} - \tilde{V}_{nn})\, dt'\right] \tag{6.179}$$

with

$$\tilde{V}_{kj} = \langle \varphi_j F_j |\, H \,| F_k \varphi_k \rangle, \quad k, j = 0, n. \tag{6.180}$$

The main contribution to the amplitude $c_n(+\infty)$ comes from the regions of t where the diabatic potentials V_{00} and V_{nn} are degenerate. In the other regions, the oscillations of the exponential in (6.179) are strong and $c_n(t)$ is vanishingly small.

If one defines the φ_k as eigenfunctions of the atomic Hamiltonians H_0^{A} and H_0^{B},

$$H_0^{\mathrm{A}} \phi_k^{\mathrm{A}} = \varepsilon_k^{\mathrm{A}} \phi_k^{\mathrm{A}}, \quad H_0^{\mathrm{B}} \phi_k^{\mathrm{B}} = \varepsilon_k^{\mathrm{B}} \phi_k^{\mathrm{B}}, \tag{6.181}$$

and if F_k, F_j are the plane-wave translational factors, the probability $|c_n(+\infty)|^2$ for a direct $0 \mapsto n$ transition is

$$P_{n,\mathrm{A}}^{\mathrm{DWBA}} = \left| \int_{-\infty}^{\infty} dt\, V_{0n}^{\mathrm{B}} \exp\left[-\mathrm{i} \int_{-\infty}^{t} (V_{00}^{\mathrm{B}} - V_{nn}^{\mathrm{B}} + \varepsilon_0^{\mathrm{A}} - \varepsilon_n^{\mathrm{A}})\, dt\right] \right|^2 \tag{6.182}$$

where

$$V_{kj}^{\mathrm{B}} = \langle \phi_j^{\mathrm{A}} |\, V^{\mathrm{B}} \,| \phi_k^{\mathrm{A}} \rangle, \quad k, j = 0, n, \tag{6.183}$$

and $V^{\mathrm{B}} = H - H_0^{\mathrm{A}}$. An expression analogous to (6.182) can be written for the probability $P_{n,\mathrm{B}}^{\mathrm{DWBA}}$ with the potential $V^{\mathrm{A}} = H - H_0^{\mathrm{B}}$ instead of V^{B}.

Equation (6.182) gives the probability of a direct process (excitation or ionization) in the distorted-wave Born approximation (DWBA) [6.81]. V_{00}^{B} and V_{nn}^{B} play the role of distortion potentials in the initial and final channels, respectively. If these potentials are neglected, one obtains the first Born (B1) approximation for the direct process

$$P_{n,\mathrm{A}}^{\mathrm{B1}} = \left| \int_{-\infty}^{+\infty} dt\, V_{0n}^{\mathrm{B}} \exp\left[-\mathrm{i}\, (\varepsilon_0^{\mathrm{A}} - \varepsilon_n^{\mathrm{A}})\, dt\right] \right|^2, \tag{6.184}$$

and an analogous expression for $P_{n,\mathrm{B}}^{\mathrm{B1}}$.

For the rearrangement (electron capture) channels, the probabilities $P_{n,\mathrm{cx}}^{\mathrm{DWBA}}$ and $P_{n,\mathrm{cx}}^{\mathrm{BI}}$ have to be obtained directly from (6.114) to account for the non-orthogonality of the travelling atomic orbitals $F^{\mathrm{A}}\phi_0^{\mathrm{A}}$ and $F_n^{\mathrm{B}}\phi_n^{\mathrm{B}}$. This gives [6.81, 82]

$$P_{n,\mathrm{cx}}^{\mathrm{DWBA}} = \left| \int_{-\infty}^{+\infty} dt\, \tilde{V}_{0n}^{\mathrm{AB}} \exp\left[-\mathrm{i} \int_{-\infty}^{t} (V_{00}^{\mathrm{B}} - V_{nn}^{\mathrm{A}} + \varepsilon_0^{\mathrm{A}} - \varepsilon_n^{\mathrm{B}})\, dt' \right] \right|^2 , \tag{6.185}$$

$$P_{n,\mathrm{cx}}^{\mathrm{BI}} = \left| \int_{-\infty}^{+\infty} dt\, \tilde{V}_{0n}^{\mathrm{AB}} \exp\left[-\mathrm{i}\, (\varepsilon_0^{\mathrm{A}} - \varepsilon_n^{\mathrm{B}})\, t \right] \right|^2 , \tag{6.186}$$

where

$$\tilde{V}_{0n}^{\mathrm{AB}} \approx V_{0n}^{\mathrm{AB}} - S_{0n}^{\mathrm{AB}} V_{00}^{\mathrm{B}} , \tag{6.187}$$

$$V_{0n}^{\mathrm{AB}} = \langle \phi_n^{\mathrm{B}} |\, \exp(\mathrm{i}\, \boldsymbol{v} \cdot \boldsymbol{r})\, V^{\mathrm{A}}\, | \phi_0^{\mathrm{A}} \rangle , \tag{6.188 a}$$

$$S_{0n}^{\mathrm{AB}} = \langle \phi_n^{\mathrm{B}} |\, \exp(\mathrm{i}\, \boldsymbol{v} \cdot \boldsymbol{r})\, | \phi_0^{\mathrm{A}} \rangle , \tag{6.188 b}$$

and terms of the order of $(S_{0n}^{\mathrm{AB}})^2$ have been neglected.

The orthogonalized form of $P_{n,\mathrm{cx}}$, given by (6.186), is termed the Bates-Born approximation. The usual first Born approximation $P_{n,\mathrm{ex}}$ is obtained by setting $S_{0n}^{\mathrm{AB}} = 0$ in (6.188). The equivalence of the first Born approximation in the semi-classical and wave-mechanical forms has been proved by many authors (see, e.g. [6.40]).

Let us note that the adiabatic form of the coupled equations [e.g. the PSS equations (6.103)] is inconvenient for a perturbation treatment. Even if the coupling matrix element $\langle \Phi_n | \, \partial/\partial t_r \, | \Phi_0 \rangle$ is small everywhere, the non-vanishing energy difference $(E_0 - E_n)$ in the adiabatic phase leads to strong oscillations of the amplitude and to a vanishingly small result. A correct result can be obtained only if the integration is carried out along a contour in the complex t-plane, passing through the complex zero of the difference $(E_0 - E_n)$ [6.83].

6.3.5 Second-Order and Related Approximations

Since, in the Born and distorted-wave series for the T matrix of the $\mathrm{A} + \mathrm{B}^{Z+}$ collision processes, the expansion parameter is $(Z v_0)/v$, the validity condition for the first-order approximation $v \gg Z v_0$ is satisfied for high values of Z only at very high collision energies. In the lower energy region, where collision experiments are most frequently done, the calculations ought to be performed beyond the first-order approximation. Moreover, for the electron-transfer processes, the perturbation series do not converge towards the first-order result, but rather towards the second-order one, meaning that the transitions via intermediate states play a crucial role in these processes.

The second Born approximation for the transition matrix elements of direct and rearrangement processes is given by the first two terms in the expansions (6.169a, b) respectively, i.e.

$$T_{if}^{B2} = T_{if}^{(1)} + T_{if}^{(2)}, \tag{6.189}$$

where we have omitted the superscripts "d" and "r" in T_{if}. The difficulties in the calculations of T_{if}^{B2} come substantially from the term $T_{if}^{(2)}$, which contains an infinite number of intermediate states. These difficulties are resolved by using various approximations either within the expression for $T_{if}^{(2)}$ itself, or by devising procedures which account for the effects of some parts of the interaction to an infinite order. The most successful of the latter approximations are the continuum distorted-wave (CDW) approximation [6.84], the impulse approximation (IA) [6.85, 86] and the eikonal approximation (EA) [6.41]. These approximations have been developed mainly in connection with the electron-transfer problem and in this context are thoroughly discussed elsewhere [6.87, 88]. We shall discuss them briefly.

Within the semi-classical approximation, the rearrangement amplitude $a_{if}^{r,dw}(\varrho)$ in the distorted-wave formalism in the "post"- and "prior-interaction" form is given by [compare with (6.177)]

$$a_{if}^{r,dw}(\varrho) = \mathrm{i} \int_{-\infty}^{\infty} dt \, \langle \chi^- \,|\, V^A - U_f \,|\, \Psi^+ \rangle \,, \quad (\text{"post"}) \tag{6.190a}$$

$$= \mathrm{i} \int_{-\infty}^{\infty} dt \, \langle \Psi^- \,|\, V^B - U_i \,|\, \chi_i^+ \rangle \,, \quad (\text{"prior"}) \tag{6.190b}$$

where V^B and V^A are the ion-atom interactions in the entrance and exit channels, U_i and U_f are the respective distorting potentials in these channels, χ_i^+ and χ_f^- are the corresponding distorted waves and Ψ_i^+ (Ψ_f^-) is the exact electron wave function satisfying an outgoing-(incoming-)wave boundary condition. The distorted waves χ_i^+, χ_f^- satisfy the boundary conditions

$$\lim_{t\to\infty} \langle \chi_i^+ \mid \Psi_f^- \rangle = \lim_{t\to-\infty} \langle \chi_f^- \,|\, \Psi_i^+ \rangle = 0 \,. \tag{6.191}$$

In the continuum distorted-wave (CDW) approximation, the functions χ_i^+ and χ_f^- are chosen in the form

$$\chi_i^+ = F_i^A \psi_i^A \mathcal{L}_i^+ \,, \quad \chi_f^- = F_f^B \psi_f^B \mathcal{L}_f^- \,, \quad \text{with} \tag{6.192a}$$

$$\psi_i^A = \phi_i^A \exp(-\mathrm{i}\, \varepsilon_i^A t) \,, \quad \psi_f^B = \phi_f^B \exp(-\mathrm{i}\, \varepsilon_f^B t) \,, \tag{6.192b}$$

where ϕ_i^A, ϕ_f^B are atomic orbitals and F_i^A, F_f^B are the corresponding plane-wave translational factors.

The distorting potential U_i is chosen in such a way that $\mathcal{L}_i^+$ is a continuum Coulomb function which represents the interaction of the electron in the initial state with the projectile. For the (Z_A, e, Z_B) system with (Z_A, e) being the target, $\mathcal{L}_i^+$ has the form

$$\mathcal{L}_i^+(\boldsymbol{r}_B, t) = \mathcal{N}_B \exp[i(Z_A Z_B/v) \ln(vR - v^2 t)] \times F(i Z_B/v; 1; i v r_B + i \boldsymbol{v} \cdot \boldsymbol{r}_B), \quad (6.193)$$

$$\mathcal{N}_B = \exp(\pi Z_B/v)\, \Gamma(1 - i Z_B/v),$$

where $F(\alpha; \beta; x)$ is the confluent hypergeometric function.

The function $\mathcal{L}_f^-(\boldsymbol{r}_A, t)$ has an analogous meaning for the exit channel,

$$\mathcal{L}_f^-(\boldsymbol{r}_A, t) = \mathcal{L}_i^+(i \to -i, A \leftrightarrow B). \quad (6.194)$$

The CDW approximation consists in replacing Ψ_i^+ and Ψ_f^- in (6.190) by χ_i^+ and χ_f^-, respectively. The final result for the transition amplitude in the CDW approximation is [6.88] ("post" form)

$$a_{if}^{CDW}(\varrho) = i \mathcal{N}_A \mathcal{N}_B (\varrho v)^{2i Z_A Z_B/v} \int_{-\infty}^{\infty} dt \int d\boldsymbol{r} \exp[-i(\boldsymbol{v} \cdot \boldsymbol{r} + \Delta\varepsilon\, t)]\, S_{if}^{(+)}, \quad (6.195)$$

$$S_{if}^{(+)} = \phi_i^A(\boldsymbol{r}_A)\, F(i Z_B/v; 1; i v r_B + i \boldsymbol{v} \cdot \boldsymbol{r}_B)\, \nabla_{r_B} \phi_f^B(\boldsymbol{r}_B) \times \nabla_{r_A} F(i Z_A/v; 1; i v r_A + i \boldsymbol{v} \cdot \boldsymbol{r}_A), \quad (6.196)$$

where $\Delta\varepsilon = \varepsilon_i^A - \varepsilon_f^B$, and $\boldsymbol{r}$ is the electron position vector with respect to the mid-point of the internuclear distance. An analogous expression for a_{if}^{CDW} can be written in the "prior" form.

An essential feature of the CDW approximation is that the asymptotic conditions for the problem are strictly satisfied by the distorted waves χ_i^+ and χ_f^-, given by (6.192). If further approximations are made on these functions (i.e. if the distorting potentials are chosen in some other way), the resulting distorted waves will have an incorrect asymptotic behaviour, leading to loss of accuracy in the transition amplitude. For example, if either of $\mathcal{L}_i^+$ and $\mathcal{L}_f^+$ is approximated by a Coulomb phase factor only (which corresponds to the choice of U_i or U_f in the form of the internuclear interaction), one obtains the continuum intermediate state (CIS) approximation [6.89]

$$a_{if}^{cis} = i \mathcal{N}_B (\varrho v)^{2i Z_A Z_B/v} \int_{-\infty}^{+\infty} dt \int d\boldsymbol{r} \exp[-i(\boldsymbol{v} \cdot \boldsymbol{r} + \Delta\varepsilon\, t)]\, \Lambda_{if}^{(+)}, \quad (6.197)$$

$$\Lambda_{if}^{(+)} = \phi_i^A(\boldsymbol{r}_A)\, \phi_f^{B*}(\boldsymbol{r}_B)\, V^A F(i Z_B/v; 1; i v r_B + i \boldsymbol{v} \cdot \boldsymbol{r}_B). \quad (6.198)$$

Similarly to the CIS approximation, the impulse approximation (IA) also includes a Coulomb continuum state in one of the arrangement channels. The transition amplitude in this approximation is given (in the "post" form) by [6.88]

$$a_{\mathrm{if}}^{\mathrm{IA}}(\varrho)=\mathrm{i}\,\mathcal{N}_{\mathrm{B}}(k)\int_{-\infty}^{+\infty} dt\int d\boldsymbol{k}\int d\boldsymbol{r}_{\mathrm{B}}\exp[\mathrm{i}\,\boldsymbol{k}\cdot\boldsymbol{\varrho}+\mathrm{i}\,(k^2/2+v^2/2-\boldsymbol{k}\cdot\boldsymbol{v}-\Delta\varepsilon)\,t]$$

$$\times\exp(\mathrm{i}\,\boldsymbol{k}\cdot\boldsymbol{r}_{\mathrm{B}}-\mathrm{i}\,k^2\,t/2)\,\mathcal{T}_{\mathrm{if}}^{(+)}\,, \qquad (6.199)$$

$$\mathcal{T}_{\mathrm{if}}^{(+)}=(2\,\pi)^{-3/2}\,\tilde{\phi}_{\mathrm{i}}^{\mathrm{A}}(\boldsymbol{k}-\boldsymbol{v})\,\phi_{\mathrm{f}}^{\mathrm{B}*}(\boldsymbol{r}_{\mathrm{B}})\,V^{\mathrm{A}}F(\mathrm{i}\,Z_{\mathrm{B}}/k;1;\mathrm{i}\,k\,r_{\mathrm{B}}+\mathrm{i}\,\boldsymbol{k}\cdot\boldsymbol{r}_{\mathrm{B}})\,, \qquad (6.200)$$

where $\tilde{\phi}_{\mathrm{i}}^{\mathrm{A}}(\boldsymbol{q}=\boldsymbol{k}-\boldsymbol{v})$ is the Fourier transform of $\phi_{\mathrm{i}}^{\mathrm{A}}(\boldsymbol{r}_{\mathrm{A}})$ and $\mathcal{N}_{\mathrm{B}}(k)=\mathcal{N}_{\mathrm{B}}(v=k)$. The IA amplitude is usually evaluated in the "peaking" approximation $k\approx v$, when the k-integration in (6.199) can be carried out analytically. A detailed review of the impulse approximation is given by *Coleman* [6.90].

While the CDW, CIS and impulse approximations can be considered as truly second-order approximations, the eikonal approximation (EA) may be regarded only as an extension of the first Born approximation. Starting from the general expression for the amplitude of the rearrangement channel in the semi-classical approximation (in the "prior" form),

$$a_{\mathrm{if}}=-\mathrm{i}\int_{-\infty}^{+\infty} dt\,\langle F_{\mathrm{f}}^{\mathrm{B}}\Psi_{\mathrm{f}}^{-}|\,V^{\mathrm{B}}\,|F_{\mathrm{i}}^{\mathrm{A}}\,\psi_{\mathrm{i}}^{\mathrm{A}}\rangle\,, \qquad (6.201)$$

the EA is obtained, approximating Ψ_{f}^{-} by

$$\Psi_{\mathrm{f}}^{-}=\psi_{\mathrm{f}}^{\mathrm{B}}\exp\left(\mathrm{i}\int_{t}^{\infty} V^{\mathrm{A}}\,dt'\right), \qquad (6.202)$$

i.e.

$$a_{\mathrm{if}}^{\mathrm{EA}}=-\mathrm{i}\int_{-\infty}^{+\infty} dt\,\langle F_{\mathrm{f}}^{\mathrm{B}}\,\psi_{\mathrm{f}}^{\mathrm{B}}|\,V^{\mathrm{B}}\exp\left(-\mathrm{i}\int_{t}^{\infty} V^{\mathrm{A}}\,dt'\right)/F_{\mathrm{i}}^{\mathrm{A}}\,\psi_{\mathrm{i}}^{\mathrm{A}}\rangle\,. \qquad (6.203)$$

If the eikonal phase is set equal to unity, $a_{\mathrm{if}}^{\mathrm{EA}}$ reduces to the first Born approximation. For certain types of potentials V^{A} and V^{B}, the amplitude $a_{\mathrm{if}}^{\mathrm{EA}}$ can be calculated analytically. If, in calculating $a_{\mathrm{if}}^{\mathrm{EA}}$, one neglects the longitudinal momentum transfer, then the Glauber approximation is obtained.

Although the second-order approximations mentioned above have been demonstrated for the rearrangement channels, they can also be easily formulated for the direct channels. For these channels however, the first-order approximation of the standard Born or distorted-wave perturbation series provides a correct high-energy limit, and the second-order approximations are required to improve the accuracy of the calculations in the intermediate energy region only. The use of the atomic-state based close-coupling methods in this region is a more adequate approach.

6.4 Classical Descriptions

The use of the concepts of classical mechanics in atomic collision physics has a long history [6.91] and is reviewed in several articles [6.92–94]. The relative success of the classical methods in describing the ion-atom collision processes, at collision velocities comparable to or greater than the orbital velocity of the initial-state electron, is probably based on the specific nature of the Coulomb interaction. It is known that in systems with pure Coulomb interactions (the hydrogen atom and the one-electron two-Coulomb-centre system), additional (dynamical) symmetries are present, which make some properties of these systems the same in classical and quantum mechanics. For example, the dynamical symmetry of the hydrogen atom (the conservation of the Runge-Lentz vector) leads to identical momentum distributions of the bound electron in both quantum (after summation over the degeneracies) and classical mechanics [6.95]. Similarly, the higher symmetry of the (Z_1, e, Z_2) system is reflected in the possibility of separating the variables in the equations of motion both in classical and quantum mechanics (using prolate spheroidal coordinates). Finally, the differential (Rutherford) cross section for scattering in a Coulomb field is the same in classical and quantum collision theory. Of course, when highly excited states are involved in the collision, the use of classical methods is justified by the Bohr and Heisenberg correspondence principles [6.96].

There exist three general methods for treating the inelastic ion-atom collisions in classical mechanics: the classical impulse (or binary-encounter) approximation (BEA), classical perturbation theory, and direct numerical integration of the classical equations of motion by a random selection of the initial conditions and the trajectories (classical trajectory Monte Carlo – CTMC – method). Of these, only the BEA and CTMC methods have been extensively used for practical cross section calculations. They are described briefly below.

6.4.1 Binary-Encounter Approximation (BEA)

The classical impulse approximation treats ion-atom scattering within the following assumptions: (I) the incident ion (assumed to be structureless) interacts with only one of the target electrons, and (II) the interaction of the target electrons with each other and with the nucleus can be neglected during the collision. Assumption (II) requires that the ion velocity $\boldsymbol{v}$ be much larger than the velocity $\boldsymbol{u}$ of the target electron. The BEA model does not require knowledge of the electron position in its initial state but only its velocity distribution. The model is reasonably justified for ionization, and less well founded for excitation and electron transfer, due to the discrete character of the final state.

If ΔE is the energy transferred from the ion to the electron then, after averaging over the random distribution of the angles between the vectors $\boldsymbol{v}$ and $\boldsymbol{u}$, the differential cross section for this energy transfer is [6.93]

$$\frac{d\sigma}{d\Delta E} = \frac{2\pi Z^2}{v^2 (\Delta E)^3} F(v, u, \Delta E), \tag{6.204}$$

where Z is the ion charge and $F(v, u, \Delta E)$ is given by

$$F(v, u, \Delta E) = \Delta E + 2u^2/3, \quad \Delta E \le 2v(v-u) \tag{6.205a}$$

$$= (u/6)\,[4v^2 - (u' - u)^3/2], \quad 2v(v-u) \le \Delta E \le 2v(v+u) \tag{6.205b}$$

$$= 0, \quad 2v(v+u) \le \Delta E, \tag{6.205c}$$

where

$$u'^2 = (2\Delta E + u^2). \tag{6.205d}$$

The ionization cross section is obtained by integrating $d\sigma/d\Delta E$ over ΔE in excess of the ionization potential I_{A_i} of a particular ith shell (or sub-shell) and over the velocity distribution $f(u)$ of the electron,

$$\sigma_{\text{ion}} = \int_0^\infty du\, f(u) \int_{I_{A_i}}^{\Delta E_{\max}} \left(\frac{d\sigma}{d\Delta E}\right) d\Delta E = \int_0^\infty du\, f(u)\, \sigma_i(v, u, I_{A_i}) \tag{6.206}$$

where

$$\sigma_i(v, u, I_{A_i}) = \frac{2\pi Z^2}{v^2} G(v, u, I_{A_i}), \tag{6.207}$$

$$G(v, u, I_{A_i}) = \frac{1}{I_{A_i}} + \frac{u^2}{3 I_{A_i}^2} - \frac{1}{2(v^2 - u^2)}, \quad I_{A_i} \le 2v(v-u) \tag{6.208a}$$

$$= \frac{1}{I_{A_i}} + \frac{1}{3 I_{A_i}^2}[2v^3 + u^3 - (u^2 + 2I_{A_i})^{3/2}] + \frac{1}{2u(v+u)}, \quad 2v(v-u) \le I_{A_i} \le 2v(v+u) \tag{6.208b}$$

$$= 0, \quad 2v(v+u) \le I_{A_i}. \tag{6.208c}$$

The velocity (or momentum) distribution $f(u)$, needed in (6.206) to evaluate σ_{ion}, is given by the square of the quantum-mechanical wave function of the bound electron in momentum space. For the hydrogen atom, $f(u)$ can be calculated classically and coincides with the quantum result,

$$f(u)=\frac{8\,u_c^5}{\pi^2\,(u^2+u_c^2)^4}\,,\quad u_c=(2\,I_H)^{1/2}\,. \tag{6.209}$$

For complex atoms $f(u)$ is usually taken in the form of a delta function,

$$f(u)=\delta\,(u-u_c)\,,\quad u_c=(2\,I_{A_i})^{1/2}\,, \tag{6.210}$$

which, for $v \gg u_c$, is proved to be a sufficient approximation.

For the excitation and electron-transfer processes in which discrete changes in the electron energy and angular momentum are involved, the use of BEA cannot be rigorousely justified, except for the highly excited states. For these processes, *Gryzinski* [6.97] proposed to proceed as in the case of the cross section calculation for ionization, but in the integral over ΔE the limits have to be adequately determined. The excitation and charge-exchange cross sections can be respectively written as

$$\sigma_{ex}=\int_0^\infty du\,f(u)\int_{E_{ex,n}}^{E_{ex,n+1}}\left(\frac{d\sigma}{d\Delta E}\right)d\Delta E\,, \tag{6.211}$$

$$\sigma_{cx}=\int_0^\infty du\,f(u)\int_{E_0-I_B}^{E_0+I_B}\left(\frac{d\sigma}{d\Delta E}\right)d\Delta E\,, \tag{6.212}$$

where $E_{ex,n}$ is the excitation energy of the nth electronic shell, $E_0=(I_A+u^2/2)$ and $-I_B$ is the binding energy of the captured electron in the product ion.

A different two-step BEA model for the electron-capture process has been proposed by *Thomas* [6.98]. In this model, the target electron first interacts with the incident ion and acquires a velocity $\boldsymbol{u}_1$. The scattered electron then undergoes a binary interaction with the target nucleus and acquires a new velocity $\boldsymbol{u}_2$. If $\boldsymbol{u}_2$ has the same magnitude and direction as the velocity $\boldsymbol{v}$ of the incident ion (assumed to be approximately unaltered during the collision), then capture is considered to have taken place. The kinematics of this model requires that in each of the two successive binary encounters, the electron is scattered at 60°. If the target electron is initially in a circular orbit of radius r_0, the electron capture is given by

$$\sigma_{cx}^{Th}=\frac{4\pi\,2^{1/2}}{3}\,\frac{S_{eB}\,(v,60°)\,S_{eA}\,(v,60°)}{r_0^2\,v^3}\,|\mathcal{U}_B\,(r_0)\,|^{3/2}\,, \tag{6.213}$$

where $S_{eA(B)}\,(v,60°)$ is the 60° differential elastic cross section in the BEA and $\mathcal{U}_B\,(r_0)$ is the potential energy of the electron in the field of the projectile, when they are a distance r_0 from each other. It is important to note that for electron capture by a proton from a hydrogen atom (ground-to-ground-state transition), $\sigma_{cx}^{Th} \sim v^{-11}$. The same velocity dependence of the capture cross section is found in the second (double scattering) term of the second Born approximation for this process.

The applicability of the classical impulse approximation is restricted by the condition $v \gg u$. With decreasing collision velocity, quantum interference effects become increasingly important and the basic assumptions of the binary interaction model fail.

6.4.2 Classical Trajectory Monte Carlo (CTMC) Method

Another approach to the problem of inelastic processes in ion-atom collisions would be to undertake a numerical solution of the classical equations of motion for the particles (the electrons and the two nuclei) involved in the collision, with all interactions in the system included. Such an approach has been pioneered by *Hirschfelder* et al. [6.99] for treating chemical reaction dynamics, and in the context of ion-atom collisions was first applied by *Abrines* and *Percival* [6.100] for the three-body Coulomb system. The essential point of this approach is to find an adequate classical representation of the initial quantum-mechanical state of the target atom. For the hydrogen-like atom, the initial bound state of the electron can be represented by a microcanonical distribution of classical orbits [6.100]. The initial conditions for the classical equations of motion are determined by picking members of this microcanonical ensemble at random by a Monte Carlo technique. The cross sections of particular processes are obtained by a statistical analysis of the state of the system long after the collision.

Let us consider in more detail the CTMC method for the (Z_1, e, Z_2) system, with (Z_1, e) being the target and Z_2 (bare nucleus) the projectile. The motion of this three-particle system in classical mechanics is described by the Hamilton equations

$$\frac{\partial H}{\partial p_i} = \dot{q}_i \,, \quad \frac{\partial H}{\partial q_i} = -\dot{p}_i \,, \tag{6.214}$$

where q_i and p_i are coordinates and their conjugate momenta, respectively. Let q_i, $i = 1, 2, 3$ be the Cartesian coordinates of the electron with respect to the charge Z_1, q_i, $i = 4, 5, 6$ the coordinates of the projectile Z_2 with respect to the centre of mass of the target (Z_1, e), and q_i, $i = 7, 8, 9$ the coordinates of the centre of mass of the entire system. The quantities p_i $(i = 1, 2, \dots, 9)$ are the corresponding conjugate momenta. By subtracting the centre of mass motion, the system (5.214) is reduced to twelve coupled equations for

q_i $(i = 1, \ldots, 6)$ and p_i $(i = 1, \ldots, 6)$ [6.101]:

$$\dot{q}_i = (M_1^{-1} + 1)\, p_i\,, \qquad i = 1, 2, 3\,, \tag{6.215a}$$

$$\dot{q}_i = [M_2^{-1} + (1 + M_1)^{-1}]\, p_i\,, \quad i = 4, 5, 6\,, \tag{6.215b}$$

$$\dot{p}_i = (1 + M_1)^{-1} [(1 + M_1)^{-1} q_i + q_{i+3}] \frac{Z_2}{R_1^3} + q_i \frac{Z_1}{R_2^3} + M_1 (1 + M_1)^{-1} [M_1 (1 + M_1)^{-1} q_i - q_{i+3}] \frac{Z_2}{R_3^3}\,, \quad i = 1, 2, 3, \tag{6.215c}$$

$$\dot{p}_i = [(1 + M_1)^{-1} q_{i-3} + q_i] \frac{Z_1 Z_2}{R_1^3} - [M_1 (1 + M_1)^{-1} q_{i-3} - q_i] \frac{Z_1}{R_3^3}\,, \quad i = 4, 5, 6\,, \tag{6.215d}$$

where M_1 and M_2 are the masses of the nuclei Z_1 and Z_2, and R_1, R_2, R_3 are the $(Z_1 - Z_2)$, $(e - Z_1)$, $(e - Z_2)$ separations, respectively. The R_i $(i = 1, 2, 3)$ are given by

$$R_1^2 = \sum_{i=1}^{3} [(1 + M_1)^{-1} q_i + q_{i+3}]^2\,, \tag{6.216a}$$

$$R_2^2 = \sum_{i=1}^{3} q_i^2\,, \tag{6.216b}$$

$$R_3^2 = \sum_{i=1}^{3} [M_1 (1 + M_1)^{-1} q_i - q_{i+3}]^2\,. \tag{6.216c}$$

In order to solve (6.215), the initial conditions q_i^0 and p_i^0 for q_i and p_i must be specified. If the z axis of the Cartesian coordinate system is chosen in the direction of the initial relative velocity vector $\boldsymbol{v}$, then

$$p_4^0 = p_5^0 = 0\,, \quad p_6^0 = [(1 + M_1)^{-1} + M_2^{-1}]\, v\,. \tag{6.217}$$

Further, if the coordinate system is oriented so that the projectile and the centre of mass of the hydrogen atom lie in the $y-z$ plane, then

$$q_4^0 = 0\,, \quad q_5^0 = \varrho\,, \quad q_6^0 = -(R_*^2 - \varrho^2)^{1/2}\,, \tag{6.218}$$

where ϱ is the impact parameter and R_* is the initial distance of Z_2 from the centre of mass of the target atom. (In the calculations R_* is usually taken to be about $10-20$ times Z_2.)

The initial conditions for the electronic coordinates are taken to be

$$\begin{aligned} q_1^0 &= r_e \sin\theta \cos\phi \,, \\ q_2^0 &= r_e \sin\theta \sin\phi \,, \\ q_3^0 &= r_e \cos\phi \,, \end{aligned} \tag{6.219}$$

where (r_e, θ, ϕ) are the spherical coordinates of the electron. Introducing the angle Ω between the plane of the orbit and the plane containing both the z axis and the major axis of the orbit, the initial conditions for the electron momenta can be written as

$$\begin{aligned} p_1^0 &= -p_e (\cos\theta \cos\phi \cos\Omega + \sin\phi \sin\Omega) \,, \\ p_2^0 &= -p_e (\cos\theta \sin\phi \cos\Omega - \cos\phi \sin\Omega) \,, \\ p_3^0 &= p_e \sin\theta \cos\Omega \,, \end{aligned} \tag{6.220}$$

where p_e is the magnitude of the electron momentum in the hydrogen atom. The values r_e and p_e can be expressed in terms of the eccentricity ε of the orbit and the eccentric angle χ by

$$r_e = (Z/2\,U)\,(1 - \varepsilon \cos\chi) \,, \tag{6.221 a}$$

$$p_e = (2\,U)^2\,(1 - \varepsilon^2 \cos^2\chi)^{1/2}/(1 - \varepsilon \cos\chi) \,, \tag{6.221 b}$$

where $-U$ is the electron binding energy in the target atom. The orbit itself is defined by the Kepler equation,

$$\theta_n = \chi - \varepsilon \sin\chi \,, \tag{6.222}$$

where θ_n is the angle swept by the radius vector $\boldsymbol{r}_e$ along the orbit.

Since the energy of the electron in its initial state is fixed, only five of the variables specifying the electronic state in phase space are free. For fixed binding energy, the quantities $\cos\theta$, ϕ, Ω, θ_n and ε^2 define a microcanonical ensemble of classical orbits, the members of which are uniformly distributed in the intervals: $-1 \leq \cos\theta \leq 1$, $0 \leq \phi \leq 2\pi$, $0 \leq \Omega \leq 2\pi$, $0 \leq \varepsilon^2 \leq 1$, $0 \leq \theta_n \leq 2\pi$ [6.100]. In the calculations, numbers from these intervals are selected by the Monte Carlo method. Additionally, the initial condition q_6^0 is also randomly determined by selecting the impact parameter from the interval $0 \leq \varrho \leq \varrho_{max}$. The maximum impact parameter ϱ_{max} is chosen in such a way that the region $\varrho > \varrho_{max}$ gives a contribution to the cross section smaller than the statistical error of the calculations.

With the specified initial conditions, the Hamilton equations (6.214) are solved numerically for a large ($\gtrsim 2000$) number of trajectories. If at the end of the calculation the electron is found near the target nucleus, the process is catalogued as elastic scattering; if it is found in the vicinity of the projectile, then it is concluded that an electron transfer has taken place, and if the electron is far from both the target nucleus and the projectile, the event is interpreted as ionization. Let N_{el}, N_{ion}, and N_{cx} be the number of trajectories leading to elastic scattering, ionization and charge transfer, respectively, and $N = N_{el} + N_{ion} + N_{cx}$ their total number. The cross sections for these processes are defined as

$$\sigma_{el} = (N_{el}/N)\, \pi \varrho_{max}^2 (1 \pm \Delta_{el}) , \tag{6.223}$$

$$\sigma_{ion} = (N_{ion}/N)\, \pi \varrho_{max}^2 (1 \pm \Delta_{ion}) , \tag{6.224}$$

$$\sigma_{cx} = (N_{cx}/N)\, \pi \varrho_{max}^2 (1 \pm \Delta_{cx}) , \tag{6.225}$$

where

$$\Delta_\lambda = [(N - N_\lambda)/N N_\lambda]^{1/2} , \quad \lambda = \text{el, ion, cx} \tag{6.226}$$

is the standard deviation of the results for each particular process. From a computational point of view, an attractive feature of the CTMC method is that all three cross sections, σ_{el}, σ_{ion} and σ_{cx}, can be calculated from one set of numerical integrations.

The above expressions for σ_λ represent the total cross sections for the corresponding processes. For the electron capture, a further analysis of the electron states after the collision is possible. Continuous classical "principal" and "angular quantum numbers" n_c and l_c can be associated with the classical energy E and momentum J by the usual quantum relations

$$E = -Z^2/2n_c^2 , \quad J = [l_c (l_c + 1)]^{1/2} . \tag{6.227}$$

If, further, one performs a "quantization" of n_c and l_c according to the rules

$$n - 1/2 \leq n_c \leq n + 1/2 , \quad l \leq l_c \leq l + 1 , \tag{6.228}$$

the capture into specific n, l quantum substates can also be described by the CTMC method.

The velocity region in which the CTMC method has been demonstrated to provide fairly accurate cross sections is restricted to $v_0 \lesssim v \lesssim 3v_0$, where $v_0 = (2U)^{1/2}$.

7. Collisions of Atoms (Ions, Molecules) with Highly Charged Ions: Charge-Transfer Processes

We start the consideration of specific inelastic processes that occur in the collisions of multiply charged ions with neutral atomic particles by looking at the class of charge-transfer (or electron-capture, or charge-exchange) processes. The change of the charge state of the collision partners, the creation of excited reaction products and the extremely high cross sections bring to this class of processes a great practical importance. The fact that the projectile ion is highly charged makes possible the capture of more than one electron from the target in a single collision. In this chapter we shall consider both single- and many-electron-capture processes in the energy range from a few eV to several MeV/amu. Our main attention, however, will be devoted to the single-charge-transfer processes, for which the theoretical and experimental information at present is most complete. Emphasis in our considerations will be given to the basic physical mechanisms governing the charge-transfer processes and their theoretical description.

The general theoretical methods, described in the preceding chapter, will be a basis for further developments of the theory of charge-transfer processes. However, we shall also describe other methods which have been devised to reflect some specific aspects of the electron-capture collisions involving highly charged ions. In addition to ion-atom systems, the charge-transfer processes occurring in ion-ion and ion-molecule collisions will also be treated briefly. The electron-capture processes, accompanied by other inelastic electron transitions either within the target spectrum or into the continuum will be discussed in the next two chapters.

The basic information gained thus far on the charge-transfer processes of multiply charged ions with atoms and molecules is contained in several review articles [7.1–9].

7.1 General Considerations

One of the main features of the single-electron-capture process in atom-highly charged ion collisions,

$$\mathrm{A} + \mathrm{B}^{Z+} \rightarrow \mathrm{A}^{+} + B^{(Z-1)+}, \quad Z \gg 1, \tag{7.1}$$

is that many electronic states are always involved in the collision dynamics. For low relative velocities ($v \ll 1$) the molecular picture of the collision system is appropriate, and the nonadiabatic transitions at pseudocrossings of the initial-state potential energy curve with the multitude of the ionic potential energy curves, corresponding to different excited product-ion final states, are the dominant mechanisms in the processes. At intermediate collision velocities ($v \sim 1$), the collision dynamics includes strong interactions of many discrete atomic states, and at still higher velocities, the role of continuum states becomes increasingly important. As was stressed in Sect. 6.3.5, the second-order effects (i.e. transitions *via* intermediate states) in (7.1) remain essential even in the asymptotic velocity region, thus invalidating (rigorously speaking) any two-state or first-order perturbation treatment of this process.

Another essential feature of the process (7.1) is that, except for very high collision energies, the electron is always preferentially captured into a relatively small group of excited states of the product ion. This is a consequence of two circumstances: (I) the electron-capture process has the highest probability when an energy resonance takes place between the initial and final states of the reaction, and (II) when Z is high, the initial-state energy level is in resonance with the energy levels of excited product-ion states. Creation of excited states characterizes not only the single-electron-capture process (7.1) but also some other multi-electron transition processes in which transfer of one or more target electrons into the projectile ion is involved.

Besides the relative velocity v, two other parameters of the colliding system influence the collision dynamics: the charge Z of the multiply charged ion and the number of electrons in the system. The ionic charge has the role of scaling the velocity regions in which the quasi-molecular and atomic expansion methods are valid and, therefore, appears in combination with the velocity in the expansion parameters for the theory, both in the adiabatic and in the high-energy region.

The number of electrons in the system influences the collision dynamics by determining at low collision energies the complexity of the interaction paths for different reaction channels. At high collision energies, capture from the inner electronic shells becomes dominant, thus coupling the dynamics of process (7.1) with other, usually multi-electron, transition processes.

From a theoretical point of view, the hydrogen atom-fully stripped ion system (we shall denote this system as $H + Z$) is of particular importance. The electronic structure for this system can be considered known (Sect. 6.1) and, due to its high intrinsic symmetry, the number of interaction couplings involved in the charge-exchange process for a given collision energy is smaller than in any other colliding system. The study of the electron-capture process in the $H + Z$ system thus provides a possibility of comparing the accuracy of different theoretical approximations to the charge-exchange scattering problem.

We shall take the relative collision velocity v as the most relevant dynamical parameter for characterizing the process (7.1), and with respect to it, we shall distinguish three broad collision regions: low-energy (or adiabatic) region ($v \ll 1$), intermediate-energy region ($v \sim 1$) and high-energy ($v \gg 1$) region. The theoretical descriptions of the process can also be classified accordingly.

7.2 Charge Exchange at Low Energies

There are three general theoretical methods for treating the charge-exchange problem of multiply charged ions at low energies: the molecular-orbital close-coupling (MO-CC) method, the multichannel Landau-Zener theory, and various decay models. The close-coupling method provides the best description of the process, provided the expansion basis is sufficiently large. With increasing ionic charge and complexity of the electronic structure of the collision system, the application of the MO-CC method encounters considerable practical difficulties. A simplification of the MO-CC approach can be reached if the coupling between different exit channels is neglected and the transitions are considered as localized at the pseudo-crossings of adiabatic potential energy curves, corresponding to the initial and multitude of final ionic states (Fig. 7.1). In this approximation, the collision dynamics can be described by a multicurve-crossing Landau-Zener

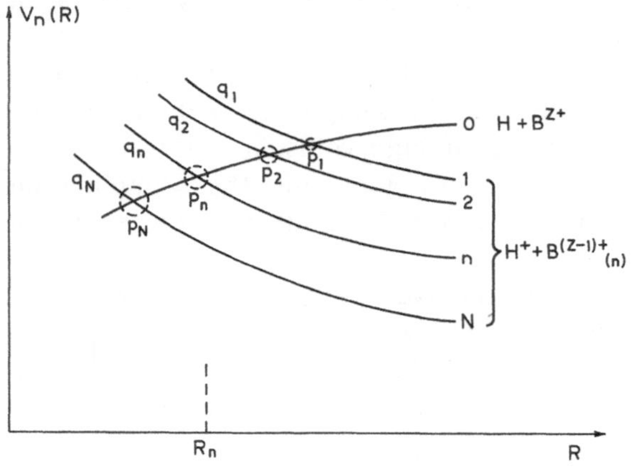

Fig. 7.1. Schematic diagram of diabatic potentials of hydrogen atom-fully stripped ion system at large distances

theory. Finally, when the number of open reaction channels becomes very large (as in the case of ions with very large Z and/or many-electron target atoms, for example), the electron capture can be represented as the decay of the initial electronic state into a quasi continuum of final ionic states. These three approaches have proved to be useful for describing different aspects of the charge-exchange process at low collision energies. We shall now elaborate them in more detail.

7.2.1 Decay Models

The basic ideas of the decay models for the electron-capture process (7.1) can be derived from the concept of the decay of a quasi-stationary state in the field of a multiply charged ion [7.10]. If the final ion product states available for the reaction are so dense that they can be considered as forming a continuum (or quasi continuum), then the initial electronic state interacting with this continuum of states becomes quasi-stationary, and its energy is characterized by a decay width Γ. The probability for electron capture per unit time is identical (in atomic units) with Γ.

Let us consider first the one-electron system (Z_1, e, Z_2) with (Z_1, e) being the target, and $Z_2 \gg Z_1$. Due to this condition, the collision system possesses a pronounced axial symmetry and the variables in the Schrödinger equation for the (Z_1, e, Z_2) system can be approximately separated in parabolic coordinates:

$$\begin{aligned} &\mu = r + z, \qquad \nu = r - z, \qquad \phi = \arctan(y/x), \\ &0 \leqq \mu < \infty, \quad 0 \leqq \nu < \infty, \quad 0 \leqq \phi < 2\pi \end{aligned} \tag{7.2}$$

where $\boldsymbol{r}(x, y, z)$ is the position vector of the electron with respect to the centre Z_1. The electron capture is pictured as electron tunelling from the atomic into the ionic potential well through a barrier between these wells, created during the collision. The main contribution to the electron tunelling flux comes from a relatively narrow cylindrical region around the internuclear axis (along which we orient the z axis), defined in the (μ, ν, ϕ) coordinates by the conditions $\mu \ll \nu < 2R$, $(\mu\,\nu)^{1/2} \ll R$, where R its the internuclear distance. Representing the electron wave function in the form

$$\Psi = \frac{X(\mu)\; Y(\nu)}{(2\pi\,\mu\,\nu)^{1/2}}\, \mathrm{e}^{\pm i m \phi}, \tag{7.3}$$

the Schrödinger equation (6.1) for the (Z_1, e, Z_2) system is transformed in the form [7.10]

– near centre Z_1:

$$X_1''(\mu) + \left[\frac{1}{2}\tilde{E} + \frac{\beta_1}{\mu} + \frac{(1-m^2)}{4\mu^2} - \frac{1}{4}\tilde{F}\mu\right] X_1(\mu) = 0, \tag{7.4a}$$

$$Y_1''(\nu) + \left[\frac{1}{2}\tilde{E} + \frac{\beta_2}{\nu} + \frac{(1-m^2}{4\nu^2} + \frac{1}{4}\tilde{F}\nu\right] Y_1(\nu) = 0 \tag{7.4b}$$

– in the cylindrical region $\mu \ll \nu < 2R$, $(\mu\nu)^{1/2} \ll R$ and $R \gg 1$:

$$X_2''(\mu) + \left(\frac{1}{2}\tilde{E} + \frac{\beta_1}{\mu} + \frac{(1-m^2)}{4\mu^2}\right) X_2(\mu) = 0, \tag{7.5a}$$

$$Y_2''(\nu) + \left(\frac{1}{2}\tilde{E} + \frac{\beta_2}{\nu} + \frac{(1-m^2)}{4\nu^2} + \frac{2Z_2\nu}{2R(2R-\nu)}\right) Y_2(\nu) = 0, \tag{7.5b}$$

$$\tilde{E} = E + Z_2/R, \quad \tilde{F} = Z_2/R^2, \tag{7.6}$$

where E is the electron energy and β_1, β_2 are the separation constants, satisfying the relation $\beta_1 + \beta_2 = Z_1$. The electron tunelling takes place only along the ν-direction, so that the boundary conditions associated with (7.4, 5) are

$$X_{1,2}(0) = 0, \quad X_{1,2}(\mu) \underset{\mu\to\infty}{\longrightarrow} 0, \tag{7.7a}$$

$$Y_{1,2}(0) = 0, \quad Y_2(\nu) \underset{\nu>\nu_1}{\longrightarrow} \text{outgoing wave} \tag{7.7b}$$

where ν_1 is the turning point of (7.5b).

The spectral parameters β_1 and β_2 are determined from the boundary-value problems (7.4–7), while the energy is obtained from the equation

$$\beta_1(E) + \beta_2(E) = Z_1. \tag{7.8}$$

Due to the second of the boundary conditions (7.7b), the separation constant β_2 is complex, which results in a complex solution of (7.8),

$$E = E_0(R) - \tfrac{1}{2}\mathrm{i}\,\Gamma_{n_1'n_2'm}(R), \tag{7.9}$$

$$E_0(R) = -\frac{Z_1^2}{2n_0^2} - \frac{Z_2}{R} + \frac{3}{2}\frac{Z_2 n_0(n_1'-n_2')}{Z_1 R^2} + O(Z_2/R), \tag{7.10}$$

$$\begin{aligned}\Gamma_{n_1'n_2'm}(R) = {} & \frac{Z_1^2}{n_0^3\, n_2'!\,(n_2'+m)!}\left(\frac{4Z_1^3R^2}{Z_2 n_0^3}\right)^{2n_2'+1+m} \\ & \times \exp\{-[Z_1^3R^2/Z_2n_0^3 - (9/2)(n_1'-n_2')]f(\alpha) \\ & - 2(2n_2'+1+m)\,g(\alpha)\}\,[1+O(Z_2/R)],\end{aligned} \tag{7.11}$$

where $n_0 = n_1' + n_2' + m + 1$ and n_1', n_1', m are the parabolic quantum numbers of the electron centred on Z_1. The functions $f(\alpha)$ and $g(\alpha)$ are defined by

$$f(\alpha) = (1/\alpha)\,[1 - g(\alpha)/\alpha]\,, \quad \alpha = Z_1^2 R/2 n_0^2 Z_2\,,$$
$$g(\alpha) = \alpha^{1/2} \ln [\alpha^{1/2} + (1+\alpha)^{1/2}]/(1+\alpha)^{1/2}\,. \quad (7.12)$$

The expression (7.11) for $\Gamma_{n_1' n_2' m}(R)$ is valid under the condition $R \gg n_0^2 (2 Z_2/Z_1^3)^{1/2}$.

$\Gamma_{n_1' n_2' m}(R)$ determines the decay probability of the Stark state $[n_0\, n_1'\, n_2'\, m]$. The electron transition probability $W_{n_0 l_0 m}(R)$ in the angular momentum representation $(n_0\, l_0\, m)$ is given by

$$W_{n_0 l_0 m}(R) = \sum_{n_1' + n_2' = n_0 - 1 - m} |\langle l_0 | n_1'\, n_2' \rangle|^2\, \Gamma_{n_1' n_2' m}(R) \quad (7.13)$$

where $\langle l_0 | n_1'\, n_2' \rangle$ are the Clebsch-Gordan coefficients. The leading term in (7.13) is [7.10]

$$W_{n_0 l_0 m}(R) = \frac{Z_1^2 (2 l_0 + 1)(l_0 + m)!}{n_0^3\, m!\, (n_0 + l_0)!\, (n_0 - l_0 - 1)!\, (l_0 - m)!} \left(\frac{4 Z_1^3 R^2}{n_0^3 Z_2}\right)^{2n_0 - 1 - m}$$
$$\times \exp\{-[Z_1^3 R^2/n_0^3 Z_2 + (9/2)(n_0 - 1 - m)]\, f(\alpha)$$
$$- 2(2n_0 - 1 - m)\, g(\alpha)\}\, [1 + O(Z_2/R^2)]\,. \quad (7.14)$$

The above expression for $W_{n_0 l_0 m}(R)$ is valid only for the (Z_1, e, Z_2) system. It can be easily generalized to the case of an arbitrary $\mathrm{A}(n_0\, l_0\, m) + \mathrm{B}^{Z+}$ colliding system, treating the atom in the independent-electron approximation and considering B^{Z+} as a structureless ion. In that case $(Z_1 = 1, Z_2 = Z \gg 1)$, dropping the linear Stark term in (7.14), one obtains

$$W_{n_0 l_0 m}(R) = A_c^2 \left(\frac{n_0}{2}\right)^{2n_0} \frac{(2 l_0 + 1)(l_0 + m)!}{m!\,(l_0 - m)!} \left(\frac{4R^2}{n_0^3 Z}\right)^{2n_0 - 1 - m}$$
$$\times \exp\left[-\frac{R^2}{n_0^3 Z} f(\alpha) - 2(2n_0 - 1 - m)\, g(\alpha)\right] [1 + O(Z/R^2)]\,, \quad (7.15)$$

$$\alpha = R\,(2 Z n_0^2)^{-1}, \quad n_0 = (2 I_\mathrm{A})^{-1/2}, \quad (7.16\mathrm{a})$$

$$A_c = (1/n_0)(2/n_0)^{n_0} [\Gamma(n_0 + l_0 + 1)\, \Gamma(n_0 - l_0)]^{-1/2}, \quad (7.16\mathrm{b})$$

with $-I_\mathrm{A}$ being the initial-state electron binding energy. The constant A_c is the Coulomb "normalization" constant of the asymptotic form of the initial-state atomic wave function. If I_A is not sufficiently small A_c has to be replaced with another constant, A, determined from the equation

$$\psi_\mathrm{HF}(r)\big|_{r=r_0} = A\, r^{n_0 - 1}\, \mathrm{e}^{-r/n_0}\big|_{r=r_0} \quad (7.17)$$

Table 7.1. The values of the atomic initial-state parameters $(1/n_0) = (2I_A)^{1/2}$ and A appearing in (7.15) for a number of ground-state atoms

Atom	$1/n_0$	A	Atom	$1/n_0$	A
H	1	2	K	0.565	0.52
He	1.344	2.87	Ca	0.670	0.95
Li	0.630	0.82	Ti	0.708	1.16
Be	0.828	1.62	Cr	0.705	1.13
B	0.781	0.88	Fe	0.761	1.40
C	0.910	1.30	Co	0.760	1.42
N	1.034	1.50	Ni	0.749	1.42
O	1	1.30	Cu	0.754	1.29
F	1.132	1.59	Zn	0.831	1.69
Ne	1.259	1.76	Kr	1.014	2.22
Na	0.615	0.74	Rb	0.554	0.48
Mg	0.750	1.32	Sr	0.647	0.86
Al	0.663	0.61	Zr	0.709	1.15
Si	0.774	1.10	Mo	0.722	1.23
P	0.878	1.65	Ag	0.746	1.18
S	0.873	1.11	Xe	0.944	2.4
Cl	0.976	1.78	Cs	0.535	0.41
Ar	1.076	2.11			

where $\psi_{HF}(r)$ is the radial part of the wave function of the active electron in the Hartree-Fock approximation and r_0 is the distance at which ψ_{HF} enters the asymptotic region and still accurately represents the electronic motion. The values of the constant A and $n_0^{-1} = (2I_A)^{1/2}$ for a number of ground state atoms are given in Table 7.1 [7.11].

In two limiting cases, the expression (7.15) for $W_{n_0 l_0 m}(R)$ can be simplified. For $\alpha \gg 1$ (i.e. $I_A \gg Z/R$) one has

$$W^{(0)}_{n_0 l_0 m}(R) = A^2 \left(\frac{n_0}{2}\right)^{2n_0} \frac{(2l_0+1)\,(l_0+m)!}{m!\,(l_0-m)!} \times \left(\frac{2R}{n_0}\right)^{2n_0-1-m} \exp\left(-\frac{2R}{n_0}\right), \quad R \gg 2n_0^2 Z\,. \tag{7.18}$$

$W^{(0)}_{n_0 l_0 m}$ does not depend on the ionic charge Z, which means that the electron transition has a properly resonant character. In the opposite case, $\alpha \ll 1$ (i.e. $I_A \ll Z/R$), $W_{n_0 l_0 m}(R)$ becomes

$$W^{(F)}_{n_0 l_0 m}(R) = \frac{A^2}{n_0}\left(\frac{n_0}{2}\right)^{2n_0} \frac{(2l_0+1)\,(l_0+m)!}{m!\,(l_0-m)!} \left(\frac{4R^2}{n_0^3 Z}\right)^{2n_0-1-m} \times \exp\left(-\frac{2}{3}\frac{R^2}{n_0^3 Z}\right), \quad n_0(2Z)^{1/2} \ll R \ll 2n_0^2 Z\,. \tag{7.19}$$

The expression $W^{(F)}_{n_0 l_0 m}(R)$ coincides with the decay probability of an atomic state in a homogeneous external electric field $F = Z/R^2$, and has been considered by *Chibisov* [7.12].

For the system of a ground-state hydrogen atom and a multicharged ion Z the above expressions for $W_{n_0 l_0 m}$ have the form ($n_0 = 1$, $l_0 = m = 0$, $A = 2$)

$$W_H(R) = \frac{4R^2}{Z} \exp\left[-\frac{R^2}{Z} f(\alpha_H) - 2g(\alpha_H)\right], \quad \alpha_H = R/2Z, \tag{7.20}$$

$$W_H^{(0)}(R) = 2R\, e^{-2R}, \quad R \gg 2Z, \tag{7.21}$$

$$W_H^{(F)}(R) = (4R^2/Z)\, e^{-2R^2/3Z}, \quad (2Z)^{1/2} \ll R \ll 2Z. \tag{7.22}$$

Expressions for $W_H(R)$, analogous to those given by (7.20–22) can be derived by approximately solving the Schrödinger equation for the $(H+Z)$ system in the cylindrical region around the internuclear axis, using the prolate spherical coordinates (ξ, η, ϕ) and an "outgoing-wave" boundary condition along the η coordinate [7.13].

In the adiabatic energy region, the probability P_{cx} for the electron-capture process is related to the decay probability $W(R)$ by

$$P_{cx} = 1 - \exp\left[-\int_{-\infty}^{\infty} W(R)\, dt\right], \tag{7.23}$$

where $R = R(t)$ is the classical trajectory for the nuclear motion. Using a rectilinear trajectory, $R^2 = \varrho^2 + (v\,t)^2$, the charge-exchange cross section in the decay model can be written in the form

$$\sigma_{DM} = 2\pi \int_0^{\infty} \varrho\, d\varrho\, \{1 - e^{-\chi(\varrho, v)}\} \quad \text{with} \tag{7.24}$$

$$\chi(\varrho, v) = 2\int_{\varrho}^{\infty} \frac{W(R)\, R\, dR}{v\,(R^2 - \varrho^2)^{1/2}}. \tag{7.25}$$

Since $W(R)$ depends strongly on R, $\chi(\varrho, v)$ also strongly varies with ϱ. Using this fact, the cross section σ_{DM} can be represented as [7.10, 12]

$$\sigma_{DM} = \pi \varrho_0^2, \quad \chi(\varrho_0, v) = e^{-C} = 0.56, \tag{7.26}$$

where $C = 0.577\ldots$ is the Euler constant. Introducing further a new variable

$$x_0 = \varrho_0^2 / n_0^3 Z \tag{7.27}$$

and estimating $\chi(\varrho_0, v)$ with the expression (7.15) for $W(R)$, one obtains [7.10]

$$\sigma_{DM} = \pi Z n_0^3 x_0, \tag{7.28}$$

where x_0 is determined from the equation

$$x_0^{2n_0-1-m} \exp\left[-x_0 f(\alpha_0)\right] = \frac{v}{B}\left(\frac{f(\alpha_0)}{\pi n_0^3 Z}\right)^{1/2} \exp\left[\varphi(\alpha_0) - C\right] \qquad (7.29\,\mathrm{a})$$

with

$$B = \frac{A^2}{n_0}\left(\frac{n_0}{2}\right)^{2n_0} 2^{2(2n_0-1-m)} \frac{(2l_0+1)(l_0+m)!}{m!\,(l_0-m)!}\,, \qquad (7.29\,\mathrm{b})$$

$$\varphi(\alpha_0) = 2(2n_0 - 1 - m)\, g(\alpha_0)\,, \quad \alpha_0 = (1/2)(x_0/n_0 Z)^{1/2}\,. \qquad (7.29\,\mathrm{c})$$

The function $x_0 = x_0(n_0, v, Z)$ is a slowly varying (logarithmic) function of v and n_0, and almost independent of Z. Thus, the general dependence of the cross section on v, Z and n_0 in the decay model is [7.10]

$$\sigma_{\mathrm{DM}} \sim a - b \ln v\,, \quad \sigma_{\mathrm{DM}} \sim Z\,, \quad \sigma_{\mathrm{DM}} \sim n_0^3\,, \qquad (7.30)$$

where a and b are constants. The specific form of the decay model described above is frequently called the "electron tunnelling model".

The charge-exchange problem in a colliding system with a large number of reaction channels can be alternatively considered as a limiting case of the multichannel Landau-Zener theory (see below). In a dense potential energy curve-crossing system, one can define an internuclear distance R_s such that for $\varrho > R_s$, the pseudocrossings can be considered as being passed completely diabatically (i.e. without nonadiabatic transitions between the interacting states), and for $\varrho \leq R_s$ the reaction probability can be taken equal to unity. In such an absorbing sphere model (ASM), the charge-exchange cross section is given by

$$\sigma_{\mathrm{ASM}} = \pi R_s^2\,. \qquad (7.31)$$

The critical distance R_s can be determined by using different theoretical arguments. *Olson* and *Salop* [7.14] proposed to determine R_s from the condition that at $\varrho = R_s$, the Landau-Zener cross section for an equivalent single-crossing problem attains its maximum. This condition is

$$\frac{\pi \Delta^2(R_s)}{2v\,|\Delta F(R_s)|} = 0.15\,, \qquad (7.32)$$

where $\Delta F = (Z-1)/R^2$ and $\Delta(R)$ is the one-electron exchange interaction. An ad hoc generalization of the expression (6.53) for the H $(1s) + Z$ system to the case of an arbitrary atom A, having an ionization potential $I_{\mathrm{A}} = 1/(2n_0^2)$, gives [7.14]

$$\Delta(R) = 18.26\, Z^{-1/2} \exp(-1.324\, R/n_0 Z^{1/2})\,. \qquad (7.33)$$

A first-order iterative solution of (7.32) with the above expression for $\Delta(R_s)$ gives $R_s \approx n_0 Z^{1/2} \ln(c/vZ)$, where c is constant. The absorbing sphere model predicts, thus, the form of the charge-exchange cross section as

$$\sigma_{ASM} \simeq \pi n_0^2 Z \ln^2\left[\frac{c}{vZ}\ln^2\left(\frac{c}{vZ}\cdots\right)\right]. \tag{7.34}$$

Similarly to σ_{DM}, σ_{ASM} depends logarithmically on v, but its dependence on n_0 and Z is somewhat different from that predicted by the electron tunnelling model. It should be noted, however, that the dependence of σ_{ASM} on these quantities is determined entirely by the form of the combination in which R_s, n_0 and Z appear in the exponent of the expression for $\Delta(R_s)$. In (7.33) this combination has been established on semi-empirical grounds. In Fig. 7.2 the Z-dependence of σ_{DM} and σ_{ASM} is presented for a collision velocity $v = 7 \times 10^7$ cm/s.

The electron tunnelling model predicts that only a very narrow band of final-state ionic levels is populated by the electron-capture process. This is a consequence of the basic assumptions of the model: the quasi-continuous character of the ionic energy spectrum and the quasi-resonance character of the tunnelling mechanism. The ionic energy level, which is preferentially populated by the electron-capture process, can be determined on the basis of the energy resonance between the shifted initial- and final-state energy levels $[I_A = I_B - (Z-1)/R,\ I_A = 1/(2n_0)^2,\ I_B = Z^2/(2n_B^2)]$, taken at the internuclear distance R_0 at which the process dominantly takes place. The existence of this distance follows from the exponential behaviour of $W(R)$ at large values of R. For high values of Z, the equation $\chi(R_0, v) = 0.56$ gives (to logarithmic accuracy) $R_0 \simeq 2n_0^2(2Z)^{1/2}$. With this value of R_0, the

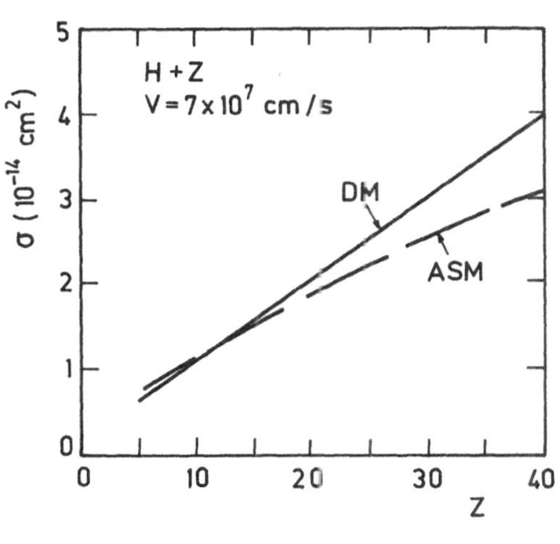

Fig. 7.2. Z-dependence of the electron-capture cross section in H+Z collisions at $v = 7 \times 10^7$ cm/s in the decay model (DM) and the absorbing sphere model (ASM)

above-mentioned resonant condition gives, for the value of the principal quantum number of the preferentially populated final-state ionic level [7.15],

$$n_m^{\mathrm{DM}} = n_0 Z\,[1 + (Z-1)\,(2Z)^{-1/2}]^{-1/2}, \tag{7.35a}$$

$$\xrightarrow[Z^{1/2} \gg 1]{} 2^{1/4} n_0 Z^{3/4}. \tag{7.35b}$$

Numerous cross section calculations have been performed by using the decay models [7.10, 12–15]. Both one- and many-electron targets have been considered. For the H+Z system, the electron-capture cross section in the decay model can be represented in the form (within a 10% accuracy) [7.13]

$$\sigma_{\mathrm{DM}} = 1.7\pi Z \ln(314/v) \tag{7.36}$$

for $Z \geq 4$ and $v \leq 1$.

7.2.2 Multichannel Landau-Zener Model

If the number of final ionic states of the system $A + B^{Z+}$ interacting with the initial covalent state is not excessively large, and if the corresponding pseudocrossing regions are mutually well separated, the electron-capture process can be treated by the multichannel Landau-Zener theory [7.16]. The Landau-Zener model for a single two-state pseudocrossing requires that the conditions

$$a = 2\delta R_c/n_0 \ll 1, \quad b = (\pi/2)\,(\delta R_c/R_c)^{1/2} \ll 1 \tag{7.37a}$$

be fulfilled, where R_c is the diabatic crossing point and $\delta R_c \simeq \Delta(R_c)/[2\,|F(R_c)|]$ is the width of the transition zone. With the expression for $\Delta(R_c)$ in the form of (7.33), $a \sim b \sim Z^{-1/2}$, so that for high values of Z, the two-state Landau-Zener model is applicable. The fulfillment of the condition for non-overlapping of the transition zones,

$$\Delta R_{c,n}^{c,n+1} = R_{c,n+1} - R_{c,n} \gg \delta R_{cn}, \tag{7.37b}$$

where $R_{c,n}$ and $R_{c,n+1}$ are two successive crossings, depends on the colliding system. Although $\delta R_c \sim Z^{-1/2}$, $\Delta R_{c,n}^{c,n+1}$ for complex atoms may be comparable and even smaller than δR_c. For the $H(1s) + Z$ system, however, $\Delta R_{c,n}^{c,n+1} \approx 2(Z-1)(2n+1)/(Z^2 - n^2)^2 \sim Z^{-1/4}$ (when $n \sim n_m \sim Z^{3/4}$) and, for sufficiently high Z, the condition (7.37b) is satisfactorily fulfilled.

Let us consider a collision system with N reaction channels. Let $R_1 > R_2 > \ldots > R_n > \ldots > R_N$ be the crossing points of the final-state diabatic potentials $V_n(R)$ and the potential of the initial state $V_0(R)$ (see Fig. 7.1). If p_n $(n = 1, 2, \ldots, N)$ is the two-state transition probability at the crossing R_n, then, neglecting interference effects, the probability P_n for population of the nth ionic level in the electron-capture process is given by [7.17]

$$\begin{aligned} P_n = & p_1 p_2 \dots p_n (1-p_n) [1 + (p_{n+1} p_{n+2} \dots p_N)^2 \\ & + (p_{n+1} p_{n+2} \dots p_{N-1})^2 (1-p_N)^2 \\ & + (p_{n+1} p_{n+2} \dots p_{N-2})^2 (1-p_{N-1})^2 \\ & \vdots \\ & p_{n+1}^2 (1-p_{n+2})^2 + (1-p_{n+1})^2] . \end{aligned} \tag{7.38}$$

The single-crossing probability p_n is given by the Landau-Zener formula [see (6.139)]

$$p_n = \exp\left[-\frac{\pi \Delta^2(R)}{2 v_R |\Delta F(R)|} \right]_{R=R_n} , \tag{7.39}$$

where in the straight-line-trajectory approximation the radial velocity is

$$v_R = v [1 - \varrho^2/R^2 - V_0(R)/E]^{1/2} \approx v (1 - \varrho^2/R^2)^{1/2} , \tag{7.40}$$

and $\Delta F = (d/dR) [V_0(R) - V_n(R)] \approx (Z-1)/R^2$. The partial cross section for capture into a final state n is given by

$$\sigma_n = 2\pi \int_0^{\varrho_n} \varrho \, d\varrho \, P_n(\varrho, v) , \tag{7.41}$$

$$\varrho_n = R_n [1 - V_0(R_n)/E]^{1/2} \approx R_n , \tag{7.42}$$

and the total charge-exchange cross section is

$$\sigma = \sum_{n=1}^{N} \sigma_n . \tag{7.43}$$

The multichannel Landau-Zener (MLZ) model presented here has an important limitation: it completely neglects the electron transitions caused by the rotation of the internuclear axis. This leads to an underestimation of the cross section in the low-energy region (below the cross section maximum) by a factor of two, or even more.

The inclusion of rotational transitions in the MLZ model can be easily done for the H+Z collision system [7.18, 19]. In this system, the initial molecular state [1000] interacts by radial coupling in the pseudocrossing region δR_n only with one of the degenerate ionic molecular states $[n\, n_1\, n_2\, m]$, namely the one with parabolic quantum numbers $n_1 = 0$, $n_2 = n-1$, $m = 0$. The transitions between the two molecular states [1000] and $[n, 0, n-1, 0]$ are described by the Landau-Zener formula (7.39). In the region $R < R_n$, the populated $[n, 0, n-1, 0]$ Stark state interacts with the other n^2-1 Stark states through rotational coupling. If the collision velocity is not too small, the radial coupling (at $R \approx R_n$) and the rotational coupling (at $R < R_n$) may be considered independently. The probability q_n of the rotational decay of the $[n, 0, n-1, 0]$ state to all other n^2-1 Stark states (taken to be degenerate) in the region $R < R_n$ is given by [7.18]

$$q_n = 1 - (1 - \sin^2\beta \sin^2\alpha)^{2(n-1)}, \qquad (7.44)$$

$$\beta = \arctan(2Z\varrho v/3n), \quad \alpha = (\Delta\chi/2)[1 + (3n/2Z\varrho v)^2]^{1/2}, \qquad (7.45)$$

where $\Delta\chi$ is the angle of rotation of the internuclear axis,

$$\Delta\chi = 2\arccos(\varrho/R_n). \qquad (7.46)$$

For a system of N open reaction channels, the probability for the population of a given product-ion level n including rotational transitions is [7.19]

$$\begin{aligned} P_n = p_1 p_2 \dots p_n (1-p_n) [1 &+ (p_{n+1} p_{n+2} \dots p_N)^2 \\ &+ (p_{n+1} p_{n+2} \dots p_{N-1})^2 (1-p_N)^2 (1-q_N) \\ &+ (p_{n+1} p_{n+2} \dots p_{N-2})^2 (1-p_{N-1})^2 (1-q_{N-1}) \\ &\vdots \\ &+ p_{n+1}^2 (1-p_{n+2})^2 (1-q_{n+2}) + (1-p_{n+1})^2 (1-q_{n+1})] \\ &+ p_1 p_2 \dots p_{n-1} (1-p_n)^2 q_n . \end{aligned} \qquad (7.47)$$

Setting $q_k = 0$ ($k = 1, \dots, N$), (7.47) is reduced to (7.38). The number of curve-crossing regions which contribute to P_n is limited. The distant crossings are passed by the system diabatically, whereas the ones with small R_k and, therefore, large $\Delta(R_k)$ are avoided adiabatically. The partial (σ_n) and the total (σ) cross sections are again given by (7.41) and (7.43), respectively. With respect to the ordinary MLZ model (with only radial-coupling-induced transitions taken into account), the inclusion of the rotational transitions leads to a shift of the maximum cross section towards lower energies and to a considerable increase in the cross section in the region below the maximum.

In the calculations of the partial cross sections σ_n, the MLZ model is limited to $n < Z$, since the ionic energy levels with $n \geq Z$ do not cross the initial-state potential curve. This follows from the expression for the crossing point [see (6.52)],

$$R_n \simeq 2n^2(Z-1)/(Z^2-n^2), \quad n < Z. \qquad (7.48)$$

However, for high values of Z this limitation is not too essential, since for low collision energies the electron capture goes dominantly into the levels around $n_m \approx 2^{1/4} Z^{3/4}$, and far from n_m the cross sections σ_n drastically decrease.

The validity of the MLZ model is restricted to the adiabatic energy region. The adiabatic condition $\tau_c \gg \tau_t$ for the H+Z system, with $\tau_c \sim \varrho_0/v \sim Z^{1/2}/v$ and $\tau_t \approx (\Delta E_{n,n+1})^{-1} \sim (n_m^3/Z^2) \sim Z^{1/4}$, now gives

$$v \ll Z^{1/4} . \tag{7.49}$$

Extensive total cross section calculations have been performed for the H(1s) + Z system ($5 \leq Z \leq 54$) in the region $10^{-2} \lesssim v \lesssim 1$ using the MLZ model [7.19, 20]. Some results are shown in Fig. 7.3. For $Z \geq 16$ and $0.04 \leq v \leq 1$, the MLZ total cross sections for this system can be represented in the form (within an accuracy of 5%)

$$\sigma_{\text{MLZ}} = 2.90\, \pi\, a_0^2 \ln(15/v) , \quad Z \geq 16, \quad 0.04 \leq v \leq 1 . \tag{7.50}$$

For $Z < 16$, σ_{MLZ} as a function of Z exhibits oscillations in the low-velocity region.

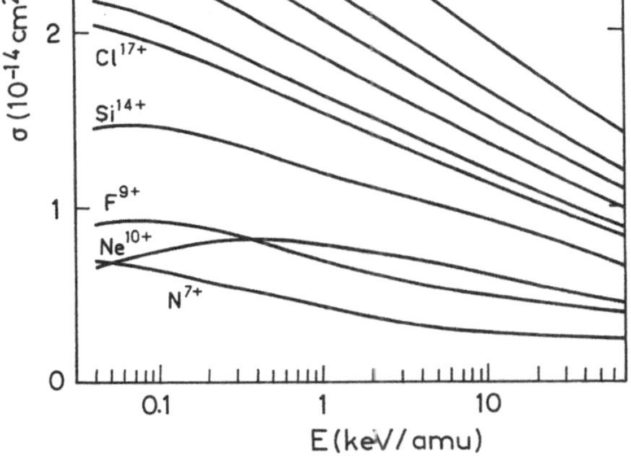

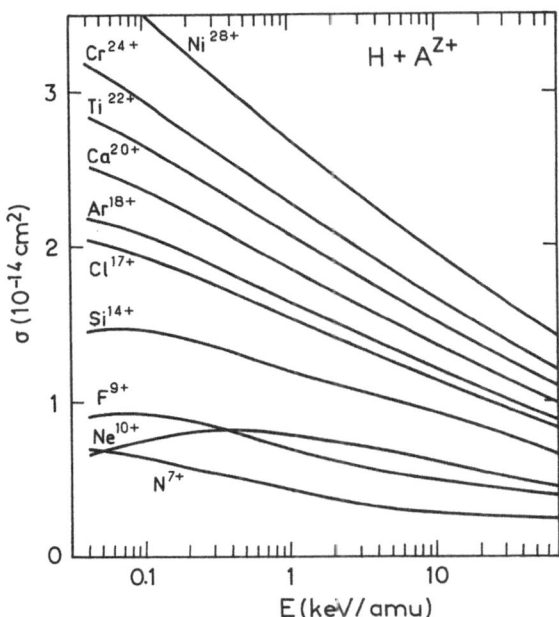

Fig. 7.3. Total electron-capture cross sections for a number of hydrogen atom-fully stripped ion colliding systems in the multichannel Landau-Zener approximation allowing for rotational coupling [7.19]

7.2.3 Molecular-Orbital Close-Coupling Calculations

The application of the molecular-orbital close-coupling (MO-CC) methods to the charge-exchange problem of highly charged ions on atoms consists of two steps. In the first step, electronic structure calculations are carried out for the collision system to provide molecular orbitals, energies and coupling matrix elements; in the second step, the coupled equations themselves are solved. From the point of view of molecular structure calculations, the hydrogen atom-fully stripped ion system is the easiest one to deal with. The eigenvalue problem for this system has been discussed in Sect. 6.1. Extensive numerical computations of the radial and rotational coupling matrix elements, Γ_{kj}^R and Γ_{kj}^L, for different H+Z pairs, have come to the following conclusions [7.21]. If the ionic charge Z is not too high, the most intense of the radial couplings between the states [1000] and $[n, 0, n-1, 0]$ is the one for $n = Z-1$. With increasing Z, the strength of the radial couplings for the states with $n = Z$ and $n = Z-2$ increases rapidly, and as Z further increases, other Γ_{kj}^R also become significant. It is important to note that with increasing Z, the radial coupling $\Gamma_{kj}^R(R)$ becomes increasingly close to that of the Landau-Zener model,

$$\Gamma_{\mathrm{LZ}}(R) = \frac{1}{2}\,\frac{\xi(R_n)}{\xi^2(R_n) + (R - R_n)^2}\,, \tag{7.51}$$

$$\xi(R_n) = \Delta(R_n)/|\Delta F(R_n)| \simeq R_n^2 \Delta(R_n)/(Z-1)$$

where R_n is the crossing point and $\Delta(R)$ is the exchange interaction.

The strength of the rotational $\sigma-\pi$ coupling is a maximum for those states which, within a given principal quantum number N, have the same value of the angular quantum number l (in the united-atom-limit notation of molecular states). The relative importance of rotational coupling with respect to the radial coupling decreases with increasing Z.

Thus, from the molecular structure calculations one obtains an insight into the collision dynamics and an idea of which molecular states have to be included in the truncated basis set. For a given energy range, the size of the minimal basis is determined by the required accuracy of the results. The number of important radial couplings increases rapidly with increasing Z. In the calculations of partial capture cross sections (capture into a specific final state), the required basis (to satisfy a prescribed accuracy) is usually much larger than the one needed for total cross section calculations of the same accuracy.

An important issue in the MO-CC calculations is the function of electron translational factors (ETF's). For relative velocities below $v \approx 0.1$, ETF's influence the cross section results very little, but for $v \gtrsim 0.3$ their role becomes significant. With a basis of $N \approx Z$ terms, different forms of the electron translation factors may change the total capture cross section by about 50%, or even more. Of course, with increasing basis size, the results become less sensitive to the choice of ETF's.

Calculations of MO-CC total capture cross sections in the PSS approximation (translational factors ignored) have been performed for the $H(1s)+Z$ system with $Z=3$ [7.22] (5 MO states), $Z=4, 5, 8$ [7.23] (3 MO states), $Z=6, 8, 26$ [7.24–26] (6, 8 and 17 MO states, respectively). As discussed in Sect. 6.2.2, the neglect of ETF's in calculations of the PSS total cross sections can be justified only if the coordinate origin is placed on the target atom (the initial-state boundary conditions are then trivially satisfied) and if the number of basis functions correlating with the ionic states at infinity is sufficiently large ("near-completeness" condition). Except for the $Z=3$ [7.22], $Z=6$ [7.24] and $Z=8$ [7.25] cases, these two conditions are not satisfied in the above-mentioned PSS calculations. More elaborate MO-CC calculations for the $H-Z$ system, with ETF's of different forms included, have been reported for $Z=2$ ([7.27]: 10 states, plane-wave ETF's; [7.28]: 12 states, switching factor; [7.29]: 5 states, optimized ETF's), $Z=6$ ([7.30]: 33 states, optimized ETF's) and $Z=8$ ([7.31]: 33 states, optimized ETF's). In the lower energy part, the integration of the coupled equations is usually carried out along non-linear trajectories. The total charge-exchange cross section for the $H+C^{6-}$ low-energy collisions, calculated by the MO-CC method with a basis of 33 states [7.30] is shown in Fig. 7.4. It is compared with the available experimental data for this system [7.32] and other theoretical results.

The application of the MO-CC method to the charge-exchange problem in many-electron collision systems, encounters serious technical difficulties

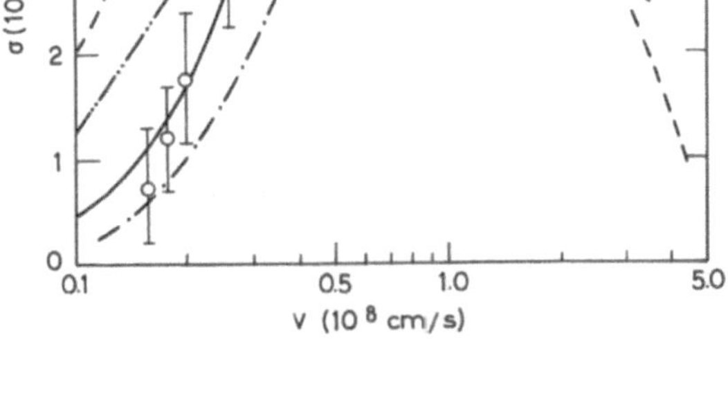

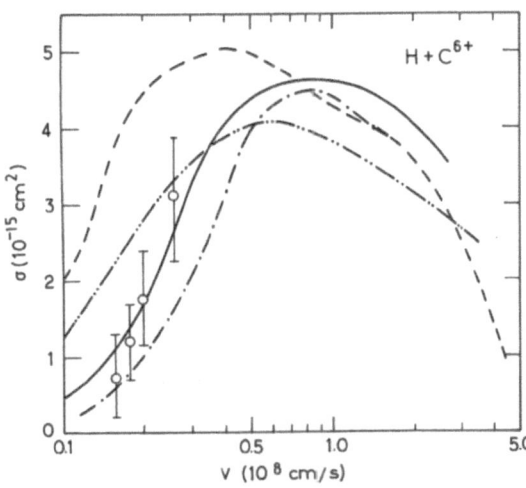

Fig. 7.4. Total cross section for the $H+C^{6+} \rightarrow H^{+}+C^{5+}$ electron-capture reaction. The curves represent the results of theoretical calculations. Solid curve: 33-state MO-CC [7.30], dashed curve: unitarized distorted-wave approximation [7.33], dot-dashed curve: analytical model for diabatic MO-CC method [7.34], two dots-dashed curve: M-LZ with rotational coupling [7.19]. The experimental data are due to *Phaneuf* et al. [7.32]

in dealing with the electron structure calculation. Two approaches are usually pursued: (I) standard quantum-chemistry methods are used to solve the multi-electron eigenvalue problem (configuration interaction method [7.35, 36], LCAO valence bond method [7.37], etc.), and (II) use of the pseudopotential method [7.38]. Most of the calculations have been performed by the PSS method and for the thermal energy region. PSS calculations for hydrogen atom-incompletely stripped ions, involving B^{3+}, C^{4+}, C^{5+}, N^{5+} and O^{6+}, have also been reported at energies above 10 eV/amu [7.37, 39].

7.2.4 Analytic Treatment of Diabatic MO-CC Equations

We shall now consider an analytic multichannel model for the charge-exchange $A+B^{Z+}$ collision problem, within which the corresponding coupled-channel equations of the MO-CC method in a diabatic basis can be solved approximately. The results of the decay and multichannel Landau-Zener models will be obtained as limiting cases of the analytic multichannel close-coupling (AMCC) model. If the diabatic molecular basis includes one state which, at $t \to -\infty$, correlates with the initial channel, and all the states correlating at $t \to \infty$ with the charge-transfer channels, the coupled-channel equations (6.118) of the diabatic MO-CC method can be put in the form

$$\mathrm{i}\frac{da_0}{dt} = \sum_n b_n(t)\, V_{0n}^* \exp\left(\mathrm{i}\int_{-\infty}^{t} \omega_{0n}\, dt'\right), \tag{7.52a}$$

$$\mathrm{i}\frac{db_n}{dt} = a_0(t)\, V_{n0} \exp\left(-\mathrm{i}\int_{-\infty}^{t} \omega_{0n}\, dt'\right), \tag{7.52b}$$

$$V_{0n} = \langle n|\, H\, |0\rangle\,, \quad \omega_{0n} = V_{00} - V_{nn}\,, \tag{7.53}$$

where $a_0(t)$ and $b_n(t)$ are, respectively, the amplitudes of the initial- and final-channel molecular states, $|0\rangle$ and $|n\rangle$. We assume that the basis is sufficiently large and ignore the electron translational factors in $|0\rangle$ and $|n\rangle$. We also assume that the interaction matrix is Hermitian ($V_{0n}^* = V_{n0}$) and that unitarity is preserved,

$$|\, a_0(t)\,|^2 + \sum_n |\, b_n(t)\,|^2 = 1\,. \tag{7.54}$$

The initial conditions for the system (7.52) are

$$|\, a_0(-\infty)\,| = 1\,, \quad b_n(-\infty) = 0\,. \tag{7.55}$$

Let us further assume that the functions $\omega_{0n}(R)$ and $V_{0n}(R)$ (R being the internuclear distance) have the properties (we specify the states $|0\rangle$ and $|n\rangle$

by the separated-atom quantum numbers, $|0\rangle = |n_0, l_0, m_0\rangle$, $|n\rangle = |n, l, m\rangle$)

$$\omega_{n_0 l_0 m_0, nlm} = \omega_n(R), \tag{7.56a}$$

$$V_{n_0 l_0 m_0, nlm} = \delta_{m m_0} f(n, l)\, V_n(R). \tag{7.56b}$$

Thus, for a given initial state, the difference in diabatic energies depends only on the principal quantum number of the final state, while the angular dependence of the coupling matrix element is separated from its R-dependence. Referring to the quasi-classical expression (6.56) for $\Delta(R)$ ($\approx 2 V_{0n}$), we see that the above assumptions are fulfilled for the (Z_1, e, Z_2) system in the asymptotic region of R and when $Z_2 \gg Z_1$.

The quasi-classical expression (6.56) for $V_{0n} = (1/2)\Delta(R)$ specifies $f(n, l)$ and $V_n(R)$, appearing in (7.56b), in the form

$$f(n, l) = f(l) = (2l+1)^{1/2} \exp[-l(l+1)/2Z], \tag{7.57a}$$

$$V_n(R) = n^{-3/2}\, U(R), \quad U(R) = (2Z/\pi)^{1/2} R\, e^{-R^2/3Z}. \tag{7.57b}$$

With the assumptions (7.56), the l-dependence can be eliminated from (7.52) and, assuming further that the capture takes place into a broad group of levels around $n = n_0$, the remaining system of coupled equations can be reduced to a system of two coupled equations for the $0 \to n_0$ transition, which effectively (through a complex energy width) accounts for the interaction of the initial state with all final states. The corresponding system of coupled equations is ($n_0 = n$) [7.34]

$$\mathrm{i}\frac{da_0}{dt} = -\mathrm{i}\,\Gamma(t)\, a_0 + V_n(t) \exp\left[-\mathrm{i}\int_{-\infty}^{t} \omega_n(t')\, dt'\right] b_n(t), \tag{7.58a}$$

$$\mathrm{i}\frac{db_n}{dt} = V_n(t) \exp\left[\mathrm{i}\int_{-\infty}^{t} \omega_n(t')\, dt'\right] a_0(t), \tag{7.58b}$$

$$\Gamma(t) = (\pi/Z^2)\, V^2(t), \quad a_0(-\infty) = 1, \quad b_n(-\infty) = 0. \tag{7.58c}$$

The system of coupled equations (7.58) describes the interaction of a discrete $|n\rangle$ and a quasi-stationary $|0\rangle$ state [7.40, 41]. In the limiting case $\Gamma \gg |V_n|$ it describes the decay of the initial state due to its interaction with all the final states. Using the initial conditions in (7.58a), one obtains for the decay probability

$$W_{\mathrm{dec}} = 1 - |a_0(+\infty)|^2 = 1 - \exp\left[-2\int_{-\infty}^{\infty} \Gamma(t)\, dt\right]. \tag{7.59}$$

In the general case, using the VPS method (Sect. 6.2.4) to solve approximately the system of coupled equations (7.58), one obtains

$$W_n = |b_n(+\infty)|^2 = \left| \int_{-\infty}^{\infty} V_n(t) \exp\left[i \int_{-\infty}^{t} \Omega(t')\, dt' - \int_{-\infty}^{t} \Gamma(t')\, dt' \right] dt \right|^2, \tag{7.60}$$

where

$$\Omega(t) = [\omega_n^2(t) + 4 V_n^2(t)]^{1/2}, \quad \omega_n \simeq (Z-1)/R + I_A - Z^2/2n^2 .$$

In the case $|V_n| \ll \omega_n$, (7.60) reproduces the perturbation solution of the coupled equations (7.58) [7.42].

The region $\varrho \sim R_n$ (R_n being the crossing point) gives the main contribution to the charge-exchange cross section. In this region, the energy splitting ω_n can be written as

$$\omega_n(t) = \frac{(Z-1)}{2R_n^3} [(R_n^2 - \varrho^2) - v^2 t^2],$$

i.e. at $\varrho \sim R_n$, $\omega_n(t) \sim t^2$. The integration in (7.60) can now be carried out analytically in the complex t-plane, along a contour which passes through the complex zeroes of $\Omega(t)$ that lie closest to the real axis, by expanding $\Omega(t)$ in the vicinity of its zeroes [7.34]. The probability W_n for capture into a specific level n is expressed in terms of the Airy functions $Ai(x)$ and $Bi(x)$. It peaks at a value of the impact parameter ϱ which is close to the diabatic crossing R_n, decreases exponentially for $\varrho > R_n$, and oscillates (with a strong damping) for $\varrho < R_n$. The integration of $W_n(\varrho, v)$ over ϱ with an exponential accuracy leads to an expression for the partial cross section σ_n [7.34],

$$\sigma_n = 2\pi \int_0^{\infty} W_n \varrho\, d\varrho \sim R_n^2 \delta_n \exp[-(2\delta_n + \gamma_n)], \tag{7.61 a}$$

$$\delta_n = \frac{\pi V_n^2(R_n)}{v \Delta F(R_n)}, \quad \gamma_n = \frac{(6\pi Z)^{1/2}}{v} \left(\frac{R_n^2}{Z} + \frac{3}{4} \right) \exp\left(-\frac{2R_n^2}{3Z} \right). \tag{7.61 b}$$

The total cross section $\sigma = \sum_n \sigma_n$ for the charge-transfer reaction in the collison system $H + C^{6+}$, calculated using the expression (7.60) for W_n, is given in Fig. 7.4.

This approximate treatment of diabatic coupled equations (7.52) can be extended beyond the assumptions (7.56), i.e. to non-Coulomb systems. An appropriate approximation of the coupled equations in this case is

$$\frac{da_0}{dt} = -\Gamma(t)\, a_0, \tag{7.62 a}$$

$$\mathrm{i}\,\frac{db_{nl}}{dt} = V_{nl}(t)\exp\left[\mathrm{i}\int_{-\infty}^{t}\omega_{nl}(t')\,dt'\right]a_0(t), \tag{7.62b}$$

which is valid under the condition

$$\delta_{nl} = \pi\, V_{nl}^2(R_{nl})/v\,\Delta F(R_{nl}) \ll 1. \tag{7.62c}$$

This condition, however, does not imply that $\delta_n = \sum_l \delta_{nl}$ is small. From (7.62), one obtains immediately

$$W_{nl} = \left|\int_{-\infty}^{\infty} V_{nl}(t)\exp\left[\mathrm{i}\int_{-\infty}^{t}\omega_{nl}(t')\,dt' - \int_{-\infty}^{t}\Gamma(t')\,dt'\right]dt\right|^2, \tag{7.63a}$$

$$W_n = \sum_l W_{nl}. \tag{7.63b}$$

Since, however, the system of coupled equations (7.62) is not Hermitian, the unitarized transition probabilities are given by

$$P_{nl} = W_{nl}/(\sum_m W_m - W_{\mathrm{dec}} + 1), \tag{7.64a}$$

$$P_n = \sum_l P_{nl}, \tag{7.64b}$$

where W_{dec} is given by (7.59). Using (7.64a, b), one can calculate respectively the partal cross sections σ_{nl} and σ_n for state-selective electron capture from an arbitrary atomic (or molecular) target by a highly charged ion. Using this method, the total cross sections $\sigma = \sum_n \sigma_n$ have been calculated for the low-energy collisions of H_2 with fully stripped ions with $Z = 6, 8, 10$ and 18 [7.34].

7.3 Electron Capture at Intermediate Energies

The electron-capture problem at intermediate collision energies ($v \sim 1$) is characterized by the strong interaction of a large number of states and by the fact that no suitable small expansion parameter exists. In this situation, the collision dynamics of the process can best be described by a coupled-channel formalism. Since for most of the collision time the electron is bound to one of the colliding particles, the atomic basis is an appropriate one. Attachment of electron translation factors (which in this case are the plane-wave ETF's) to the basis functions is an essential requirement, since the electron momentum transfer cannot be ignored in this energy region. For practical reasons, the basis is always restricted to a finite (and relatively small) number of terms. In order to account for the neglected discrete or continuum states, the basis could be supplemented by pseudostates, or even replaced by a Sturmian basis. On the other hand, the pronounced role of the multistate coupling and the continuum states, makes it possible to use classical mechanics in the description of the process.

The coupled-channel equations of the atomic-orbital close-coupling (AO-CC) method can be treated either numerically or using some suitable approximation. We shall consider below these approaches in more detail.

7.3.1 AO-CC Method: Numerical Calculations

In general, the AO basis is constructed by orbitals representing the unperturbed electronic states around each of the centres (two-centre expansion). The coupled-channel equations obtained from such a basis (with plane-wave translational factors included) are given by (6.122), and have to be solved numerically subject to the appropriate initial conditions (6.125). The unitarity condition $M_{kj} - M_{jk}^{*} + \mathrm{i}\,(d/dt)\,S_{kj} = 0$ where M_{kj} and S_{kj} are the interaction-coupling and overlap matrix elements, respectively, usually serves as a useful check on the accuracy of the computations. It should be emphasized that the application of the two-centre expansion AO-CC method to the charge-exchange problem for highly charged ions requires a large number of terms in the basis. The reason is that for large values of Z, a considerable range of highly excited levels is populated by the electron-capture process, and each of these levels contains a high angular and magnetic multiplicity. For this reason two-centre expansion AO-CC calculations have been performed only for charge exchange in hydrogen atom-low-charged ion systems: $\mathrm{H{-}He^{2+}}$ [7.43] and $\mathrm{H{-}Li^{3+}}$ [7.44], with all orbitals with $n \leqq 3$ included in the calculation. In these calculations, the electron capture into higher n-shells has been accounted for by Oppenheimer's n^{-3} rule, a procedure which in the region $v \lesssim 1$ is not quite appropriate. Nevertheless, the two-centre expansion AO-CC calculations with a large basis give results which for $v \sim 1$ agree closely with those of the MO-CC method, if the electron momentum-transfer effects are adequately treated.

For the $\mathrm{H} + Z$ $(Z \gg 1)$ system, a reasonably good choice of the restricted two-centre expansion basis would be to take one AO centred on H^{+}, representing the initial state, and a sufficiently large number of AO's centred on the projectile charge, which represent the ionic states around the resonance energy level. With such a basis for the $\mathrm{H} + \mathrm{C}^{6+}$ system, containing only 11 AO's, agreement with the 33-state. MO-CC calculations [7.30] has been obtained in the total charge-exchange cross section down to 100 eV/amu [7.45]. For lower energies, the integration of the coupled-

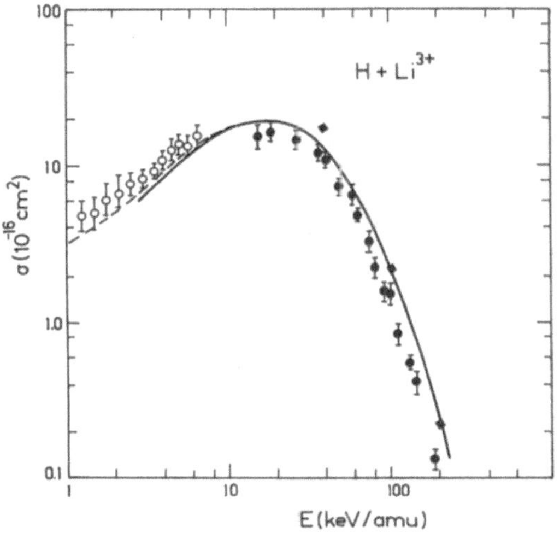

Fig. 7.5. Total cross section for the $H+Li^{3+} \rightarrow H^{+}+Li^{2+}$ charge-exchange reaction. Solid curve represents the results of 22-state AO-CC calculations [7.44], dashed curve is the result of 24 AO+UA pseudostate close-coupling calculations [7.46], and solid squares are the results of the OHCE method [7.47]. The experimental points are due to *Shah* et al. [7.48] (●) and *Seim* et al. [7.49] (○)

channel equations of the AO-CC method is usually carried out along non-linear trajectories.

To improve the calculations within a restricted AO-CC method, appropriate pseudostates are added to the basis (see Sect. 6.2.3). The pseudostates may be chosen either to represent the molecular effects in the collision system (united atom pseudostates [7.46]), or the continuum states (as in the "one and a half centre expansion", OHCE method [7.47]). The results for the $H+Li^{3+}$ system, obtained from the large-size coupled-channel calculations with 22 AO's [7.44], 24 AO's + united-atom pseudostates [7.46], and by the OHCE method [7.47] are shown in Fig. 7.5 and compared with the available experimental data [7.48, 49]. Elaborate 35 AO + united atom pseudostate CC calculations have also been performed for the $H+C^{6+}$ system [7.47]. For the $H+He^{2+}$ system, accurate exchange cross section calculations have been carried out using a Sturmian expansion with 24 functions [7.50].

7.3.2 Approximate Treatments of Coupled-Channel Equations

Under certain simplifying assumptions, the coupled-channel equations of the AO expansion method allow approximate treatments, leading to a significant reduction of the computational work. These treatments are usually based on the neglect of transitions via intermediate states, i.e. on the assumption that the states representing the exit reaction channels are strongly coupled only with the initial state. Within such an approximation the probability for populating a given exit channel is expressed in terms of two-state transition probabilities. Further approximations are made in the calculations of the two-state close-coupling problem. Such an approximate approach to the multichannel charge-exchange problem in atom-highly charged ion collisions has been adopted by *Presnyakov* and *Ulantsev* [7.51], who calculated the two-state transition probability in the VPS approximation (6.146), and by *Ryufuku* and *Watanabe* [7.33], who used the distorted-wave Born (DWB) approximation (6.185) for the two-state transition matrix elements. These two approximate treatments of the coupled-channel equations can be respectively considered as generalizations of the VPS and DWB approximations to a multichannel situation. In both treatments orthogonalized atomic orbitals with plane-wave ETF's have been used and a unitarization of the S matrix has been accomplished. Below, we shall give a brief account of these two methods.

A. Multichannel VPS Approximation

If the basis set is constructed from one function representing the target initial state, and a complete set of functions representing the electronic states of the product ion (including the continuum), all of them being centred on the multiply charged ion, and orthogonalized, then the system of coupled equations for the expansion amplitudes is Hermitian. It coincides with the diabatic CC equations (7.52), in which an integral over the continuum should be added to the sum in (7.52a), and all matrix elements are calculated in the one-centre atomic (diabatic) basis, defined above. Introducing the notation

$$V_n(t) = V_{0n} \exp\left(-\mathrm{i} \int\limits_{-\infty}^{t} \omega_{0n}\, dt'\right), \tag{7.65}$$

the system (7.52) can be written as

$$\mathrm{i}\frac{da_0}{dt} = \sum_n V_n^* b_n, \tag{7.66 a}$$

$$\mathrm{i}\frac{db_n}{dt} = V_n a_0. \tag{7.66 b}$$

Introducing, further, the new function $K_n(t)$ by the relation

$$K_n(t) = b_n(t)/a_0(t), \tag{7.67}$$

the system of coupled equations (7.66) can be transformed into a system of non-linear Riccati equations

$$\frac{dK_n}{dt} + \mathrm{i}(V_n - V_n^* K_n^2) = Q_n(t)\, K_n(t), \tag{7.68}$$

where

$$Q_n(t) = -\mathrm{i} \sum_{m \neq n} K_m(t)\, V_m^*(t). \tag{7.69}$$

The system of equations (7.68) is a multichannel generalization of (6.144). From (7.69, 54), it follows that

$$|a_0(t)|^2 = \left[1 + \sum_m |K_m(t)|^2\right]^{-1}, \quad |b_n(t)|^2 = \frac{K_n(t)^2}{1 + \sum_m |K_m(t)|^2}. \tag{7.70}$$

This approach, analogous to the K-matrix formalism in general scattering theory, ensures that unitarity is preserved independent of the approximation to which $K_n(t)$ is calculated. In particular, the first-order approximation to (7.68) gives

$$K_n(t) = -\mathrm{i} \int_{-\infty}^{t} V_{0n} \exp\left(-\mathrm{i} \int_{-\infty}^{t''} \omega_{0n}\, dt'\right) dt'' \tag{7.71}$$

which, together with (7.70), gives the normalized distorted-wave approximation.

The term on the right-hand side of (7.68) describes transitions via intermediate states. If they can be regarded as small and neglected in a zero-order approximation, the system (7.68) reduces to a set of uncoupled Riccati equations. $|K_n(+\infty)|^2$ can then be approximately calculated in the form [7.51]

$$|K_n(+\infty)|^2 = p_{0n}(1 - p_{0n})^{-1}, \tag{7.72}$$

where p_{0n} is given by (6.146), the VPS approximation. Using (7.70), one obtains the channel probabilities

$$P_0 = |a_0(\infty)|^2 = T^{-1} q, \quad q = \prod_m (1 - p_{0m}), \tag{7.73}$$

$$P_n = |b_n(\infty)|^2 = T^{-1} p_{0n} \bar{q}_n, \quad \bar{q}_n = \prod_{m \neq n} (1 - p_{0m}), \tag{7.74}$$

with

$$T = q + \sum_n p_{0n} \bar{q}_n. \tag{7.75}$$

The validity of (7.73, 74) is restricted by the conditions $\ln|\bar{q}_n| \ll 4$, $p_{0n}^{1/2} \ln|\bar{q}_n| \ll 2$.

For the system $A + B^{Z+}$ the "frequency" function $\Omega(R) = (\omega_{0n}^2 + 4 V_{0r}^2)^{1/2}$ in the complex R-plane has the following singular points (all matrix elements being evaluated with hydrogenic wave functions) [7.51]:

$$R^{(1)} \simeq 1.2\, n^2/Z + \mathrm{i}\,[(\pi/2\,\gamma) + 4\,\pi\, k], \quad k = 0, 1, 2, \ldots, \tag{7.76}$$

$$R^{(2)} \simeq R_n + \mathrm{i}\,(2\, R_n/\Delta\,\varepsilon_{0n})\; V_{0n}\,(R_n), \tag{7.77}$$

$$\gamma = \frac{1}{2}\left[\left(\frac{Z^2}{n^2} + \frac{v^2}{4}\right)^{1/2} + \left(2\, I_A + \frac{v^2}{4}\right)^{1/2}\right], \tag{7.78}$$

$$R_n \simeq (Z-1)/\Delta\,\varepsilon_{0n}, \tag{7.79}$$

where n is the principal quantum number of the final ionic state and $\Delta\,\varepsilon_{0n} = \varepsilon_0^{\mathrm{A}} - \varepsilon_n^{\mathrm{B}}$ is the energy defect. Transitions take place predominantly around $R^{(1)}$ and $R^{(2)}$. The singular point $R^{(2)}$ corresponds to a pseudo-crossing; the corresponding transitions are of the Landau-Zener type. The singular point $R^{(1)}$ is related to the underbarrier transitions, described by (6.150). For all ionic levels with $(Z/2) < n < Z$, the transition regions $R^{(1)}$ and $R^{(2)}$ are well separated and p_{0n} can be written as $p_{0n} = p_{0n}^{(1)} + p_{0n}^{(2)}$. Consequently, the electron-capture probability P_n can be represented in the form

$$P_n = P_n^{(1)} + P_n^{(2)}, \tag{7.80}$$

leading to a similar structure for the partial charge-exchange cross section,

$$\sigma_n = \sigma_n^{(1)} + \sigma_n^{(2)}. \tag{7.81}$$

For $v \ll 1$, $\sigma_n^{(1)}$ is exponentially small, while the Landau-Zener cross section $\sigma_n^{(2)}$ reaches its maximum in this region. In the region $v \sim 1$ where $\sigma_n^{(2)}$ is small, the cross section $\sigma_n^{(1)}$ attains its maximum. Thus, the multichannel VPS approximation predicts the existence of two maxima in the charge-exchange cross section in the region $v \lesssim 1$. In a broad velocity region, the total cross section $\sigma = \sum_n \sigma_n$ can be represented in the form

$$\sigma = \pi (Z/I_{\mathrm{A}})^2\, Q\,(v, Z), \tag{7.82}$$

where I_{A} is the ionization potential of the target and $Q\,(v, Z)$ is a function which, for $v \lesssim 1$, is almost constant and, for $v \gg 1$, falls off as v^{-12}. The quadratic dependence of the cross section on the ionic charge is a characteristic feature of the multichannel VPS approximation.

B. Unitarized Distorted-Wave Approximation

Unitarization of the approximate solution of the AO coupled-channel equations can be achieved not only within the K-matrix formalism, but also using the interaction representation of the S matrix,

$$S^{\text{int}} = T \exp\left[-\mathrm{i} \int_{-\infty}^{\infty} H^{\text{int}}(+)\, dt\right], \tag{7.83}$$

$$H^{\text{int}}(t) = \exp\left(\mathrm{i} \int_{-\infty}^{t} H^0\, dt'\right)(H - H^0) \exp\left(-\mathrm{i} \int_{-\infty}^{t} H^0\, dt'\right), \tag{7.84}$$

where T is the chronological operator and H^0 is the diagonal part of the matrix H. If $|0\rangle$ is the initial electronic state, the probability that, at $t \to +\infty$, the electron will be in the state $|n\rangle$ is given by

$$P_{0n} = |\langle n| S^{\text{int}} |0\rangle|^2. \tag{7.85}$$

In the calculations of $\langle n| S^{\text{int}} |0\rangle$, one usually expands the exponential functions containing H^{int} and takes a time-ordered product. If the time ordering is neglected (which is justifiable at $v \gtrless 1$) and only the coupling matrix elements connecting the initial with other states are retained, one obtains [7.33]

$$\langle 0| S^{\text{int}} |0\rangle = \cos p^{1/2}, \tag{7.86}$$

$$\langle n| S^{\text{int}} |0\rangle = \mathrm{i}\, t_{0n}\, p^{1/2} \sin p^{1/2}, \tag{7.87}$$

where

$$p = \sum_n |t_{0n}|^2, \quad t_{0n} = \int_{-\infty}^{\infty} \langle n| H^{\text{int}} |0\rangle\, dt. \tag{7.88}$$

As can be seen from (7.86–38), the unitarity relation $\sum_n |\langle n| S^{\text{int}} |0\rangle|^2 = 1$ is satisfied, independent of the fact that the higher-order terms in the expansion of $\langle n| S^{\text{int}} |0\rangle$ have been neglected. Neglecting these terms, which represent transitions via intermediate states, may lead, however, to significant errors in the computed p_{0n} at high collision velocities.

If the AO basis $\{\varphi\}$ contains terms representing the electronic states about each of the Coulomb centres A^+ and B^{Z+}, then t_{0n} in (7.88) may describe an excitation transition ($|n\rangle = \varphi_n^A$), electron capture into a specific final ionic state ($|n\rangle = \varphi_n^B$), or ionization ($|n\rangle = \varphi_k^B$, $k^2/2$ being the energy of the free electron, and the continuum wave function φ_k^B being centred on B^{Z+}). The sum $p = \sum_n |t_{0n}|^2$ can, thus, be decomposed into three parts,

$$p = p_{\text{exc}} + p_{\text{ion}} + p_{\text{cx}} \tag{7.89}$$

related to the excitation, ionization and charge-exchange processes. The two-state matrix elements t_{0n} can be calculated for each of these processes by using the orthogonalized travelling AO's $\varphi_n^\lambda(\lambda = \mathrm{A, B})$, centred on each of the centres λ. Calculating them in the distorted-wave Born approximation, one has

$$t_{0n}^{\mathrm{exc}} = \int_{-\infty}^{\infty} dt\, U_{0n}^{\mathrm{AA}} \exp\left[-\mathrm{i} \int_{-\infty}^{t} (U_{00}^{\mathrm{AA}} - U_{nn}^{\mathrm{AA}} + \varepsilon_0^{\mathrm{A}} - \varepsilon_n^{\mathrm{A}})\, dt'\right], \tag{7.90}$$

$$t_{0n}^{\mathrm{cx}} = \int_{-\infty}^{\infty} dt (U_{0n}^{\mathrm{AB}} - S_{0n}^{\mathrm{AB}}\, U_{00}^{\mathrm{AA}}) \exp\left[-\mathrm{i} \int_{-\infty}^{t} (U_{00}^{\mathrm{AA}} - U_{nn}^{\mathrm{BB}} + \varepsilon_0^{\mathrm{A}} - \varepsilon_n^{\mathrm{B}})\, dt'\right], \tag{7.91}$$

$$t_{0k}^{\mathrm{ion}} = \int_{-\infty}^{\infty} dt (U_{0k}^{\mathrm{AB}} - S_{0k}^{\mathrm{AB}}\, U_{00}^{\mathrm{AA}}) \exp\left[-\mathrm{i} \int_{-\infty}^{t} (U_{00}^{\mathrm{AA}} - U_{kk}^{\mathrm{BB}} + \varepsilon_0^{\mathrm{A}} - k^2/2)\, dt'\right], \tag{7.92}$$

where

$$U_{0n}^{\mathrm{AB}} = H_{0n}^{\mathrm{AB}} = \langle \varphi_n^{\mathrm{B}} |\, H\, | \varphi_0^{\mathrm{A}} \rangle, \quad U_{jj}^{\lambda\lambda} = H_{jj}^{\lambda\lambda} - \varepsilon_j^\lambda, \quad j = 0, n, k; \quad \lambda = \mathrm{A, B},$$

and terms of order S_{0n}^2 have been neglected. The probability for electron capture into a specific final state $|n\rangle$ is, thus, given by

$$P_{0n}^{\mathrm{cx}} = p^{-1}\, |t_{0n}^{\mathrm{cx}}|^2 \sin^2 p^{1/2} \tag{7.93}$$

and the total capture probability is

$$P_{\mathrm{cx}} = p^{-1} \sum_n |t_{0n}^{\mathrm{cx}}|^2 \sin^2 p^{1/2} = (p_{\mathrm{cx}}/p) \sin^2 p^{1/2}. \tag{7.94}$$

Analogous expressions can be written for P_{0n}^{exc}, P_{0k}^{ion}, P_{exc} and P_{ion}. The above procedure of calculating the probabilities for excitation, ionization and charge exchange is referred to as the unitarized distorted-wave approximation (UDWA). In view of the particular approximations made in the calculations of t_{0n}, it may be expected that the UDWA is valid in the region $1 \lesssim v \lesssim 3-4$. Charge-exchange cross section calculations with UDWA have been performed for the H + Z system, including $Z = 1-6, 8$, 10, 14 and 20 [7.33, 52, 53]. Most of these calculations have been extended beyond the above validity limits of the UDWA. The use of the DWBA for t_{0n}^{cx} and the neglect of the second- (and higher-)order transitions mean that the UDWA considerably overestimates the charge-exchange cross section both at $v < 1$ and $v > 2-3$.

The UDWA total cross section can be adequately represented in a scaled form [7.53],

$$\sigma(E) = Z^\alpha\, \tilde{\sigma}(\tilde{E}), \quad \tilde{E} = E/Z^\beta, \quad \alpha = 1.07, \quad \beta = 0.35, \tag{7.95}$$

where E is the collision energy in units of keV/amu.

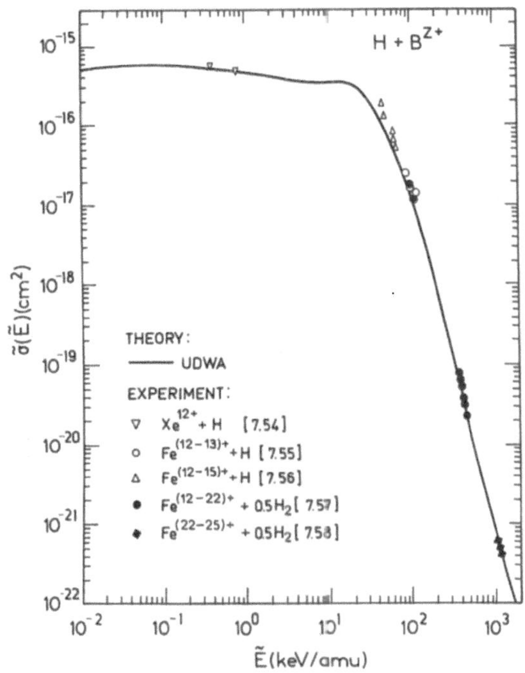

Fig. 7.6. Scaled electron-capture cross section $\tilde{\sigma}$ vs. scaled impact energy $\tilde{E}$ in the UDWA approximation for the H+Z ($Z \gtrsim 14$) system [7.53], compared with experimental results [7.54–58]

For $\tilde{E} \gtrsim 25$ keV/amu, the reduced cross sections $\tilde{\sigma}(\tilde{E})$ differ very little from each other. For $\tilde{E} \lesssim 25$ keV/amu, the difference between $\tilde{\sigma}(\tilde{E})$ for different values of $Z(\geqq 3)$ is within a factor of two. The scaling law given by (7.95) gives a good representation of all cross sections for the $H + Z$ system with $Z \geqq 14$. Figure 7.6 shows the reduced cross section $\tilde{\sigma}(\tilde{E})$ as a function of the reduced energy $\tilde{E}$. Comparison is made with the scaled experimental cross sections of $H + B^{Z+}$ and $H_2 + B^{Z+}$ (the latter multiplied by a "structure" factor of 0.5).

Starting from the expression for P_{0n}^{cx} (7.93), the UDWA theory allows one also to calculate the partial cross sections σ_{nlm}, σ_{nl} and σ_n for electron capture into a specific final state, electronic subshell or shell, respectively. State-selective electron capture will be discussed in more detail in Sect. 7.5.2.

7.3.3 Classical Treatments

a. Classical Trajectory Monte Carlo (CTMC) Method

Systematic investigations [7.59–61] of the charge-exchange process in hydrogen atom-fully stripped ion collisions have been performed by the

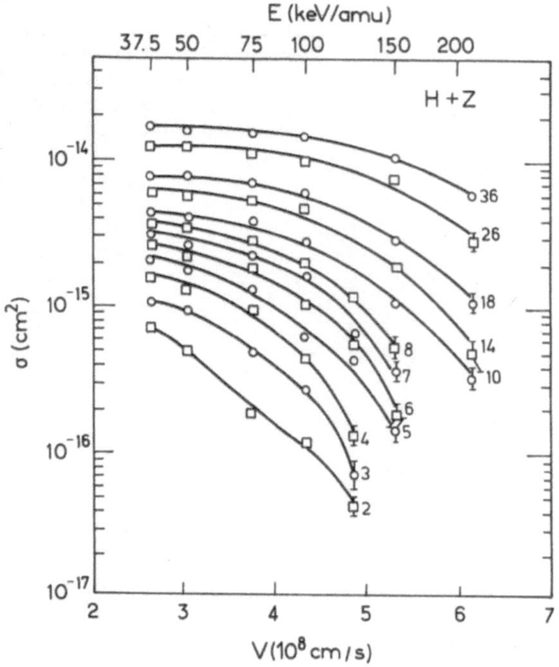

Fig. 7.7. Electron-capture cross sections for the $H+Z$ collision system (Z is shown on the curve) in the CTMC model

CTMC method, described in Sect. 6.4.2. Total (σ) and partial (σ_n and σ_{nl}) cross sections have been calculated for fully stripped ions with Z ranging from $Z=1$ to $Z=36$, in the energy range from 30 to $\sim 250-300$ keV/amu. Some of the results for the total cross sections are shown in Fig. 7.7.

b. Analytic Models

Within the concepts of classical mechanics, two analytic models have been developed for describing the electron-capture process in atom-highly charged ion collisions at intermediate energies. One of them, developed by *Bohr* and *Lindhard* [7.62], uses very simple classical relations between the forces and energies involved in the collision system. The other is based on the concept of overbarrier (classically allowed) transitions in the system [7.63, 64]. The Bohr-Lindhard model defines two critical distances, R_1 and R_2, by the relations

$$Z/R_1^2 = u/a, \quad Z/R_2 = (1/2)\, v^2, \tag{7.96}$$

where u and a are the orbital velocity and radius of the atomic electron. R_1 defines the distance at which the binding force of the electron in the atom equals the ionic Coulomb force attracting the electron. For $R \leqq R_1$, the electron is released. The second relation in (7.96) expresses the equality of the potential energy of the electron in the ionic field with the kinetic energy in the projectile-ion frame. Capture is possible when $R < R_2$. If $R_1 < R_2$, the condition for capture is automatically satisfied and (assuming a unit probability), the cross section is

$$\sigma_1 = \pi R_1^2 = \pi Z a/u . \tag{7.97}$$

When $R_1 > R_2$, whether a released electron will be captured or not depends on the probability for release (u/a), considered as a gradual process, and the time R_2/v during which the capture can take place. The cross section is now given by

$$\sigma_2 = \pi R_2^2 \left(\frac{u}{a} \frac{R_2}{v} \right) = 8\pi Z^3 u/a v^7 . \tag{7.98}$$

The cross section σ_1 holds for $v \leqq 2^{1/2} (Z a/u)^{1/4} = v_c$ whereas σ_2 holds for $v \geqq v_c$. Thus, in the region $v \leqq v_c$, the cross section predicted by the Bohr-Lindhard model is velocity independent and scales linearly with Z, while for $v \geqq v_c$, $\sigma \sim Z^3/v^7$. This Z- and v- dependence of σ in the intermediate velocity range $(0.8 \lesssim v \lesssim 2)$ has been experimentally confirmed, and is in accordance with the more elaborate theoretical models. Further developments of the Bohr-Lindhard model can be found elsewhere [7.65].

The overbarrier electron-capture model assumes that during the collision, the colliding system attains the distance R_0 at which the top of the potential barrier (created between the ionic potential wells) becomes equal to the initial-state electron binding energy. For $R < R_0$, the motion of the electron takes place in the combined potential well of the two Coulomb centres. Since the potential well of the highly charged ion is much deeper and wider than that of the target ionic core, the electron capture can be considered to have taken place (with a unit probability) once the distance $R = R_0$ is attained. The capture cross section is thus given by

$$\sigma_{\mathrm{cl}} = \pi R_0^2 . \tag{7.99}$$

In the $\mathrm{H} + Z$ system, the overbarrier transitions are possible only along the elliptic $\eta\,[=(r_1 - r_2)/R]$ coordinate. The distance R_0 in this case is $R_0 = 2(2Z - 1)^{1/2}$ [7.63, 64], so that

$$\sigma_{\mathrm{cl}} \simeq 8\pi Z . \tag{7.100}$$

If the classical density distribution of the transferred electron is taken into account, the capture cross section for the $\mathrm{H} + Z$ system is given by [7.64]

$$\sigma_{\mathrm{cl}} = \pi R_0^2 f(Z, v) , \tag{7.101}$$

where the function $f(Z, v)$ ($\leqq 1$) weakly depends on Z and decreases slowly with increasing v. The classical overbarrier transition model predicts total cross sections which for $v \lesssim 2$ are close to those calculated by the CTMC method. The overbarrier electron transition model, however, can be extended down to $v \approx 0.5$. Since recapture by the target ionic core is ignored, the validity of this model is restricted to ions with $Z \lesssim 15$.

7.4 Electron Capture at High Energies

The electron-capture process in the velocity region above $v \sim 3-4$ is characterized by a strong coupling of the rearrangement channels with the ionization ones. In order to describe accurately the influence of the continuum channels on electron capture within the coupled-channel formalism, many pseudostates have to be included in the basis. Such a treatment is highly impractical, and simpler methods, based on the perturbation approach, are evoked instead. In spite of the existence of a suitable expansion parameter in this region (the ratio Z/v), its validity for highly charged ions starts at very large values of v, where the role of the coupling with the continuum (and other higher-order effects) is so strong that the first-order perturbation results for the capture process are, rigorously speaking, never correct. Nevertheless, these results are useful in providing an elementary and physically transparent picture of the process, and in giving the general dependence of the cross section on the parameters of the collision system (except the velocity) and moderately accurate predictions of the partial cross section ratios. Another important feature of the high-energy charge-changing collisions involving complex atoms is the possibility and prevalence of electron capture from the inner electronic shells. This feature is a consequence of the fact that the cross section maximum appears at the velocity $v_m \sim (I^{\mathrm{A}}_{n_0} - I^{\mathrm{B}}_{n})^{1/2} \sim (I^{\mathrm{A}}_{n_0})^{1/2}$, $I^{\mathrm{A}}_{n_0}$ and I^{B}_{n} being the ionization potentials of the n_0th shell (or subshell) of the target atom A and the nth shell of the product ion $\mathrm{B}^{(Z-1)+}$. With increasing v, the above condition is satisfied for the electrons of increasingly deep electronic shells. Below, we shall give a brief account of the high-energy methods for treating the charge-exchange problem, emphasizing those aspects and recent developments which are pertinent to the case when highly charged ions are involved. Detailed accounts of these methods can be found in [7.66, 67]; see also [6.39–41].

7.4.1 The Brinkman-Kramers Approximation and Its Extensions

a) Brinkman-Kramers Approximation

The Brinkman-Kramers (BK) approximation [7.68, 69] consists in the neglect of the internuclear potential in the exchange matrix element of the first Born approximation [see (6.186–188)]. As discussed in Sect. 6.22, the internuclear interaction can be removed from the Schrödinger equation for the scattering problem by a phase transformation and, therefore, it has no effect (at least for linear trajectories) on the total cross section result. However, the differential cross section, which we shall not consider here, is affected by the internuclear potential.

Let us consider the problem of electron capture from a complex atom by a structureless multiply charged ion,

$$A(n_0\, l_0^N) + B^{Z+} \rightarrow A^+(n_0\, l_0^{N-1}) + B^{(Z-1)+}(n,\, l)\,, \tag{7.102}$$

where we have chosen a specific $(n_0\, l_0^N)$ shell of the atom. The quantum-mechanical scattering amplitude for the charge-exchange process in the BK approximation is (apart from an unessential mass factor)

$$f(\boldsymbol{q}) = \int d\boldsymbol{r}\, d\boldsymbol{r}'\, \phi_1^{B*}(\boldsymbol{r}')\, e^{-i\boldsymbol{\varkappa}\cdot\boldsymbol{r}'}\, V^{BK}\, e^{i\boldsymbol{\nu}\cdot\boldsymbol{r}}\, \phi_0^A(\boldsymbol{r}) \tag{7.103}$$

where $\boldsymbol{q} = \boldsymbol{k}_0 - \boldsymbol{k}_1$ is the momentum transfer, $\boldsymbol{k}_0$ and $\boldsymbol{k}_1$ are the relative momenta before and after the collision, and ϕ_0^A and ϕ_1^B are the corresponding one-electron wave functions. The vectors $\boldsymbol{\varkappa}$ and $\boldsymbol{\nu}$ are related to $\boldsymbol{q}$ and the relative velocity $\boldsymbol{v}$ by (within an accuracy of the order of $1/M_A$)

$$\boldsymbol{\varkappa} = \boldsymbol{q} + \boldsymbol{v}\, M_A/(M_A + M_B)\,, \tag{7.104a}$$

$$\boldsymbol{\nu} = \boldsymbol{q} - \boldsymbol{v}\, M_B/(M_A + M_B)\,, \tag{7.104b}$$

M_A and M_B being the masses of A and B^{Z+}. With the same accuracy, the conservation of energy in the process yields

$$\varkappa^2 - \nu^2 = I(n_0\, l_0) - Z^2/(2\, n^2) \equiv \omega \tag{7.105}$$

where $I(n_0\, l_0)$ is the ionization potential of the $(n_0\, l_0^N)$ subshell. In (7.103), V^{BK} can be used either in its "prior" (V_i^{BK}) or "post" (V_f^{BK}) form, representing the interactions in the initial and final channels of the reaction (7.102), respectively.

The cross section for capture of one of the N electrons is given by [cf. (6.164 a and 7.103)]

$$\sigma = \frac{N}{2\pi^2 v^2} \int\limits_{k_0-k_1}^{k_0+k} q\, dq\, |f(q)|^2. \tag{7.106}$$

Inserting (7.103) into (7.106) and summing over the magnetic quantum numbers, one finds [7.70, 71]

$$\sigma_{nl}(n_0\, l_0) = \frac{8\,\pi N(2\,l+1)}{v^2} \int\limits_{x_0}^{\infty} x\, dx\, T^2(x)\, Q^2(x,\omega)\,, \tag{7.107}$$

$$x_0 = \left| \frac{v}{2} + \frac{\omega}{v} \right| . \tag{7.108}$$

The quantities $T(x)$ and $Q(x, \omega)$ in the "prior" and "post" form are given by

$$T_i(x) = \int\limits_0^{\infty} R_{nl}(r)\, V_i^{\mathrm{BK}}(r)\, j_l(x\,r)\, r^2\, dr\,, \tag{7.109a}$$

$$Q_i(x,\omega) = \int\limits_0^{\infty} R_{n_0 l_0}(r)\, j_{l_0}[r(x^2 - 2\,\omega^2)^{1/2}]\, r^2\, dr\,, \tag{7.109b}$$

$$T_f(x) = \int\limits_0^{\infty} R_{nl}(r)\, j_l(x\,r)\, r^2\, dr\,, \tag{7.110a}$$

$$Q_f(x,\omega) = \int\limits_0^{\infty} R_{n_0 l_0}(r)\, V_f^{\mathrm{BK}}(r)\, j_{l_0}[r(x^2 - 2\,\omega^2)^{1/2}]\, r^2\, dr\,, \tag{7.110b}$$

where R_{nl}, $R_{n_0 l_0}$ are the radial wave functions, normalized by the condition

$$\int\limits_0^{\infty} R_{nl}^2(r)\, r^2\, dr = 1\,, \tag{7.111}$$

and j_l, j_{l_0} are the spherical Bessel functions. Since, in general, $R_{n_0 l_0}$ and R_{nl} are not orthogonal, the use of the "post" and "prior" forms of $T(x)$ and $Q(x, \omega)$ yields different results for σ. With increasing v, the "post-prior" discrepancy becomes unimportant. For the reaction (7.102), the interactions $V_i^{\mathrm{BK}}(r)$ and $V_f^{\mathrm{BK}}(r)$ are given by

$$V_i^{\mathrm{BK}}(r) = -\,Z/r\,, \tag{7.112}$$

$$V_f^{\mathrm{BK}}(r) = -\,\zeta(r)/r\,, \tag{7.113}$$

where $\zeta(r)$ is the effective charge of the atomic core, given by

$$\zeta(r) = Z_0 - \int\limits_0^{r} r'\, \varrho(r')\, dr' - r \int\limits_r^{\infty} \frac{\varrho(r')}{r'}\, dr'\,, \tag{7.114}$$

Z_0 is the nuclear charge of atom A and $\varrho(r)$ is the electronic density of the atomic core.

If B^{Z+} is a fully stripped ion, then using the well-known properties of the hydrogenic wave functions [7.72], one can sum $\sigma_{nl}(n_0\, l_0)$ over the angular quantum number l to obtain

$$\sigma_n(n_0\, l_0) = \sum_{l=0}^{n-1} \sigma_{nl}(n_0\, l_0) = \frac{8\pi N}{n^3 v^2} \int_{x_0}^{\infty} T_n^2(x)\, Q^2(x, \omega)\, x\, dx\,, \tag{7.115}$$

where $Q(x, \omega)$ is as before, and $T_n(x)$ is

$$T_{n,\mathrm{i}}(x) = (x^2 + n^{-2})^{-1}\,, \tag{7.116a}$$

$$T_{n,\mathrm{f}}(x) = 4(x^2 + n^{-2})^{-2}\,. \tag{7.116b}$$

For large values of n, $T_n(x)$ becomes independent of n, and the cross section $\sigma_n(n_0\, l_0)$ becomes proportional to n^{-3}. If n_c is the value of n above which the n-dependence of $T_n(x)$ can be neglected (to a prescribed accuracy), then for all $n > n_c$, the following relation holds:

$$\sigma_n(n_0\, l_0) = \left(\frac{n_c}{n}\right)^3 \sigma_{n_c}(n_0\, l_0)\,, \quad n > n_c\,. \tag{7.117}$$

This relation is known as Oppenheimer's n^{-3} rule. It is usually used to estimate the contribution to the total cross section from the states with $n > n_c$:

$$\sigma_t(n_0\, l_0) = \sum_{n=1}^{n_c} \sigma_n(n_0\, l_0) + n_c^3\, \sigma_{n_c}(n_0\, l_0) \left[\zeta(3) - \sum_{n=1}^{n_c} n^{-3}\right], \tag{7.118}$$

where $\zeta(3) \simeq 1.202\ldots$ is the Riemann zeta function. Numerous calculations have shown that Oppenheimer's n^{-3} rule can be applied for $n > n_c \simeq 5\, n_0\, Z/Z_0\, v$ to an accuracy of about 10%.

In the case when $A(n_0\, l_0)$ is a hydrogen-like system with charge Z_0, one can also sum $\sigma_n(n_0\, l_0)$ over the quantum numbers l_0, which gives

$$\sigma_n^{\mathrm{BK}}(n_0) = \sum_{l_0} \sigma_n(n_0\, l_0) = \pi \frac{2^8 N (Z_0 Z)^5}{5\, n_0^5\, n^3 v^2 g^5(v)}\,. \tag{7.119}$$

$$g(v) = \left(\frac{v}{2} + \frac{\omega}{v}\right)^2 + \frac{Z^2}{n^2} = \left(\frac{v}{2} - \frac{\omega}{v}\right)^2 + \frac{Z_0^2}{n_0^2}\,. \tag{7.120}$$

It follows from (7.119) that $\sigma_n(n_0)$ behaves as v^{-12} as $v \to \infty$. Such behaviour also follows directly from (7.107), namely, for $v \gg \omega^{1/2}$, the lower limit of the integral in (7.107) is $x_0 \sim v/2$ and the value of the integral is determined by the rate of decrease of the integrand at large values of x. This is, in turn, determined by the properties of the wave function at small $r \sim 1/x$. The Fourier transform of the wave function [see (7.109, 110)] falls off at least as slowly as x^{-4} (for the $n_0 l_0 = 1$s state), which corresponds to a cross section behaviour $\sigma \sim v^{-k}$, $k \geqq 12$.

The analysis of (7.107) shows that for a given reaction-energy defect ω, the cross section maximum is attained at a velocity v_m given by

$$v_m \simeq 0.7\,|\omega|^{1/2}, \tag{7.121}$$

i.e. with increasing $I(n_0 l_0)$, v_m also increases. Detailed investigations of the total electron-capture cross section in the BK approximation [7.73, 74] have revealed the asymptotic behaviour

$$\sigma_t^{BK} \sim I_0^{5/2} Z^5 v^{-12}, \quad v \gg \max\{I_0^{1/2}, Z^{1/2}\}, \tag{7.122a}$$

$$\sim I_0^{5/2} Z^3 v^{-10}, \quad Z \gtrsim \{I_0, v^2\}, \tag{7.122b}$$

where $I_0 \equiv I(n_0)$. For the one-electron system $(Z_0, e) + Z$, an explicit but rather cumbersome expression for the BK cross section $\sigma_{nlm}(n_0 l_0 m_0)$ has been derived by *Omidvar* [7.75]. The BK cross section $\sigma_{nl}(n_0 l_0)$ for this system has the l_0-, l-dependence

$$\sigma_{nl}(n_0 l_0) \sim (n_0 n)^{-3} Z_0^{5+2l_0} Z^{5+2l} v^{-2(6+l_0+l)}. \tag{7.123}$$

It is also useful to note the following properties of the BK cross section $\sigma_{Z;nlm}(Z_0; n_0 l_0 m_0)$:

$$\sigma_{Z;nlm}(Z_0; n_0) = \sigma_{Z;nlm}(Z_0/n_0; 1), \tag{7.124a}$$

$$\sigma_{Z;n}(Z_0; n_0 l_0 m_0) = n^2 \sigma_{Z/n;1}(Z_0; n_0 l_0 m_0), \tag{7.124b}$$

$$\sigma_{Z;n}(Z_0; n_0; v) = \alpha^2 n^2 \sigma_{\alpha Z/n;1}(\alpha Z_0/n_0; 1; \alpha v), \tag{7.124c}$$

where absence of the quantum numbers l, m (or l_0, m_0) denotes an average over them.

The accuracy of the absolute cross section values provided by the BK approximation is rather low and cannot be rigorously specified. For capture from $n_0 s$ states by a bare nucleus, the contribution of the higher-order terms of the Born series to σ_t^{BK} introduces a factor of about 0.3 (see below). Moreover, as will be seen in Sect. 7.4.2, the correct asymptotics behaviour of the cross section at large v is not v^{-12}, but rather v^{-11}, as predicted by the second-order Born term.

Useful cross section calculations of modest accuracy can still be performed using the BK approximation for complex atoms, where more elaborate methods cannot be applied. For this, a semi-empirical factor $a_0 \simeq 0.3$ is usually introduced to simulate the second-order effects (for not too high v) and a normalization procedure is applied to the computed cross section [7.76].

b) VPS-BK Approximation

Another possibility for improving the BK approximation in the velocity region of the cross section maximum and below, is to start with the VPS approximation for the transition probability [see (6.150)]. For large values of v the sine function in (6.150) can be replaced by its argument, and the integration over impact parameters leads to the expression for the cross section [7.71, 77]

$$\sigma_{nl}(n_0\, l_0) = \exp(-\pi\,|\omega|/\gamma\, v)\; \sigma_{nl}^{\mathrm{BK}}(n_0\, l_0)\,, \tag{7.125}$$

$$\gamma = (1/2)\,[(2I_0 + v^2/4)^{1/2} + (2I_n + v^2/4)^{1/2}]\,,$$
$$I_0 \equiv I_0(n_0\, l_0)\,, \quad I_n = Z^2/(2\,n^2)\,, \tag{7.126}$$

where $\sigma_{nl}^{\mathrm{BK}}(n_0\, l_0)$ is given by (7.107).

Considering capture from the inner shells by highly stripped ions, the cross section for the production of inner-shell vacancies is of interest:

$$\sigma_t(n_0) = \sum_{n,l,l_0} \sigma_{nl}(n_0\, l_0) \tag{7.127}$$

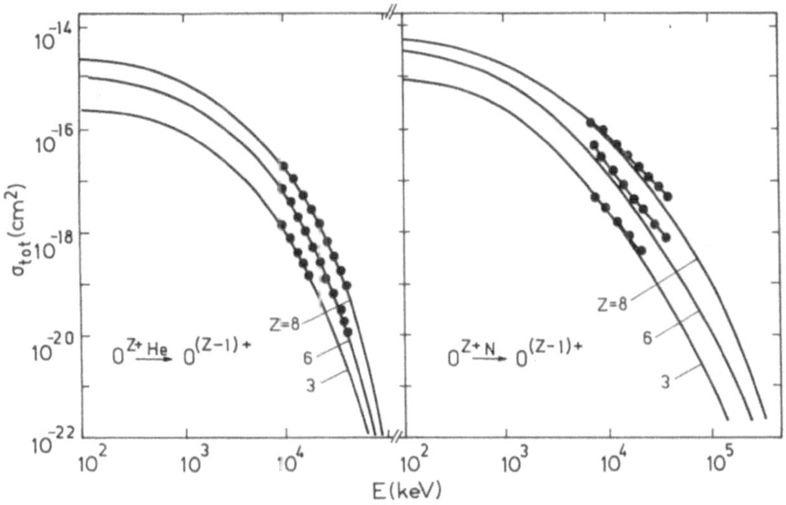

Fig. 7.8. Total charge-transfer cross sections for the O^{Z+} + He and O^{Z+} + N collision systems ($Z = 3, 6, 8$) obtained in the VPS approximation (with a 1/3 empirical correction factor) [7.71]. The points represent the experimental data [7.79]

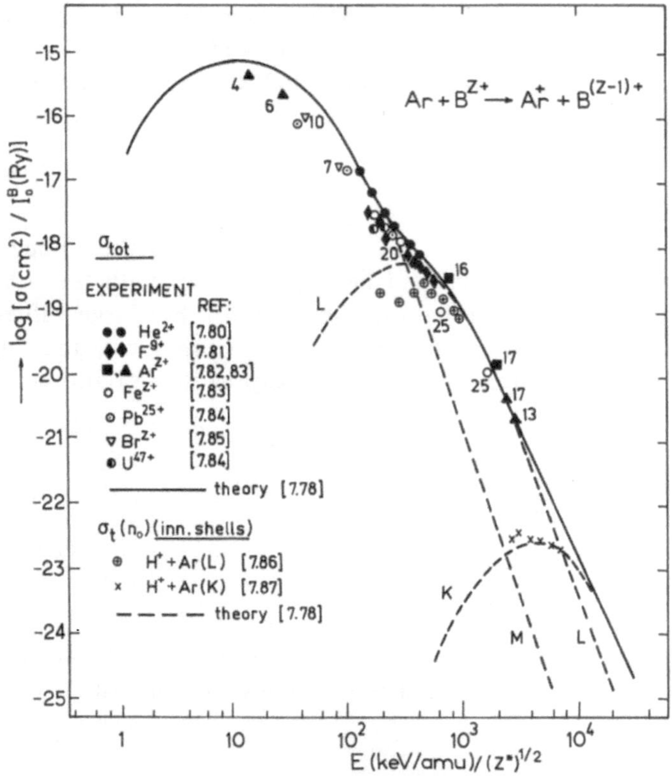

Fig. 7.9. Cross sections for electron capture from the inner shells in Ar + Br^{Z+} collisions in a reduced representation

The total capture cross section is then given by summing over the contributions of different shells:

$$\sigma_{\text{tot}} = \sum_{n_0} \sigma_{\text{t}}(n_0) . \tag{7.128}$$

Calculations of $\sigma_{\text{t}}(n_0)$ and σ_{tot} have been performed for a large number of complex atoms colliding with fully and partially stripped ions [7.71, 74, 78]. Some of these results are shown in Figs. 7.8, 9, where comparison is also made with the experimental data. The results in Fig. 7.9 are given in scaled units: $\tilde{\sigma} = \sigma / I_0^{\text{B}}$ and $\tilde{E} = E/(Z^*)^{1/2}$, where I_0^{B} is the ground-state ionization potential of the product ion $\text{B}^{(Z-1)+}$ and $Z^* = n_0^{\text{B}}(2 I_0^{\text{B}})^{1/2}$ is the effective charge of the ionic core B^{Z+}. For high Z, $Z^* \simeq Z$.

c) Eikonal-BK Approximation

A simple way to go beyond the BK approximation is to include some effects of the higher-order Born terms through an eikonal phase in the scattering function [see (6.201–203)]. If the potentials V^{A} and V^{B} appearing in the transition amplitude (6.203) of the eikonal approximation are used in the BK approximation (i.e. with the internuclear interaction omitted), one obtains an eikonal-BK (E-BK) approximation.

Let us consider a hydrogen-like atom-fully stripped ion collision system, $(Z_0, e)_{1s} + Z$. The semi-classical transition amplitude in the E-BK approximation for this system is ("prior" form)

$$a_{\mathrm{if}}^{\text{E-BK}} = -\mathrm{i} \int_{-\infty}^{+\infty} dt \left\langle F_{\mathrm{f}}^{(Z)} \phi_{\mathrm{f}}^{(Z)} \right| (-Z/r_{\mathrm{B}}) \exp\left(\mathrm{i} \int_{t}^{\infty} \frac{Z_0}{r_{\mathrm{A}}} \mathrm{d}t\right) \left| F_{\mathrm{i}}^{(Z_0)} \phi_{\mathrm{i}}^{(Z_0)} \right\rangle \tag{7.129}$$

where $F_{\mathrm{i}}^{(Z_0)}$, $F_{\mathrm{f}}^{(Z)}$ are plane-wave translational factors and $\phi_{\mathrm{i}}^{(Z_0)}$ and $\phi_{\mathrm{f}}^{(Z)}$ are hydrogenic wave functions, centred on Z_0 and Z, respectively. The above approximation has been used by *Chan* and *Eichler* [7.88, 89], who were able to perform the integrations analytically and obtain the cross section for the $1\,s \rightarrow n$ transition in the form

$$\sigma_n^{\text{E-BK}}(1\,s) = \alpha_{Z_0}(Z, n, v)\, \sigma_n^{\text{BK}}(1\,s). \tag{7.130}$$

The function $\alpha_{Z_0}(Z, n, v)$ has been obtained in closed form, and for the case $Z_0 = 1$ reads

$$\begin{aligned} \alpha_1(Z, n, v) = {} & \frac{(\pi/v)}{\sinh(\pi/v)} \exp\left[-\frac{2}{v} \arctan\left(\frac{v}{2} - \frac{\omega}{v}\right)\right] \\ & \times \left[\frac{23}{48} + \frac{1}{6}(1 + 5\,\omega)\, v^{-2} + \frac{5}{12}\,\omega^2\, v^{-4}\right], \end{aligned} \tag{7.131}$$

with $\omega = (1/2)(1 - Z^2/n^2)$. Thus, $\sigma_n^{\text{E-BK}}(1\,s)$ is just $\sigma_n^{\text{BK}}(1\,s)$, scaled by the function $\alpha_{Z_0}(Z, n, v)$. For $Z_0 = 0$, $\alpha_{Z_0}(Z, n, v) = 1$. The case $Z_0 = Z = n = 1$ has previously been treated by *Dewangan* [7.90]. The scaling function $\alpha_1(Z, n, v)$ is largely independent of Z and n, and when v varies between $v = 2$ and $v = 5$, it increases gradually from 0.15 to 0.4. The eikonal factor $\alpha_{Z_0}(Z, n, v)$ reduces the absolute BK cross section values to a level necessary for a fair agreement with the experimental data in the intermediate velocity region while leaving the Z- and n-dependence of the BK cross section practically unchanged. Since $a_{\mathrm{if}}^{\text{E-BK}}$ is only of first order in the perturbation (Z/r_{B}), it follows that the E-BK approximation is inadequate at lower velocities ($v \lesssim 1-2$). On the other hand, by expanding the eikonal phase in powers of (Z_0/r_{A}), the second-order term in the expansion is purely real, whereas the corresponding term in the Born series is complex. For this reason, the E-BK approximation must also fail at high collision velocities. The weak dependence of the scaling function $\alpha_{Z_0}(Z, n, v)$ on the charges Z_0

and Z, makes possible the extension of (7.130) to collisions involving complex atoms and/or incompletely stripped ions if the concept of effective charge is introduced [7.91]. For the $(Z_0, \mathrm{e}) + Z$ system, analytic expressions for $\sigma_{nlm}^{\text{E-BK}}(n_0\, l_0\, m)$ can also be derived [7.92].

7.4.2 Higher-Order Approximations

Of the higher-order methods (see Sect. 6.3.5), only the second Born approximation (B2) and the continuum distorted-wave (CDW) method have been applied to any extent in the studies of the charge-exchange process in high-energy collisions of atoms with highly charged ions. The only system which has been studied is the $(Z_0, \mathrm{e}) + Z$ system, the specific calculations being made for $Z_0 = 1$ and a few values of Z. Although limited, these studies are of great theoretical importance and have been recently reviewed [7.93]. Below we shall consider only the $(Z_0, \mathrm{e}) + Z$ collision system.

a) The Second Born Approximation

The charge-exchange cross section in the second Born approximation can be represented in the form [7.94, 95]

$$\begin{aligned}\sigma^{\mathrm{B2}} &= \int d\varrho \,|\, T_{\mathrm{if}}^{\mathrm{B2}}(\varrho)\,|^2 = \int d\varrho \,|\, T_{\mathrm{if}}^{(1)}(\varrho) + T_{\mathrm{if}}^{(2)}(\varrho)\,|^2 \\ &= (2\pi)^2 v^{-1} \int d\boldsymbol{q} \,|\, T_{\mathrm{if}}^{(1)}(\boldsymbol{q}) + T_{\mathrm{if}}^{(1)}(\boldsymbol{q})\,|^2\, \delta(\boldsymbol{q}\cdot\boldsymbol{v} + \omega - v^2/2)\,, \end{aligned} \tag{7.132}$$

where $T_{\mathrm{if}}^{(1)}$ and $T_{\mathrm{if}}^{(2)}$ are the first and second terms in the T-matrix expansion and $\boldsymbol{q}$ is the momentum transfer. The above cross section can be represented as

$$\sigma^{\mathrm{B2}} = \sigma^{(1)} + \sigma^{(1,2)} + \sigma^{(2)} \tag{7.133}$$

where $\sigma^{(1)} = \sigma^{\mathrm{B1}}$ is the contribution from $T_{\mathrm{if}}^{(1)}$, $\sigma^{(1,2)}$ comes from the interference term ($\sim \mathrm{Re}\,\{T_{\mathrm{if}}^{(1)*}\, T_{\mathrm{if}}^{(2)}\}$), and $\sigma^{(2)}$ is the contribution from $T_{\mathrm{if}}^{(2)}$. It has been proved [7.96, 97] that there exists a direct connection between the second-order term $\sigma^{(2)}$ and the classical double-scattering capture mechanism of *Thomas* [7.98] (see Sect. 6.4.1). Introducing the Fourier transforms of the wave functions and the potentials involved in $T_{\mathrm{if}}^{(2)}$, this amplitude can be written as [7.94]

$$\begin{aligned} T_{\mathrm{if}}^{(2)}(\boldsymbol{q}) &= \mathrm{i}\,(2\,\pi)^{-2} \int d\boldsymbol{k}\, d\boldsymbol{k}'\, \tilde{\psi}_{\mathrm{f}}(\boldsymbol{k}')\, \tilde{V}^{\mathrm{A}}(\boldsymbol{q}-\boldsymbol{k}) \\ &\quad \times D^{-1}(\boldsymbol{q}, \boldsymbol{k}, \boldsymbol{k}')\, \tilde{V}^{\mathrm{B}}(\boldsymbol{k}'-\boldsymbol{q}+\boldsymbol{v})\, \tilde{\psi}_{\mathrm{i}}(\boldsymbol{k})\,, \end{aligned} \tag{7.134}$$

where "$\sim$" denotes the Fourier-transformed quantity,

$$D^{-1}(\boldsymbol{q}, \boldsymbol{k}, \boldsymbol{k}') = (D_0 + D_1 + D_2)^{-1} \quad \text{with} \tag{7.135}$$

$$D_0 = q^2/2 - v^2/2 - (\varepsilon_i^A + i\,\eta), \; \eta \to 0^+, \tag{7.136a}$$

$$D_1 = \boldsymbol{k} \cdot (\boldsymbol{v} - \boldsymbol{q}) - \boldsymbol{k}' \cdot \boldsymbol{q}, \tag{7.136b}$$

$$D_2 = (\boldsymbol{k}_1 + \boldsymbol{k}')^2/2, \tag{7.136c}$$

and ε_i^A is the initial-state electron binding energy. V^A and V^B are the perturbations in the initial and final channels, respectively.

Direct evaluation of the cross section $\sigma^{(2)}$ using the exact amplitude (7.134) is extremely difficult even for the $1\,s \to 1\,s$ electron-capture transition [7.96, 99, 100]. The evaluation of $T_{if}^{(2)}$ can be simplified if the following approximations are made:

$$|\boldsymbol{q}| \approx v, \quad |\boldsymbol{q} - \boldsymbol{v}| \approx v, \quad \tilde{\boldsymbol{k}} \approx 0, \quad \tilde{\boldsymbol{k}}' \approx 0, \tag{7.137a}$$

$$D_2 \simeq 0. \tag{7.137b}$$

These approximations are justified for $v \gg \varepsilon_i^A, \varepsilon_f^B$ [7.101, 102].

Attaching damping exponential factors $\exp(-\alpha\,r)$ to the Coulomb potentials $V^A(r)$ and $V^B(r)$ acting in the (Z_0, e, Z) system, and using the approximations (7.137) in (7.134), one obtains [7.94]

$$T_{if}^{(2)}(\boldsymbol{q}) \simeq -2\pi\, \tilde{V}^A(\boldsymbol{q}; \alpha)\, \tilde{V}^B(\boldsymbol{q} - \boldsymbol{v}; \alpha)\, I_{if}(\boldsymbol{q}, \boldsymbol{v}), \tag{7.138}$$

$$\begin{aligned} U_{if}(\boldsymbol{q}, \boldsymbol{v}) = \int dx\, \psi_f^*(-x\,\boldsymbol{q})\, \psi_i[x(\boldsymbol{q} - \boldsymbol{v})] \\ \times \exp[-i\,x(q^2/2 - v^2/2 - \varepsilon_i^A)], \end{aligned} \tag{7.139}$$

with

$$\tilde{V}^A(\boldsymbol{q}; \alpha) = -(2/\pi)^{1/2}\, Z/(q^2 + \alpha^2), \tag{7.140}$$

and an analogous expression for $\tilde{V}^B(\boldsymbol{q} - \boldsymbol{v}; \alpha)$. The integral $I_{if}(\boldsymbol{q}, \boldsymbol{v})$ can be evaluated analytically for arbitrary hydrogenic states [7.94]. It should be noted that the approximation (7.138) for the amplitude $T_{if}^{(2)}$ tends to underestimate the exact cross section σ^{B2}. Calculations of σ^{B2} with the approximate form (7.138) of $T_{if}^{(2)}$ have been performed for the $H + C^{6+}$ system [7.95].

A remarkable feature of the "double-scattering" term $\sigma^{(2)}$ of the second Born cross section is its v^{-11} behaviour at asymptotically high velocities. Contrary to the first Born (or BK) approximation, which exhibits a $v^{-6(2+l_0+l)}$-dependence on the initial and final state angular quantum numbers, [see (7.123)], the v^{-11}-dependence of $\sigma^{(2)}$ is not dependent on l_0 and l. The velocity v_c at which the term $\sigma^{(2)}$ starts to dominate the v-dependence of σ^{B2} is strongly dependent on the final state quantum numbers n, l, m of the captured electron. For the $Z_0 = Z = 1$ system, v_c varies from $\sim 80 - \sim 5$, when $(n\,l\,m)$ is changed from $1\,s_0$ to $3\,d_{\pm 2}$ [7.95]. For the $1\,s \rightarrow 1\,s$ transition in this system, σ^{B2} is asymptotically given by [7.96]

$$\sigma_{1s}^{B2}(1\,s) = \left(0.29458 + \frac{5\,\pi\,v}{2^{12}}\right) \sigma_{1s}^{BK}(1\,s)\,. \tag{7.141}$$

The third Born (B3) approximation gives the asymptotic result [7.101, 102]

$$\sigma_{1s}^{B3}(1\,s) = \left(0.31868 + \frac{5\,\pi\,v}{2^{12}}\right) \sigma_{1s}^{BK}(1\,s)\,, \tag{7.142}$$

i.e. it changes the first term insignificantly and leaves the second v^{-11}-term unaltered.

The n^{-3} law has not been predicted by the second Born approximation, but it has been demonstrated that for $n \gg 1$ $\sigma_{nl}^{B2}(1\,s) \sim n^{-3}$ [7.97].

b) Other Second-Order Approximations

The second-order approximations discussed in Sect. 6.3.5, like CDW, CIS and IA, have not as yet been fully exploited in the studies of charge-exchange problems involving highly charged ions. CDW cross section calculations have been reported only for the systems $H + Li^{3+}$ ([7.67], total cross section), $H + F^{9+}$ ([7.103], total and partial σ_{nlm} cross sections) and $H + Z$, $Z = 1 - 6$ ([7.104], partial cross sections). With respect to the conventional formulation of the CDW amplitude (see Sect. 6.3.5), *Crothers* [7.104] has recently proposed a new formulation, which utilizes a Stark representation of the initial and final electronic states and has some computational advantages. It should be noted that the CDW total cross section has the same high-velocity asymptotic behaviour as the one predicted by the second Born approximation. However, in the impulse approximation (IA) and the continuum intermediate state (CIS) approximation, the v^{-11} asymptotic term is twice as large as in σ^{B2} or σ^{CDW}. More information about the asymptotic behaviour of the charge-exchange cross section in different second-order approximations, including capture from and to excited states, effects of backward scattering and relativistic effects, can be found in [7.67, 93].

Although σ^{B2} and σ^{CDW} have the same asymptotic behaviour (at least for $1\,s \rightarrow 1\,s$ transitions), they may differ considerably from each other at lower velocities, especially for large values of the final-state quantum numbers and high Z. An example is given in Fig. 7.10 where the relative contributions of different partial cross sections $\sigma_n(1\,s)$ to the total capture cross section of the reactions $H + Z$ $(Z = 2, 4, 6)$ at $v = 2.28$ are presented.

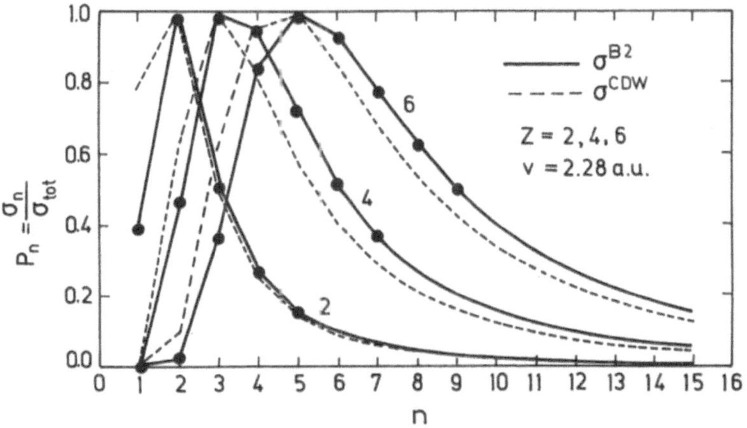

7.10. The ratio of the partial σ_r and total σ_{tot} electron-capture cross sections for the H + Z (Z = 2, 4, 6) collision systems calculated in the second Born [7.95] and CDW [7.104] approximations at a collision velocity of $v = 2.28$ a. u.

A new second-order method has recently been developed in connection with the electron capture from the inner electronic shells. This describes the situation in the $(Z_0, e) + Z$ collision system, with $Z \ll v \lesssim Z_0$ [7.105–108]. This method, termed "strong potential Born" (SPB) theory, is based on a further development of the impulse approximation to asymmetric collisions [7.105] and includes the off-energy-shell terms in the T matrix up to order $(Z_0/v)^2$ [7.107]. The SPB theory is, thus, a consistent approach which contains the weaker electron-projectile interaction to lowest order. The SPB capture amplitude is given by

$$a_{\mathrm{if}}^{\mathrm{SPB}} = (2\pi)^{3/2} \int d\boldsymbol{k}\, \tilde{\phi}_{\mathrm{f}}^{\mathrm{B}*}(\boldsymbol{k})\, \tilde{V}^{\mathrm{B}}(\boldsymbol{k}-\boldsymbol{q}) \times \langle \Psi^{-}_{\boldsymbol{k}+\boldsymbol{v},\varepsilon}(\boldsymbol{r})\, | \exp[\mathrm{i}(\boldsymbol{k}-\boldsymbol{q})\cdot\boldsymbol{r}] |\, \phi_{\mathrm{i}}^{\mathrm{A}}(\boldsymbol{r})\rangle \tag{7.143}$$

where $\phi_{\mathrm{i}}^{\mathrm{A}}$, $\phi_{\mathrm{f}}^{\mathrm{B}}$ are the initial- and final-state electron wave functions, respectively, $\tilde{V}^{\mathrm{B}}$ is the Fourier transform of the electron-projectile potential and $\varepsilon = v^2/2 + \boldsymbol{v}\cdot\boldsymbol{k} + \varepsilon_{\mathrm{f}}^{\mathrm{B}}$. The quantity $\Psi^{-}_{\boldsymbol{k}+\boldsymbol{v},\varepsilon}(\boldsymbol{r})$ is the wave function of an intermediate continuum state which satisfies the equation

$$(H_0 - \varepsilon + \mathrm{i}\,\eta)\, \Psi^{-}_{\boldsymbol{p},\varepsilon}(\boldsymbol{r}) = -(\varepsilon - p^2/2)\, \phi_{\boldsymbol{p}}^{\mathrm{A}}(\boldsymbol{r}), \tag{7.144}$$

where H_0 is the Coulomb Hamiltonian of the target (Z_0, e) and $\eta \to 0^+$. The main contribution to the integrals in (7.143) comes from the region $\boldsymbol{k} \approx \boldsymbol{q}$. The off-shell function $\Psi^{-}_{\boldsymbol{p},\varepsilon}$ can be related to the on-shell function $\Psi^{-}_{\boldsymbol{p}}$ which satisfies the homogeneous version of (7.144) [7.108]:

$$\Psi^-_{p,\varepsilon}(\boldsymbol{r}) \approx |\Delta|^{\tau}\, e^{i\pi\tau/2}\,\Gamma(1+\tau)\,\Psi^-_{p}(\boldsymbol{r}), \qquad (7.145)$$

$$\tau = Z_0[-2(\varepsilon + i\,\eta)]^{-1/2}, \quad \varepsilon \approx k^2/2,$$

$$\Delta = (p^2 + x^2)/4\,x^2, \quad x = Z_0/\tau.$$

With the function (7.145), the matrix element involved in (7.143) can be calculated analytically [7.108]. The integration over $\boldsymbol{q}$ can be performed either numerically or using the "peaking" approximation ($\boldsymbol{k} = \boldsymbol{q}$).

The transition amplitude of the impulse approximation, a^{IA}_{if}, is obtained if one replaces $\Psi^-_{p,\varepsilon}(\boldsymbol{r})$ by $\Psi^-_{p}(\boldsymbol{r})$ in (7.143).

7.4.3 Electron Capture into Continuum States

In the collisions of highly charged ions with atoms, the appearance of electrons ejected into a narrow forward cone with velocity $\boldsymbol{v}_e$, close to the projectile velocity $\boldsymbol{v}$, has been experimentally observed [7.109]. It is natural to assume that the appearance of these electrons is connected with a double-scattering mechanism, analogous to the Thomas classical model for electron capture. The process is, accordingly, termed "capture to the continuum" and its theory has been constructed within the double-scattering model [7.110–113]. The only difference from true electron capture is that a continuum wave function centred on the projectile is used for the final electron state in the calculations. For the continuum capture from a $1\,s$ state in the $(Z_0, e) + Z$ collision system, the differential cross section is [7.113]

$$\frac{d\sigma_{cc}(1\,s)}{dv_e} = \pi\, 2^5\, Z_0^4\, Z^3\, v^{-9}\, I\,(v_e, \chi), \qquad (7.146)$$

$$I(v_e, \chi) = \int d\Omega_e\, |\boldsymbol{v}_e - \boldsymbol{v}|^{-1}\, J_0(i\,\chi\, Z/Z_0)\, e^{-\chi Z/Z_0} \qquad (7.147)$$

where

$$\chi = 1 + \hat{\boldsymbol{k}}\cdot\hat{\boldsymbol{r}}^*, \quad \boldsymbol{k} = (\boldsymbol{v}_e - \boldsymbol{v}), \quad \boldsymbol{r}^* = (\pi/3)\,\boldsymbol{\varrho}, \qquad (7.148)$$

$J_0(i\,z)$ is the Bessel function and Ω_e is the solid angle around $\theta_0 \approx 0$ into which the electrons are ejected.

It is seen from (7.146, 147) that the continuum-capture differential cross section has a cusp at $\boldsymbol{v}_e = \boldsymbol{v}$. Due to the dependence of $(d\sigma_{cc}/dv_e)$ upon the direction of $\boldsymbol{k} = (\boldsymbol{v}_e - \boldsymbol{v})$, the cusp is asymmetric, the asymmetry being more pronounced for larger values of Z. Far from the cusp, the asymptotic velocity behaviour of $d\sigma_{cc}(1\,s)/dv_e$ is v^{-11}. The validity of the above expression for $d\sigma_{cc}(1s)/dv_\varepsilon$ is restricted by the conditions $v \gg Z_0$ and $v \gg (Z_0 Z)^{1/2}$.

Capture into continuum can also be treated within an approximation analogous to the Brinkman-Kramers approximation [7.89, 112]. In contrast to the double-scattering treatment, the BK version predicts a symmetric cusp in the differential cross section and a v^{-12} asymptotic behaviour.

7.5 General Features of Single-Charge-Transfer Cross Sections

The numerous experimental and theoretical investigations of the single-electron-capture process in atom-highly charged ion collisions have revealed the general dependence of the charge-transfer cross section on the main physical prameters of the colliding system (collision velocity, ionic charge and the initial- and final-state electron binding energy). For obvious reasons, the system theoretically studied most of all is the $H+Z$ system, while experimental information is more abundant for the many-electron targets and incompletely stripped ions. A critical analysis and evaluation of all the theoretical charge-exchange cross section data for the $H+B^{Z+}$ system has recently been published [7.114−116] with reference to the existing experimental data. For other collision systems, the relevant cross section information is contained in many review articles [7.2−9, 117].

7.5.1 Scaling Laws for the Total Single-Charge-Transfer Cross Section

The existence of a cross section scaling law with respect to a certain collision parameter is closely connected with the predominance of a single physical mechanism which governs the process. Since the mechanisms responsible for electron capture are different in various velocity regions, one should expect that the corresponding cross section scaling is also different. In view of their practical importance, the scaling relationships for the charge-exchange cross section have been much discussed [7.2−6, 74, 118, 119]. The most important are the scaling laws for the total charge-exchange cross section with respect to the ionic charge Z and the electron binding energy I_0 in its initial state. Various theoretical models for the process, emphasizing particular aspects of the collision dynamics in a given velocity range, predict different scaling relationships. These are presented in Table 7.2, together with some supporting experimental findings.

The knowledge of the scaling laws in different velocity regions allows theoretical and experimental cross section data to be represented in a

Table 7.2. Scaling laws for the total charge-exchange cross section in various velocity regions

Method	Z-scaling	Validity
a) $v \lesssim 1$		
Decay models:		
i) Electron tunnelling	Z	$Z \gtrsim 5$, many-electron system $Z \gtrsim 15$, $H+Z$ system
ii) Absorbing sphere	$Z \ln(a/Z)$	$Z \gtrsim 4$
Multichannel LZ with rotational coupling	Z	$Z \gtrsim 16, 0.04 \leqq v \leqq 1, H+Z$
Experiment:		
[7.3, 120]	$Z^{1.1-1.2}$	$2 \gtrsim Z \gtrsim 10$; H_2, rare-gas atom + partially stripped ions
[7.119]	Z	$5 \leqq Z \leqq 18$; H_2, $He+Z$
b) $0.8 \lesssim v \lesssim 3-4$		
UDWA	$Z^{1.07}$	$4 \leqq Z \leqq 20, v \lesssim 2, H+Z$
Classical models:		
i) Overbarrier transitions	Z	$Z \gtrsim 14; 0.5 \lesssim v \lesssim 1.5$
ii) Bohr-Lindhard	Z	$v \lesssim 2^{1/2} Z^{1/4}$
	Z^3	$v \gtrsim 2^{1/2} Z^{1/4}$
Multichannel VPS	Z^2	$Z \gtrsim 10$
BK	Z^3	$Z \gtrsim \{I_0, v^2\}$
CDW [7.73]	Z^3	$Z \geqq 2$
Experiment:		
[7.5, 117, 121]	$Z^{2.5-3}$	$2 \leqq Z \leqq 8, 1 \lesssim v \lesssim 3-4$
c) $v \gg 1$		
BK	Z^5	$v \gg I_0^{1/2}, Z^{1/2}$
Experiment:		
[7.122]	Z^5	$Z \lesssim 20, v \sim 100$

unified form. To this end, it is convenient to introduce reduced (or scaled) representations $\tilde{\sigma}$ and $\tilde{v}$ for the cross section and the collision velocity, respectively. In order to cover the velocity range from $\sim 10^{-2}$ to 10^2 a.u., we take the form $\tilde{v} = v\, Z^{-1/4}$, which follows from adiabatic considerations in the low-velocity region and, in the high-velocity region, both from classical arguments (e.g. the Bohr-Lindhard model) and the BK approximation. The corresponding reduced energy is $\tilde{E} = EZ^{-1/2}$. Since in the whole above-mentioned velocity range $\sigma \sim Z^{\alpha}$, $\alpha \geqq 1$ (see Table 7.2), an appropriate reduced from of σ would be $\tilde{\sigma} = \sigma/Z$. All the experimental cross section data for the $H+B^{Z+}$ and $He+B^{Z+}$ collisions with $Z \geqq 5$ are shown in Fig. 7.11 using such a $\tilde{\sigma}(\tilde{E})$ representation. The solid curves are the best least-squares fits to the data (see [7.118; 121]).

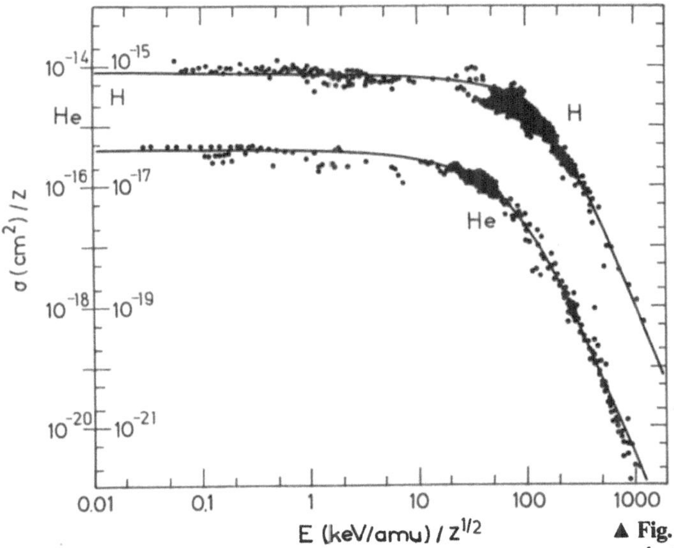

▲ **Fig. 7.11.** Reduced single-electron-capture cross sections σ/Z vs. reduced energy $E/Z^{1/2}$ for $H+B^{Z+}$ and $He+B^{Z+}$ collision systems with $Z \geqq 5$. Points represent experimental data of various authors [7.121]; solid curves are least-squares fits to the data [7.118]

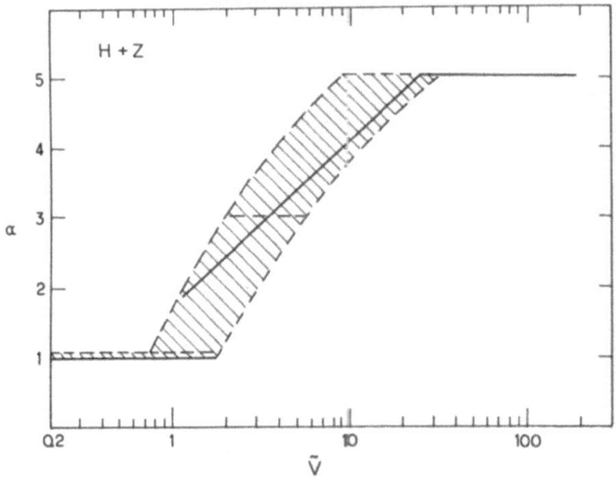

Fig. 7.12. Dependence of the exponent α in (7.149) on the scaled velocity $\tilde{v} = v/Z^{1/4}$ for the $H-Z$ collision system. The uncertainties are shown by the shaded area. The solid line within this area represents the prediction of the two-state AO-CC calculations

The scaling laws presented in Table 7.2 can be written in a unified form

$$\sigma = \sigma_0(\tilde{v})\, Z^{\alpha(\tilde{v})}, \quad \tilde{v} = v\, Z^{-1/4}. \tag{7.149}$$

For the $H+Z$ system, the function $\alpha(\tilde{v})$ in the interval $0.2 \leqq \tilde{v} \leqq 100$ behaves as shown in Fig. 7.12. The two-state coupled-channel calculations [7.74] predict an increase of α from $\alpha \approx 2$ to $\alpha \approx 5$ in the interval from $\tilde{v} \approx 1$ to $\tilde{v} \approx 25$. However, the scaling law in this region cannot be considered as well established. It should be noted that in the low-energy region, the linear

Table 7.3. I_0-dependence of the total charge-transfer cross sections

Method	Scaling
a) $v \lesssim 1$	
Decay models:	
i) Absorbing sphere	I_0^{-1}
ii) Electron tunnelling	$I_0^{-k(z)}$, $k = 3/2$, $Z \to \infty$
b) $0.8 \lesssim v \lesssim 3-4$	
Multichannel VPS	I_0^{-2}
Classical models:	
i) Overbarrier transitions	I_0^{-2}
ii) Bohr-Linhard	I_0^{-2}, $v \leqq 2^{1/2} Z^{1/4}$
	$I_0^{3/2}$, $v \geqq 2^{1/2} Z^{1/4}$
c) $v \gg 1$	
BK	$I_0^{5/2}$
B2 (double scattering term)	$I_0^{7/2}$

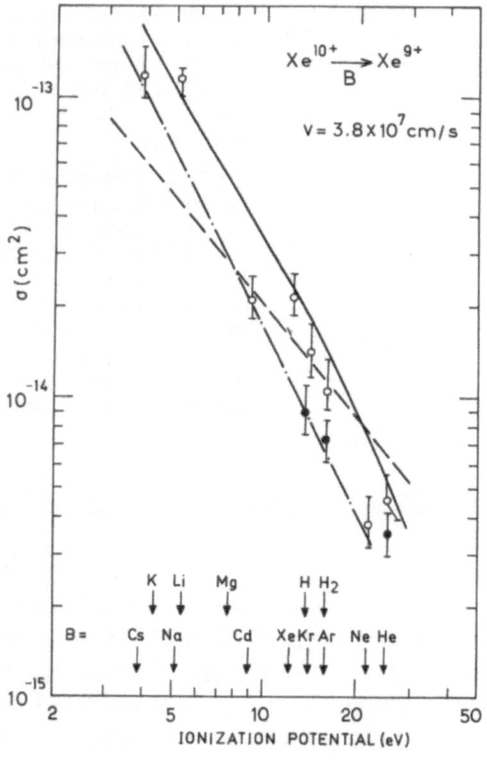

Fig. 7.13. Dependence of the capture cross section on the target ionization potential for the Xe^{10+} projectile ion at a collision velocity of $v = 3.8 \times 10^7$ cm/s. Curves represent the theoretical predictions. Solid line: electron tunnelling theory [7.10], dashed line: absorbing sphere model [7.14], dot-dashed line: the M-VPS model [7.51]. The experimental points are from [7.3] (○) and [7.54] (●)

Z-scaling is meaningful only for $Z \gg 1$. For small values of Z ($Z \lesssim 8$), the dependence of the cross section on the molecular structure effects is strongly pronounced and no scaling can be expected in this range of Z and v. In fact, as we shall see in Sect. 7.5.3, the Z-dependence of σ in this (Z, v) region has an oscillatory structure. In the high-velocity region and for incompletely stripped ions, the electronic structure of the projectile also influences the form of the scaling.

The situation at present is rather unclear regarding the scaling of the total charge-transfer cross section with respect to the binding energy I_0 of the target electron. The predictions of various theoretical models in a given velocity range may differ substantially. Table 7.3 summarizes the predicted I_0 scaling. The experimental investigations [7.54, 123, 124] of the I_0-dependence of single-charge-transfer cross sections are rather scattered and, therefore, inconclusive concerning the I_0-scaling laws. An example is given in Fig. 7.13, where some typical low-energy I_0-scaling laws are compared with experimental data.

7.5.2 Final-State Distributions of Captured Electrons

Creation of excited product states is one of the most prominent features of the charge-exchange process in atom-highly charged ion collisions (except at very high collision velocities). This feature results from the fact that the quasi-resonant energy condition, providing a high probability for the electron-capture process, is always fulfilled for highly excited ionic states when $Z \gg 1$.

The first experimental evidence for capture into specific excited states in collisions of hydrogen atoms with O^{8+} and C^{6+} ions has come from spectroscopic [7.125] and corpuscular [7.126] diagnostics of tokamak and laser-produced plasmas. Studies of state-selective electron capture in direct ion-atom collision experiments are also available, but the information gained is still rather incomplete and pertains to a restricted number of collision partners and limited ranges of variation of collision parameters. The theoretical studies are more numerous and almost entirely devoted to the hydrogen atom-fully stripped ion collision system. Currently, the state-selective electron-capture process is being investigated very actively and the subject is reviewed in several papers [7.127–130].

The main topic of interest in the study of capture into excited states is the distribution of captured electrons over the final-state quantum numbers (n, l, m). Both the absolute values of the partial cross sections σ_n, σ_{nl} and σ_{nlm} as functions of the collision velocity v, and the shapes of the n-, l- and m-distributions of captured electrons at a fixed velocity, present a sensitive test of the theoretical models for the process. Below, we shall discuss the final-state distributions of captured electrons as predicted by different models in various velocity regions.

a) *n*-Distributions

At very low collision velocities and for relatively low values of Z ($Z \lesssim 10$), electron capture takes place into a few principal electronic shells of the projectile ion, one of which is usually predominantly populated. Such a conclusion for the H+Z system follows from the adiabatic curve-crossing considerations and has been confirmed both in the multichannel Landau-Zener (MLZ) [7.19] and MO coupled-channel [7.23] calculations. The n-distribution of captured electrons is, thus, very narrow and centred at $n = n_m$. With increasing Z and collision velocity (still in the region $v \lesssim 1$), the number of final n-channels which are considerably populated increases, and the width of the n-distribution becomes larger. For the H+Z system, the shape of the n-distribution is asymmetric: due to the larger energy spacing of the states with $n < n_m$ with respect to those with $n > n_m$, the decrease of σ_n for $n < n_m$ is faster than for $n > n_m$. This picture remains unchanged as the collision velocity further increases and is illustrated in Fig. 7.14, where the CTMC cross section calculations of σ_n for the H+Z system are shown [7.61].

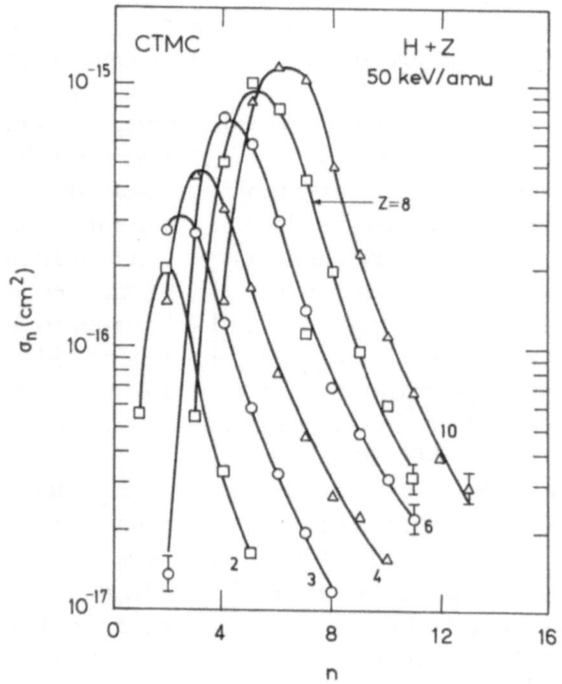

Fig. 7.14. n-distributions of captured electrons in the H+Z collisions at $E = 50$ keV/amu as predicted by the CTMC method [7.61]

An important question in connection with the n-distributions is the dependence of the quantum number n_m of the dominantly populated level on the ionic charge Z and collision velocity. In the low-velocity region ($v \lesssim 1$), the decay models, MLZ model and UDWA theory all predict that n_m is *almost* independent of v, and its Z-dependence is given either by (7.35a) obtained within the decay models, or by the best fit to UDWA results,

$$n_m^{\mathrm{UDWA}} \simeq Z^{0.768} . \tag{7.150}$$

The expressions (7.35a) and (7.150) are numerically equivalent (for $n_0 = 1$) up to $Z \approx 50$.

For $v \gtrsim 2$, n_m becomes v-dependent, reflecting the fact that electron momentum transfer starts to play an increasingly important role in the electron-capture process. The BK approximation gives for $n_m(Z, v)$ in the $(Z_0, e) + Z$ system,

$$n_m^{\mathrm{BK}} = 17^{1/2}\, Z\,[2\,(25\,v^4 + v_0\,v^2 + 25\,v_0^4)^{1/2} - 7\,(v^2 - v_0^2)]^{-13} , \tag{7.151}$$

where $v_0 = (Z_0/n_0)$ is the initial-state electron velocity. For $v \to \infty$, $n_m^{\mathrm{BK}} \to (17/3)^{1/2}\, Z/v$, which means that, independently of Z, Z_0 and n_0, the maximum populated final level is always $n = 1$.

b) *l*-Distributions

In the low-velocity region ($v \lesssim 1$), the problem of the population of different l-substates having the same principal quantum number n has been studied for the H+Z system in a general form using the multichannel Landau-Zener model [7.131] and an approximate analytical treatment of the diabatic MO-CC equations [7.34, 42]. The population mechanism of different l-substates within the M-LZ model is the mixing of Stark states due to the rotation of the internuclear axis. This mixing is weak at very low collision velocities and strong when v approaches unity. In these two limits, the relative population $W_{nl} = \sigma_{nl}/\sigma_n$ of different l-substates is respectively given by [7.19, 131]

– weak mixing:

$$W_{nl} = (2\,l + 1)\,[(n-1)!]^2/(n+l)!\,(n-1-l)!$$
$$\approx \left(\frac{2\,l+1}{n}\right) \exp\,[-\,l(l+1)/n], \quad n, l \gg 1 , \tag{7.152}$$

– strong mixing:

$$W_{nl} = (2\,l + 1)/n^2 , \tag{7.153}$$

where (7.153) represents just the statistical l-distribution. In the intermediate case W_{nl} has the form [7.131]

$$W_{nl} = [2^{2(n-1)} - 1]^{-1} \sum_{k_1, k_2 \neq |0\rangle} \begin{bmatrix} \frac{n-1}{2} & \frac{n-1}{2} & l \\ k_1 & k_1 & m \end{bmatrix} \binom{n-1}{k_1} \binom{n-1}{k_2}, \quad (7.154)$$

where [: : :] is the Clebsch-Gordan coefficient, (:) is the binomial coefficient, $k_{1,2} = [m \pm (n_2 - n_1)]/2$, $|0\rangle$ is the state with $k_1 = -k_2 = (n-1)/2$ and n_1, n_2, m are the parabolic quantum numbers of the particular product-ion state. Expression (7.154) for W_{nl} is valid for $Z^{-3/4} < v < 1$. For $n = 3, 4$ and 5, the above three l-distributions are presented in Fig. 7.15. These l-distributions have been qualitatively confirmed by the more elaborate MO-CC [7.30] and UDWA [7.52] calculations.

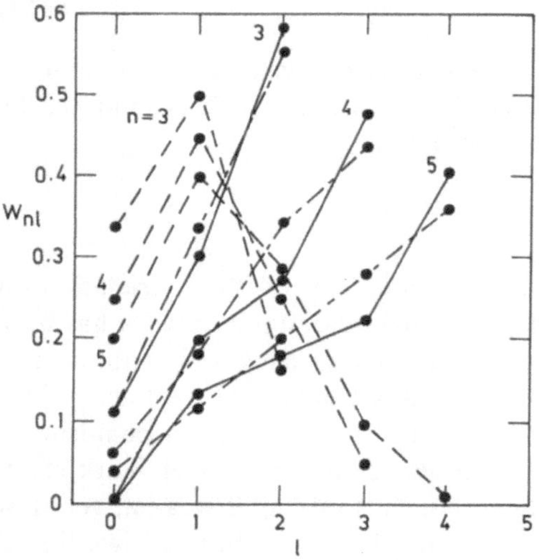

Fig. 7.15. Relative population of specific l-substates in the H+Z electron-capture collisions due to rotational mixing of Stark states within the $n = 3$, 4 and 5 manifolds. The points connected by dashed lines are for the weak mixing regime (7.152), the points connected by dot-dashed lines are for the strong mixing regime (7.153), and the points connected by solid lines represent the case of intermediate Stark mixing (7.154)

With increasing collision velocity ($v \sim 1-3$), both the UDWA and the CTMC methods predict that for $n < n_m$, the dominantly populated substate is the one with $l = l_m = n - 1$, whereas for $n > n_m$, $l_m \approx (Z-1)$ or $(Z-2)$ [7.129]. This character of the l-distributions in the intermediate-velocity region can be explained in terms of the classical picture of the process. For $n < n_m$, the capture arises primarily from small-impact-parameter collisions in which the electron is captured into an orbit with small eccentricity (classical analogue of high l). For $n > n_m$, the capture is dominated by large-impact-parameter collisions in which the electron is preferentially captured into an orbit having a large eccentricity (small l).

With a further increase in collision velocity, the l-distribution maximizes at smaller l_m values. In the $v \gg 1$ region, the BK approximation gives [see (7.123)]

$$W_{nl}^{\mathrm{BK}} \sim (Z/v)^{2l} \tag{7.155}$$

which for $v \gg Z$ predicts that the $l=0$ substate is dominantly populated. In the asymptotic region, where the v^{-11} second Born term dominates the cross section, the second-order theories (B2, CDW, CIS, IA) also predict that $l=0$ is the most-populated substate.

c) *m*-Distributions

The population of specific magnetic substates in the low-velocity collision region is governed by the requirement of conservation of the projection of electronic angular momentum on the internuclear axis and by the under-barrier (tunelling) character of the transition. If the initial magnetic substate is not specified, then the pronounced axial symmetry of the electron tunnelling mechanism predicts that substates with $m=0$ will be dominantly populated in the adiabatic energy region. This result follows from the R^{-m}-dependence of the exchange interaction $\Delta_{nlm}(R)$ (see Sect. 6.1.3) and from the similar dependence of the electron tunnelling probability (Sect. 7.2.1). The predominant population of the $m=0$ substate has also been predicted by UDWA in the region $1 \lesssim v \lesssim 5$ [7.52] for the H+Z system. The BK approximation for $v \gg 1$ leads to the same results.

The electron capture into excited states of multiply charged ions has also been investigated experimentally [7.132–135]. However, the agreement with theoretical predictions has so far been relatively modest.

7.5.3 *Z*-Oscillations of the Total Cross Sections

As we have seen, the selective character of electron capture at low collision energies results from the fact that localized nonadiabatic transitions are the underlying mechanism of the process. For low values of the ionic charge, only one final level (n_m), for which the quasi-resonance (or other equivalent maximum capture probability) condition is fulfilled, is predominantly populated. For $Z \lesssim 20$, the UDWA value for n_m, namely $n_m \simeq Z^{0.786}$, is very close to $n_m \simeq Z/2$, so that for odd values of Z the capture is shared by two neighbouring levels with a reduced probability. Thus, regarded as a function of Z, the electron-capture cross section for the H+Z system will exhibit maxima and minima for even and odd values of the ionic charge Z, respectively. In the velocity region where UDWA is not valid ($v \lesssim 0.5$),

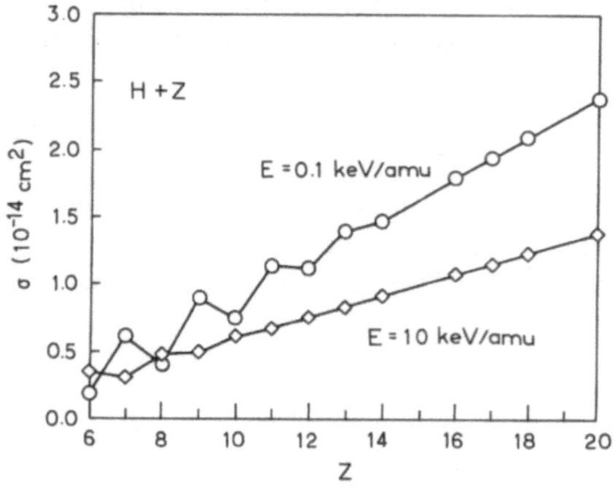

7.16. *Z*-oscillations of the total electron-capture cross section in low-energy H+*Z* collisions, predicted by the multichannel Landau-Zener theory [7.19]

other models for the electron-capture process (such as the decay and the multichannel Landau-Zener model) predict a weak (logarithmic) dependence of n_m on the collision velocity [7.130, 19], which is nevertheless sufficient to change the ordering of the extrema of $\sigma(Z)$ at lower collision velocities. Two examples of such *Z*-oscillations of the total capture cross section are given in Fig. 7.16, obtained by the multichannel Landau-Zener model with rotational coupling included [7.19]. It can be seen from the figure that above some $Z = Z_v$, the oscillations are smeared out, and that with increasing v, both Z_v and the amplitude of the oscillations decrease.

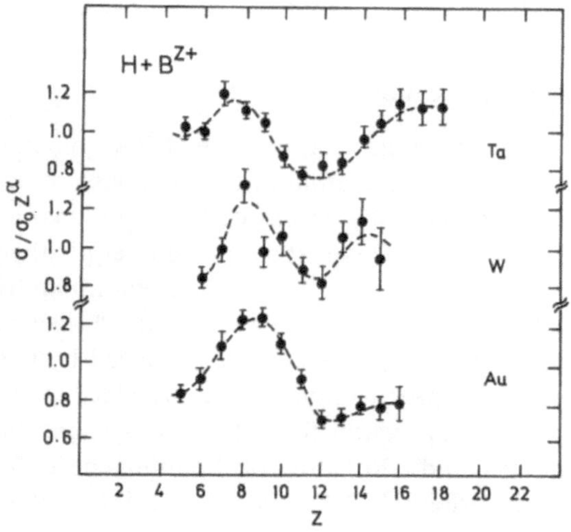

Fig. 7.17. *Z*-oscillations observed in the total electron-capture cross section in the high-energy $H+B^{Z+}$ collisions [7.140]. The collision velocities are $v = 3.4 \times 10^8$ cm/s for W^{Z+} and Au^{Z+}, and $v = 3.9 \times 10^8$ cm/s for Ta^{Z+}

Z-oscillations in the electron-capture cross section at low energies have also been obtained in a classical model of the process [7.136] as well as in the analytical treatment of the diabatic MO-CC equations [7.34, 42]. Oscillations of this type have also been experimentally observed in the charge-exchange cross sections of both fully [7.137, 138] and partially [7.124, 139] stripped ions on many-electron targets and on H atoms [7.139].

Regular deviations (up to 30%) from the monotonic Z-dependence of the capture cross section, $\sigma = \sigma_0 Z^\alpha$, have been also observed in the high-energy collisions of hydrogen atoms with incompletely stripped ions [7.56, 140]. The high-energy Z-oscillations of σ for ions with not too high atomic masses (like Fe^{Z+}, Mo^{Z+}, ...) can be directly related to the variations in the geometrical ionic cross section, $\sigma_g = \pi r_i^2$, where r_i is the ionic radius [7.56]. For heavier ions (such as Ta^{Z+}, W^{Z+} and Au^{Z+}), however, such a correlation of the Z-oscillations (see Fig. 7.17) with r_i cannot be established. An attempt [7.140] to describe the latter type of oscillations in terms of two-potential (Coulomb + screened-Coulomb) scattering of the transferring electron failed to reproduce the main features of the oscillations.

7.6 Single-Charge Transfer in Ion-Ion and Ion-Molecule Collisions

The basic mechanisms of the charge-exchange process in ion-ion and ion-molecule collisions remain the same as in the ion-atom case. However, in ion-ion and ion-molecule colliding systems some new aspects appear in the collision dynamics.

In ion-molecule collisions, the main new aspect is the vibrational motion of the nuclei in the target molecule and product molecular ion, which at low collision energies has to be explicitly included in the theoretical description. At high energies, the molecular structure also influences the electron-capture process.

The main specific feature of ion-ion collisions is the need to account for the Coulomb repulsion of the ions. This aspect, however, is important only at low collision energies.

There are relatively few theoretical studies and cross section calculations of ion-ion and ion-molecule charge-exchange processes. The experimental investigations of ion-ion electron-capture collisions are restricted to low charged ions [7.141] only, whereas the experimental cross section data for the ion-molecule charge exchange are more available [7.3–8]. We shall now give a brief account of charge-exchange investigations in these two collision systems.

7.6.1 Ion-Ion Charge Exchange

Let us consider first the resonant charge-transfer process in symmetric ion-ion systems

$$A^{(Z-1)+} + A^{Z+} \rightarrow A^{Z+} + A^{(Z-1)+}, \quad Z \geq 2. \tag{7.156}$$

Within the impact-parameter method, the cross section for reaction (7.156) is given by

$$\sigma = 2\pi \int_0^\infty \sin^2 \eta\,(\varrho, v)\, \varrho\, d\varrho\,, \tag{7.157}$$

$$\eta\,(\varrho, v) = (1/v) \int_{\varrho_1}^\infty \Delta\,(R) \left(1 - \frac{\varrho_0}{R} - \frac{\varrho^2}{R^2}\right)^{-1/2} dR\,, \tag{7.158}$$

where R is the internuclear distance, ϱ_0 and ϱ_1 are, respectively, the distances of closest approach for a head-on collision and for a collision with a given impact parameter ϱ,

$$\varrho_0 = Z\,(Z-1)/E\,, \quad \varrho_1 = \varrho_0/2 + (\varrho_0^2/4 + \varrho^2)^{1/2}, \tag{7.159}$$

and E is the centre-of-mass collision energy.

The quantity $\Delta\,(R)$ in (7.158) represents the energy separation of the symmetric (gerade) and antisymmetric (ungerade) states of the $A_2^{(2Z-1)+}$ system which at $R \rightarrow \infty$ are degenerate. For an ion A^{Z+} with closed electronic shell (or subshell), $\Delta\,(R)$ is given asymptotically by [7.142]

$$\Delta_{g-u}(R) = A_c^2\, \mathscr{I}(\gamma, l, m)\, R^{2Z/\gamma-1-m} \exp\,(-\gamma R - Z/\gamma)\,, \tag{7.160}$$

$$A_c = \gamma\,(2\gamma)^{Z/\gamma}\, Z^{-1/2}\,[\Gamma\,(Z/\gamma + l + 1)\,\Gamma\,(Z/\gamma - l)]^{-1/2}, \tag{7.161}$$

$$\mathscr{I}(\gamma, l, m) = (2\gamma)^{-2m}\,(2l+1)\,(l-m)!/(l+m)!\,, \tag{7.162}$$

where $-\gamma^2/2$ is the electron binding energy, l, m are the usual quantum numbers and the condition $\gamma R \gg 1$ is assumed. For a hydrogenic (n, l, m) state, the exchange interaction $\Delta_{g-u}(R)$ is obtained by setting $\gamma = Z/n$ in (7.160–162).

Expressions for $\Delta_{g-u}(R)$ for an ion A^{Z+} with an open electronic shell can be found elsewhere [7.142]. Assuming that the exponential dependence of $\Delta_{g-u}(R)$ is dominant, the charge-exchange phase shift $\eta\,(\varrho, v)$ can be calculated in the form [7.143]

$$\eta\,(\varrho, v) = \pi\, \varrho_1\, \Delta\,(\varrho_1)/(2\gamma)^{1/2}\, v\, (\varrho_0^2/4 + \varrho^2)^{1/2}\,. \tag{7.163}$$

For $E \ll Z(Z-1)$ one has $\varrho_0 \gg 1$, $\varrho_1 \approx \varrho_0 + \varrho^2/\varrho_0$, so that the phase shift $\eta(\varrho, v)$ is small. Replacing the sine-function in (7.157) by its argument, one obtains asymptotically

$$\sigma \approx 2\pi\varrho_0^2[\pi\Delta(\varrho_0)/2\gamma v]^2 . \tag{7.164}$$

Thus, for $E \ll Z(Z-1)$, the resonant electron-capture cross section decreases exponentially. This behaviour is different from the ion-atom resonant cross section which, according to (7.157), increases (logarithmically) with decreasing energy, until it attains the polarization capture limit. In the opposite case, $E \gg Z(Z-1)$, $\varrho_0 \ll 1$ and $\eta(\varrho, v)$ is large. The situation is similar to the case of resonant charge transfer in ion-atom collisions (straight-line trajectories), so that σ behaves as [7.142]

$$\sigma \sim \gamma^{-2}\ln(a/v) \sim (n/Z)^2\ln(a/v) , \tag{7.165}$$

where a is a constant.

The Z^{-2}-dependence of the cross section remains approximately valid also in the region of the cross section maximum, $E_m \approx Z(Z-1)$ [7.144]. Thus, with increasing ionic charge, the magnitude of the cross section for $E \gtrsim E_m$ decreases roughly as $1/Z^2$, and the position of the cross section shifts towards higher collision energies. Cross section calculations for resonant single-charge transfer have been performed for the case when $A^{(Z-1)+}$ is a ground-state hydrogen-like ion [7.144], and $2 \le Z \le 8$. In this case $\Delta_{g-u}(R)$ is

$$\Delta_{g-u}(R) = (4/e)Z^3 R\, e^{-ZR} . \tag{7.166}$$

With the above expression for $\Delta_{g-u}(R)$, the charge-transfer phase shift $\eta(\varrho, v)$ has an extremum and the cross section exhibits an oscillatory structure which is more pronounced at higher energies ($E > E_m$) and for higher values of Z. The cross section for the $He^+ + He^{2+}$ resonant charge-transfer reaction has also been measured experimentally [7.145], however, below the energy at which the oscillations of the cross section are noticeable.

Turning to the non-resonant ion-ion charge-exchange reactions

$$A^{Z_1+} + B^{Z_2+} \rightarrow A^{(Z_1+1)+} + B^{(Z_2-1)+*} + \Delta\varepsilon , \tag{7.167}$$

one has to distinguish two cases: $Z_1 \ll Z_2$ and $Z_1 \approx Z_2$. In the first case, the collision dynamics at relatively low energies is governed by a series of pseudocrossings of the initial molecular potential energy curve with those corresponding to different final (including excited) molecular states. The radial coupling is the main electron transition mechanism and a multichannel Landau-Zener model can be used to estimate the cross section. As in the atom-ion case, the electron-capture process is highly state-selective, and (7.167) can be competitive with electron impact in producing excited ion species in a hot plasma. *Bazylev* and *Chibisov* [7.146] have considered the reaction (7.167) for $Z_1 \ll Z_2$, with, B^{Z_2} being a fully stripped ion. Keeping in mind the Stark splitting of the final electron states in the field of the $A^{(Z_1+1)+}$ ion, they derived the following expression for the adiabatic energy splitting at the diabatic crossing point R_c of the parabolic state $n_1 = 0$, $n_2 = p-1$, $m = 0$ with the initial state,

$$\Delta(R_c) = D(n_0, n, Z_1, Z_2)\, R_c^{n_0 Z_1 + n Z_2 - 1} \exp\left[-\frac{R_c}{Z_1 + Z_2}\left(\frac{Z_1}{n_0} + \frac{Z_2}{n}\right)\right], \tag{7.168}$$

$$D(n_0, n, Z_1, Z_2) = \frac{A_1}{n^2}\left(\frac{2}{n}\right)^{p-1} \frac{(n_0 + n)(Z_1 + Z_2)}{(p-1)!\,(n_0 Z_1 + n Z_2)}$$

$$\times \frac{[Z_1/(Z_1+Z_2)]^{(n_0-n)Z_1}}{[Z_2/(Z_1+Z_2)]^{(n_0-n)Z_2}} \exp\left[-\frac{Z_1 Z_2 (n_0+n)}{Z_1+Z_2}\right],$$

$$n = p/Z_2, \quad p = n_1 + n_2 + m + 1, \tag{7.169}$$

where p is the principal quantum number of the final electronic state, $n_0 = (2I_0)^{-1/2}$ with I_0 being the ionization potential of A^{Z_1+}, and A_1 is the asymptotic constant of the initial-state electron wave function ($A_1 \approx A_{1,\mathrm{Coul}}$). The Landau-Zener mechanism provides capture cross sections (into specific final states) of the order of $10^{-15}\,\mathrm{cm}^2$ at collision energies of $\sim 1-5$ keV even for charges Z_2 as high as $Z_2 = 7, 8$ (for $Z_1 \sim 1-3$) [7.146].

In the case $Z_1 \approx Z_2$, apart from the Landau-Zener mechanism, a Rosen-Zener-Demkov coupling may also be responsible for electron-capture transitions in the $A^{Z_1} + B^{Z_2+}$ collision system. The exchange interaction $\Delta(R)$, necessary for calculating the transition probability (either P_{LZS} or P_{D}, see Sect. 6.2.4), is given by

$$\Delta(R) = A_1 A_2 \mathscr{G}(l_1, l_2, m)\, 2^{(Z_2 - Z_1)(1/\gamma_1 + 1/\gamma_2)} R^{Z_1/\gamma_1 + Z_2/\gamma_2 - 1 - m}$$

$$\times \exp\left[-\frac{1}{2}(\gamma_1 + \gamma_2) R - \frac{Z_1}{2\gamma_1} - \frac{Z_2}{2\gamma_2}\right], \tag{7.170}$$

$$\mathscr{G}(l_1, l_2, m) = (\gamma_1 + \gamma_2)^{-m}\left[\frac{(2l_1+1)(2l_2+1)(l_1-m)!\,(l_2-m)!}{(l_1+m)!\,(l_2+m)!}\right]^{1/2} \tag{7.171}$$

where $-\gamma_1^2/2$ and $-\gamma_2^2/2$ are the binding energies of the electron in its initial and final state, respectively. Cross section calculations for a number of small energy defect ($|\Delta E| \ll 1$) reactions with $Z_2 = Z_1 - 1$ ($2 \le Z_1 \le 8$) proceeding via a Demkov coupling mechanism have been performed [7.147]. The quasi-resonant cross section, including the peripheral transitions, is given by

$$\sigma = \delta_D + \delta_p , \tag{7.172}$$

$$\sigma_D = 2\pi \int_0^{R_0} \operatorname{sech}^2 \chi \sin^2 \eta\, \varrho \, d\varrho , \tag{7.173}$$

$$\sigma_p = 2\pi \int_{R_0}^{\infty} P_{per}(\varrho, v)\, \varrho \, d\varrho , \tag{7.174}$$

$$\chi = [\pi \Delta E/(\gamma_1 + \gamma_2)\, v_R]|_{R=R_0} , \quad \Delta E = \left| \frac{\gamma_2^2}{2} - \frac{\gamma_1^2}{2} \right| \tag{7.175}$$

$$P_{per}(\varrho, v) = \eta^2 \exp[-2\Delta E^2 (\varrho_0^2/4 + \varrho^2)^{1/2}/(\gamma_1 + \gamma_2)\, v^2] , \tag{7.176}$$

where R_0 is determined from the condition $\Delta(R_0) = \Delta E$, and P_{per} is the probability for peripheral transitions [7.147]. The computed cross sections exhibit pronounced oscillations in the energy region above the cross section maximum. The peripheral transitions contribute to the total cross section by up to $\sim 10-20\%$, but only in the region $E \lesssim E_m$.

The theoretical investigations of ion-ion charge collisions at intermediate and high energies are restricted to low-charged ions only. The one-electron $H^+ + He^+$ system has been studied the most, and cross section calculations have been carried out using the AO-CC and related expansion methods with various basis sets [7.148–150]. Experimental data are also available [7.151, 152].

The inner-shell electron-capture processes can also be regarded from the ion-ion collision point of view. We shall, however, consider them in Chap. 9, together with the other inner-shell processes.

7.6.2 Ion-Molecule Charge Exchange

The specific features of the electron-capture process in collisions of highly charged ions with molecules can be demonstrated in the simplest case of a diatomic molecule A_2. Let us consider first the low-energy collisions and restrict ourselves to collision energies which are much greater than the rotational energy of the molecule A_2. Under this condition we can consider the molecular axis fixed during the collision and average the results over its orientations with respect to the line joining the centre of mass of the molecule A_2 with the multicharged ion B^{Z+}. Regarding the vibrational motion of the molecule, two limitting situations can be distinguished.

If the effective collision time for the charge-transfer process, $\tau_c \sim \varrho_0/v$, is much larger than the characteristic time for the vibrational transitions in the molecule $\tau_v \sim \omega_{vib}^{-1}$ (ω_{vib} being the vibrational quantum), then the collision dynamics with respect to the $A + B^{Z+}$ system is modified only by

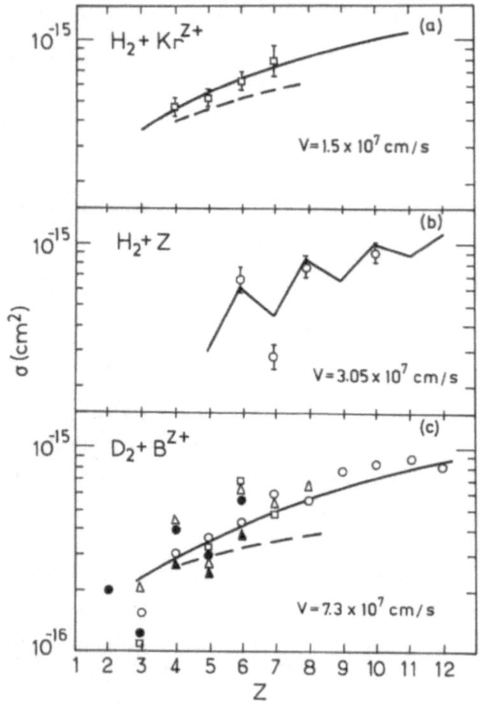

Fig. 7.18 a – c. Low-energy charge-exchange cross sections for $H_2 (D^2) + B^{Z+}$ collisions. (**a**) Theory: (——) electron tunnelling model [7.153], (- - -) absorbing sphere model [7.14]. Experimental data are from [7.120]. (**b**) Theory [7.34], experiment [7.137]. (**c**) Theory: as (**a**), experiment: (○) Ar^{Z+}, (●) C^{Z+}, (△) O^{Z+}, (□) N^{Z+} [7.154, 155], (▲) O^{Z+} [7.156]

replacement of the coupling matrix elements H_{kj} by $S_{v_0 v} H_{kj}$ [7.142], where $S_{v_0 v}$ is the vibrational overlap integral, and v_0 and v are the vibrational quantum numbers of A_2 and A_2^+. In the opposite case, $\tau_c \ll \tau_v$, the vibrational transitions are intense and, due to the sum rule for the overlap integrals $S_{v_0 v}$, the vibrational motion of A_2 has no effect on the electron-capture process. The calculations of the single-charged ion – diatomic molecule charge-transfer process have shown that cross sections obtained in these two limiting cases differ from each other by a factor of two or less [7.142].

Calculations of electron-capture cross sections for the $H_2 (D_2) + B^{Z+}$ system have been performed in the extreme situations described above by using the absorbing sphere model [7.14], the electron tunnelling model [7.153], and by approximate solution of the MO-CC equations [7.34]. These calculations have shown that in the low-energy region, the general picture of the electron-capture process in molecule-highly charged ion collisions remains the same as in the atom-ion case. In Fig. 7.18, some of the theoretical and experimental results [7.154 – 156] for the Z-dependence of the total electron cross section are presented. (Note that the data in Fig. 7.18 a, b represent the sum of single- and double-electron capture.) State-selective electron capture in molecule-highly charged ion collisions has also been studied [7.132, 133]. If very-high-resolution translational spectroscopy is used for determining the state of the reaction products, it is also possible to study the population of vibrational states of the $A_2^+(v)$ ion from the capture reaction. Such studies have recently been initiated by *Huber* [7.157] on the $Xe^{2+} + H_2$ system.

The high-energy ($v \gg 1$) charge transfer between highly charged ions and molecules has been studied only experimentally, with H_2 being the target mostly used [7.56, 158]. For this particular target molecule, the velocity behaviour of the total cross section is, generally speaking, similar to that of atomic hydrogen. In the velocity range $v = 2.2 - 6.2 \times 10^8$ cm/s, the ratio $r = \sigma(H_2)/\sigma(H)$ decreases when Z increases from $Z = 4$ to $Z = 25$, and increases from $r \approx 1$ to $r \approx 2.5$ with increasing velocity [7.56].

7.7 Multiple-Charge Transfer

Multielectron transition processes are a characteristic feature of the collisions of highly charged ions with complex atoms. At low collision energies, this feature stems from the multitude of relaxation pathways of the highly excited multicharged ion-atom complex, and at high energies it is a result of the strong potential interaction of the atomic electrons with the Coulomb field of the ion. In these processes transitions within both the discrete and the continuum spectrum are involved. We shall consider only processes leading to multiple-charge transfer in the collision. Other many-electron transition processes will be considered in the next two chapters. Obviously, the dynamics of many-electron transition processes is much more complex than for single-electron transition processes and, consequently, the corresponding theoretical methods for its description are much less developed.

7.7.1 Two-Electron-Capture Processes

The two-electron capture in atom-highly charged ion collisions proceeds via the following two basic modes:

– simultaneous capture,

$$A + B^{Z+} \to A^{2+} + B^{(Z-2)+} , \tag{7.177a}$$

– two successive one-electron captures,

$$A + B^{Z+} \to A^{+} + B^{(Z-1)+} \to A^{2+} + B^{(Z-2)+} . \tag{7.177b}$$

In the low-energy collision region, the simultaneous two-electron capture mode presumes that the two-electron exchange coupling in the system is predominant either in a broad range of internuclear distances (as in the symmetric resonance case) or in a restricted nonadiabatic transition zone (with a strong Landau-Zener, Demkov or rotational coupling). This situation is rather rare in the $A + B^{Z+}$ systems, since the two-electron coupling matrix elements are usually much smaller than those for one-electron coupling. Thus, the two-step mode (7.177b) is a much more frequent mechanism for the two-electron capture process at low collision energies. In contrast to this, the direct mode (7.177a) is a dominant capture mechanism in the high-energy region.

a) Low-Energy Two-Electron Capture

In most cases of the $A + B^{Z+}$ collision system, the basic mechanisms (7.177a, b) compete with each other, and a set of nonadiabatic regions with one- and two-electron coupling has to be considered. The simplest possible situations are shown in Fig. 7.19, where the ground-state diabatic potential

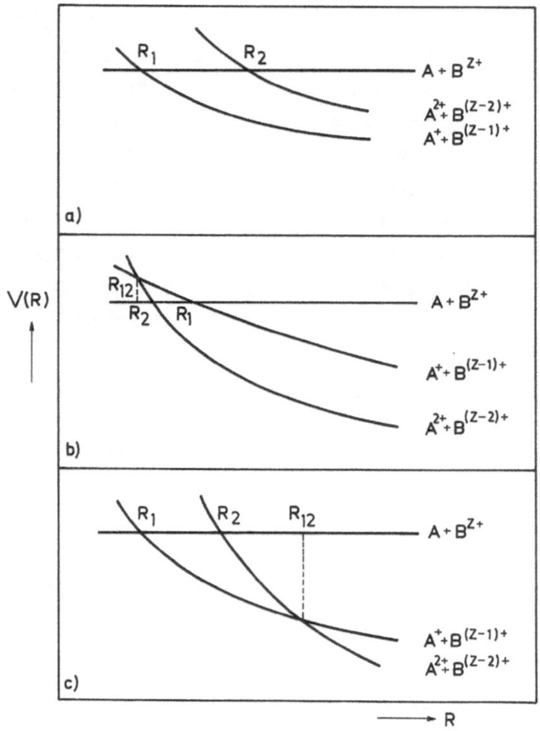

Fig. 19a–c. Schematic diagram of diabatic potentials of the ground-state $A+B^{Z+}$, $A^{+}+B^{(Z-1)+}$ and $A^{2+}+B^{(Z-2)+}$ systems, for three different relationships of ionization potentials of the particles involved

curves of the configurations $[A + B^{Z+}]$, $[A^+ + B^{(Z-1)+}]$ and $[A^{2+} + B^{(Z-1)+}]$ are shown for three different relationships between the ionization potentials of the particles involved.

If $I_A^{(1)}$ and $I_A^{(2)}$ are, respectively, the first and the second ionization potentials of atom A, and $I_B^{(Z)}$ and $I_B^{(Z-1)}$ are, respectively, the Zth and $(Z-1)$th ionization potentials of atom B, then the crossing points R_1, R_2 and R_{12} shown in the figure are given by

$$R_1 = (Z-1)/(I_B^{(Z-1)} - I_A^{(1)}) , \tag{7.178 a}$$

$$R_2 = 2\,(Z-2)/(I_B^{(Z)} + I_B^{(Z-1)} - I_A^{(1)} - I_A^{(2)}) , \tag{7.178 b}$$

$$R_{12} = (Z-3)/(I_B^{(Z)} - I_A^{(2)}) . \tag{7.178 c}$$

The mutual positions of these crossing points on the R axis determine the dynamic pathway of the two-electron capture process. The probability for two-electron capture for the situation represented by Fig. 7.19 a is

$$P^{(a)} = p_2\,(1-p_2)\,[1 + p_1^2 + (1-p_1)^2] , \tag{7.179}$$

where p_i $(i = 1, 2)$ is the Landau-Zener probability at the crossing point R_i. If $p_1 \to 0$ (when R_1 is very small, for instance), the two-electron capture probability is determined entirely by the nonadiabatic transitions around the crossing point R_2, i.e.

$$P^{(a)} \simeq 2p_2\,(1-p_2) . \tag{7.180}$$

This case corresponds to the direct two-electron transfer process (7.177 a).

The situation represented in Fig. 7.19b includes, as well as the direct two-electron capture at R_2, a pathway which corresponds to the consecutive capture of the two electrons at the crossing points R_1 and R_{12} (7.177b). For $p_2 \to 1$ (diabatic passage of the crossing point R_2) the probability for the two-step two-electron capture is given by

$$P^{(b)} \simeq 2\,(1-p_1)\,p_{12}\,(1-p_{12}) . \tag{7.181}$$

Figure 7.19c illustrates the situation when the two-electron capture (at R_2) is strongly affected by the recapture of one of the transferred electrons by the product A^{2+} ion. The two-electron probability in this case is (for $p_1 \to 0$)

$$P^{(c)} \simeq 2p_{12}\,p_2\,(1-p_2) . \tag{7.182}$$

The inclusion of excited and doubly excited states of the ionic species involved in Fig. 7.19 makes the number of pathways for the production (as well as for the destruction) of the $A^{2+} + B^{(Z-2)+}$ reaction channel very large. However, in many cases the situation is not so complex, and only a restricted number of strong coupling nonadiabatic regions determine the exit channel probability.

As we have seen from the above examples, the two-electron-transfer process always involves a nonadiabatic region in which the initial- and final-state two-electron configurations interact strongly. Therefore, for the calculation of the reaction probability, knowledge of the corresponding two-electron exchange interaction $\Delta^{(2)}(R)$ is required. For the transition of two equivalent valence electrons from a Coulomb centre of charge Z_a to another Coulomb centre having charge Z_b, the two-electron exchange interaction has the form [7.159]

$$\Delta^{(2)}(R) = DR^{(Z_a+Z_b-1)(1/\alpha_1+1/\beta_1-5)} \exp[-(\alpha_1+\beta_1)R], \tag{7.183}$$

$$\begin{aligned} D = {} & (-1)^{S+l}\mathscr{D}_l(A_1, A_2, \alpha_1; Z_b)\,\mathscr{D}_{l'}(B_1, B_2, \beta_1, Z_a) \\ & \times \left\{ \frac{1+(-1)^{L+l'}}{[l(l+1)\,l'(l'+1)]^{1/2}} \begin{bmatrix} l & l & L \\ 1 & 0 & 1 \end{bmatrix} \begin{bmatrix} l' & l' & L' \\ 1 & 0 & 1 \end{bmatrix} \mathscr{K}_{l'}(\alpha_1, \beta_2, Z_b)\,\mathscr{K}_{l'}(\beta_1, \alpha_2, Z_a) \right. \\ & \left. + \begin{bmatrix} l & l & L \\ 0 & 0 & 0 \end{bmatrix} \begin{bmatrix} l' & l' & L' \\ 0 & 0 & 0 \end{bmatrix} \mathscr{I}_{l'}(\alpha_1, \beta_2, Z_b)\,\mathscr{I}_l(\beta_1, \alpha_2, Z_a) \right\}, \end{aligned} \tag{7.184}$$

$$\mathscr{D}_l(A_1, A_2, \alpha_1, Z_b) = (1/8)\,A_1 A_2 (2l+1)(2\alpha_1/e)^{Z_b/\alpha_1}\,\Gamma(1 - Z_b/\alpha_1), \tag{7.185a}$$

$$\mathscr{D}_{l'} = \mathscr{D}_{l'}(B_1, B_2, \beta_1, Z_a), \quad A_2 = 2\alpha_2^{3/2}, \quad B_2 = 2\beta_2^{3/2}, \tag{7.185b}$$

where $\alpha_1^2/2$, $\alpha_2^2/2$ are, respectively, the first and second ionization potentials of atom A, $\beta_1^2/2$, $\beta_2^2/2$ are the corresponding potentials of the two electrons when captured on the ion B^{Z_b+}, l and l' are the angular momenta of the individual electrons before and after capture, L and L' are the total electron angular momenta of A and $B^{(Z_b-2)+}$, S is the total spin of atom A and [: : :] is the Clebsch-Gordan coefficient. The quantities A_1 and B_1 in (7.183–185) are the asymptotic normalization constants of the single-electron wave function for an electron with binding energy $-\alpha_1^2/2$ and $-\beta_1^2/2$, respectively. The quantities $\mathscr{K}_l$, $\mathscr{K}_{l'}$ and $\mathscr{I}_l$, $\mathscr{I}_{l'}$ are integrals which can be calculated analytically for any given l and l' [7.159]. They satisfy the "symmetry" relations

$$\mathscr{K}_{l'}(\alpha_1, \beta_2, Z_b) = \mathscr{K}_l(l' \to l, \alpha_1 \leftrightarrow \beta_1, \beta_2 \leftrightarrow \alpha_2, Z_b \leftrightarrow Z_a), \tag{7.186a}$$

$$\mathscr{I}_{l'}(\alpha_1, \beta_2, Z_b) = \mathscr{I}_l/l' \to l, \alpha_1 \leftrightarrow \beta_1, \beta_2 \leftrightarrow \alpha_2, Z_b \leftrightarrow Z_a). \tag{7.186b}$$

For S states ($L=0$ or $L'=0$) the first term in the expression for D has to be omitted. The values of $\mathscr{K}$ and $\mathscr{J}$ for the $l=l'=0$ and $l=l'=1$ cases are:

$$\mathscr{J}_0(\alpha_1, \beta_2, Z_b) = \frac{2^6 (Z_b - \alpha_1)(2\beta_2 - Z_b)}{(\beta_2 + \alpha_1)^6} \left(\frac{\beta_2 - \alpha_1}{\beta_2 + \alpha_1}\right)^{Z_b/\alpha_1 - 3}, \quad (7.187\text{a})$$

$$\mathscr{K}_1(\alpha_1, \beta_2, Z_b) = \frac{2^5 (\beta_2 + \alpha_1 - 2 Z_b)}{(\beta_2 + \alpha_1)^5} \left(\frac{\beta_2 - \alpha_1}{\beta_2 + \alpha_1}\right)^{Z_b/\alpha_1 - 2}, \quad (7.187\text{b})$$

$$\mathscr{J}_1(\alpha_1, \beta_2, Z_b) = \frac{2^4 [(\beta_2 - \alpha_1)^2 - 2(Z_b - \alpha_1)(\beta_2 - Z_b + \alpha_1)]}{(\beta_2 + \alpha_1)^6} \left(\frac{\beta_2 - \alpha_1}{\beta_2 + \alpha_1}\right)^{Z_b/\alpha_1 - 3} \quad (7.187\text{c})$$

The expression $\Delta^{(2)}(R)$ is valid under the conditions $\alpha_1 R \gg 1$ and $\beta_1 R \gg 1$. In the symmetric case ($\alpha_1 = \beta_1$, $\alpha_2 = \beta_2$, $Z_a = Z_b$), $\Delta^{(2)}(R)$ is equivalent to the gerade-ungerade energy splitting $\Delta^{(2)}_{g-u}(R)$ of molecular states [7.160, 161] and determines the resonant double-charge transfer in the A_2^{Z+} system. For the He + He^{2+} collision system, $\Delta^{(2)}_{g-u}(R)$ has the form

$$\Delta^{(2)}_{g-u}(R) = 24.96\, R^{-0.537}\, e^{-2.688 R}. \quad (7.188)$$

The cross section for resonant two-electron capture in the A + A^{2+} low-energy collisions is given by [7.142]

$$\sigma = \pi R_0^2, \quad (\pi R_0 / 2\alpha_1)^{1/2} \Delta^{(2)}_{g-u}(R_0) = 0.28\, v. \quad (7.189)$$

For the He + He^{2+} system, the calculated double-charge-transfer cross section is shown in Fig. 7.20, together with the experimental data [7.162–164]. Calculations have also been made for the resonant two-

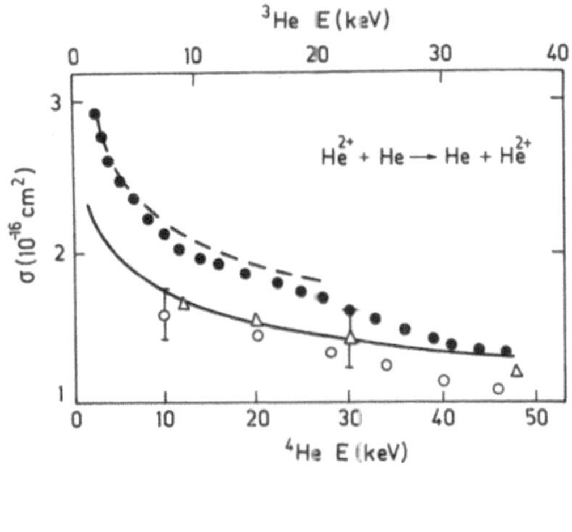

Fig. 7.20. Resonant two-capture in He + He^{2+} collisions. Theory: (—) [7.159], (- - -) [7.161]. Experimental data: (●) [7.162], (○) [7.163], (△) [7.164]

electron capture in $(Z, 2\,e) + Z$ ion-ion collisions for $3 \leqq Z \leqq 10$ using (7.157–159) [7.143]. The calculated cross sections can be represented in a scaled form,

$$\tilde{\sigma}(\tilde{E}) = Z^{1.7}\,\sigma, \quad \tilde{E} = Z^{-3}\,E. \tag{7.190}$$

The scaled cross section $\tilde{\sigma}$ has a maximum at $\tilde{E}_m \simeq 4 \times 10^{-3}$ keV/amu with a value $\tilde{\sigma}_m \simeq 1 \times 10^{-15}$ cm^2.

An example of quasi-resonant simultaneous two-electron capture is the reaction $He + C^{4+} \rightarrow He^{2+} + C^{2+}$, in which nonadiabatic Landau-Zener transitions at $R_2 \simeq 3.3$ dominate the collision dynamics in this system. Theoretical [7.159, 165] and experimental [7.165] investigations of the $He + C^{4+}$ collision system have demonstrated that in the low-energy region ($v \lesssim 4 \times 10^7$ cm/s) the double-charge-transfer cross section is higher than the cross section for single-electron capture. This circumstance is a consequence of the fact that in this system the nonadiabatic regions δR_1 and δR_2 for single- and double-electron capture are located on the internuclear axis as represented by Fig. 7.19a (i.e. $R_2 > R_1$). More frequent is, however, the opposite situation, $R_1 < R_2$, (Fig. 7.19b); then the two-electron-capture cross section is much smaller than that for single-electron capture. This is the case in, for example, $He + B^{3+}$, N^5, O^6 and Ar^{6+} collision systems [7.166, 167], the smallness of the cross section being caused by multiplication of several probability factors [see (7.181)]. When the ionic charge Z is large, both electrons may be captured into excited ion states,

$$A + B^{Z+} \rightarrow A^{2+} + B^{(Z-2)+**}(n\,l, n'\,l'), \tag{7.191}$$

producing, thus, a doubly excited (autoionizing) product ion. The MO-CC calculations for the $He + O^{8+}$ system, for instance, have shown [7.168] that both electrons are preferentially captured into the $n = 3$ or 4 and $n' = 3$ or 4 levels. The autoionizing state $B^{(Z-2)+}(n, n')$ can decay either radiatively or by an Auger (electron-ejection) mechanism. The ratio of the Auger and radiative decay probabilities can be estimated to be [7.4]

$$W_A/W_r \simeq (10/Z)^4\,(n'/n)^3, \tag{7.192}$$

whence it can be inferred that for small values of $Z\,(Z \simeq 10)$ the Auger mechanism is the main relaxation channel, whereas for high values of Z the radiative decay is predominant. Spectra of secondary Auger electrons have been observed in collisions of Ne^{4+} with Ar and Kr [7.169], $He^{2+} + Xe$ [7.170] and in other collision systems.

If each of the colliding particles A and B^{Z+} contains many electrons and/or Z is large, then one may assume that for each of the single-electron-capture steps of the two-electron-capture reaction (7.177b) the conditions for applying the electron tunnelling model are satisfied. Defining a reaction radius (R_{0i} ($i = 1, 2$) for each of the steps by

$$R_{01} = k_1 Z^{1/2}/[I_A^{(1)}]^{3/2}, \quad R_{02} = k_2 (Z-1)^{1/2}/[I_A^{(2)}]^{3/2}, \tag{7.193}$$

where $k_1 \approx k_2$ are constants, and $I_A^{(i)}$ ($i = 1, 2$) is the ith ionization potential of atom A, the single- and double-electron-capture cross section at low energies can be roughly estimated to be

$$\sigma_1 \approx \pi (R_{01}^2 - R_{02}^2), \quad \sigma_2 \simeq \pi R_{02}^2. \tag{7.194}$$

Alternatively, R_{01} and R_{02} can be determined from a classical absorbing model [7.4]

$$R_{0i} \simeq [2(Z+1-i)^{1/2}+1]/I_A^{(i)}, \quad i = 1, 2. \tag{7.195}$$

The single- and double-electron-capture cross sections for the He + Ar^{6+} collision system calculated using (7.194, 195) are shown in Fig. 7.21 and compared with experimental data [7.167].

From the relations (7.193–195), it follows that for high values of Z, the ratio $r = \sigma_2/\sigma_1 \approx R_{02}^2/R_{01}^2$ in the electron tunnelling and classical absorption models is given by

$$r_{\text{tun}} \simeq [I_A^{(1)}/I_A^{(2)}]^{3/2}, \quad r_{\text{abs}} \simeq [I_A^{(1)}/I_A^{(2)}]^2. \tag{7.196}$$

The ratios σ_2/σ_1 of the numerous experimental data on σ_1 and σ_2 for inert gas atom and ion colliding systems (see [7.3]) are scattered around r_{tun} and r_{abs} [7.4], and lie (with a few exceptions) below the adiabatic limit $r_{\text{ad}} \simeq [I_A^{(1)}/(I_A^{(1)} + I_A^{(2)})]^2$. These experimental data have shown that in the low-velocity region σ_2 is almost independent of the collision velocity and its average Z-dependence is slightly weaker than linear, $\sigma_2 \sim Z^{0.71 \pm 0.14}$ [7.171].

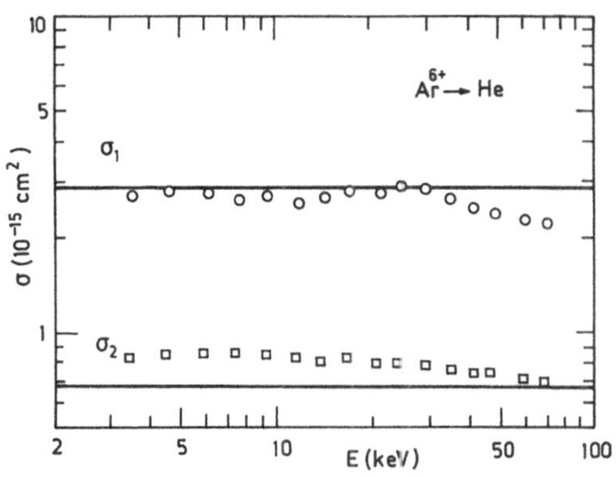

7.21. Single-(σ_1) and double-(σ_2) charge-transfer cross sections in He + Ar^{6+} collisions. (——) classical absorbing sphere model. Experimental data: [7.167]

In the case of double capture from H_2 by partially [7.124, 157] and fully stripped [7.154, 172] ions, Z-oscillations in the cross sections have been observed for charge states below $Z = 8$.

b) Medium- and High-Energy Two-Electron Capture

The theoretical investigations of double-charge-transfer processes in the medium- and high-energy regions are very limited. The main problem in the theory of two-electron transition processes is the need for an adequate description of two-electron correlation. In the high-energy region ($v \gg 1$), however, the independent particle model can be adopted as a reasonable approximation and the two-electron transition probability can be represented as a product of the electron transition probabilities for the individual electrons,

$$P_2 = p_1 \cdot p_2 . \tag{7.197}$$

If the transferred electrons are equivalent in both the initial and final states, then $p_1 = p_2$, and $P_2 = p_1^2$. The one-electron transition probabilities p_i ($i = 1, 2$) can be calculated by using any of the conventional high-energy approximations, and the two-electron-capture cross section is then evaluated from P_2 in the usual way. Such a procedure is justified only if the collision energy is sufficiently high (with respect to the interelectron correlation energy) and if, at such energies, the method applied for the calculation of p_i is highly accurate. Using this procedure, *Gayet* et al. [7.173] have performed cross section calculations for the double-charge-transfer process in $He + He^{2+}$ high-energy collisions by employing the CDW approximation for p_i.

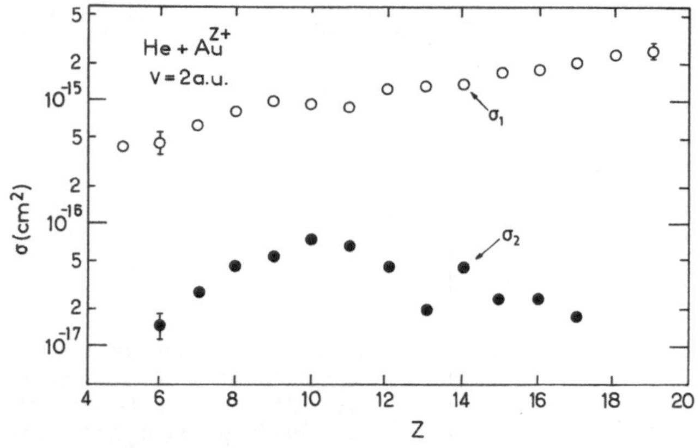

Fig. 7.22. Comparison of the Z-dependences of single-(σ_1) and double-(σ_2) charge-transfer cross sections in $He + Au^{Z+}$ collisions at the relative velocity $v = 2$ a. u. [7.176]

Systematic experimental investigations of double-charge-transfer processes at medium and high energies have been done by *Dmitriev*, *Nikolaev*, *Teplova* and their associates [7.174, 175] and by *Hvelplund* and his collaborators [7.176]. The general energy-dependence of the double-charge-transfer cross section is similar to that for single-charge transfer. However, the Z-dependences of σ_1 and σ_2 are similar (proportional to Z) only up to a certain $Z = Z_c$, after which they are drastically different. The situation is illustrated in Fig. 7.22, where the Z-dependence of σ_1 and σ_2 for the $He + Au^{Z+}$ collision at $E = 100$ keV/amu is shown. The decrease of $\sigma_2(Z)$ for $Z \gtreqless 10$ does not have a clear physical explanation. It might be connected with the Coulomb trajectory effects if the two-electron capture is governed by a two-step mechanism.

7.7.2 Multi-Electron-Capture Processes

Experimental cross section data on three- and four-electron-capture processes in atom-highly charged ion collisions were first reported by *Salzborn* and coworkers [7.177, 178] in the rare gas atom-ion systems. In the low-velocity region, the behaviour of multi-electron-capture cross sections is similar to that for single-electron-capture. An example is shown in Fig. 7.23 for the $Ar + Ar^{Z+}$ system. In this particular system, the double- and triple-charge-transfer cross sections are not very different from each other. This fact can be understood within the classical absorbing sphere model for the multi-electron-capture process. Extending the relation (7.195) for the multi-step capture radius R_{0i} to the case $i = 3$, one can estimate the single-, double-, and triple-charge-transfer cross sections by

$$\sigma_3 \approx \pi R_{03}^2, \quad \sigma_2 \approx \pi (R_{02}^2 - R_{03}^2), \quad \sigma_1 \approx \pi (R_{01}^2 - R_{02}^2 - R_{03}^2) \tag{7.198}$$

so that for $Z \gg 1$ the ratio σ_3/σ_2 is given by $\sigma_3/\sigma_2 \approx [I_A^{(2)}/I_A^{(3)}]^2$. $I_A^{(2)}$ and $I_A^{(3)}$ for Ar are not too much different from each other, which is reflected in the ratio of σ_3 and σ_2. The values of σ_1, σ_2 and σ_3, calculated from (7.198), are denoted on the right-hand side of Fig. 7.23. In the low-energy region the Z-dependence (for $Z \leqq 8$) of the cross section $\sigma_{Z,Z-k}$ for capture of k electrons obeys the empirical law [7.171]

$$\sigma_{Z,Z-k} = 10^{-12} A_k Z^{\alpha_k} (I_A\,[\mathrm{eV}])^{-\beta_k} \quad [\mathrm{cm}^2], \tag{7.199}$$

where I_A is the first ionization potential of the target atom and the values of the parameters A_k, α_k and β_k are given in Table 7.4. The Z-dependence of multi-electron-capture processes in $Xe + Ar^{Z+}$ ($Z \leqq 8$) collisions at an energy of 30 keV is shown in Fig. 7.24. We note, however, that for Z higher than a

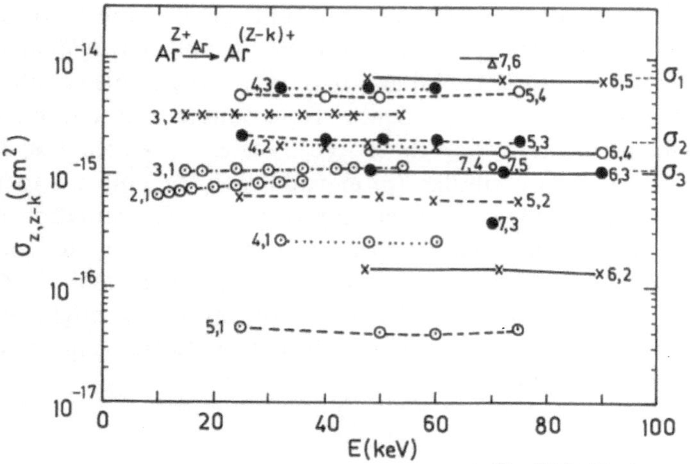

▲ **Fig. 7.23.** Single- and multiple-charge-transfer cross sections in $Ar^{Z+}+Ar$ collisions [7.177]. The right-hand ordinate shows the values of σ_1, σ_2 and σ_3 as predicted by the classical absorbing sphere model

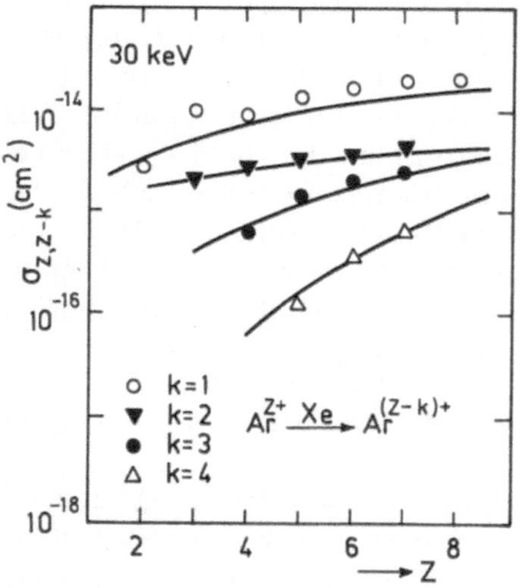

Fig. 7.24. Z-dependence of single- and multiple-charge-transfer cross sections for $Xe+Ar^{Z+}$ collisions at a collision energy of 30 keV [7.178]

Table 7.4. Values of the parameters A_k, α_k and β_k in the empirical formula (7.199) for $\sigma_{Z,Z-k}$

k	A_k	α_k	β_k
1	1.43 ± 0.76	1.17 ± 0.09	2.76 ± 0.19
2	1.08 ± 0.95	0.71 ± 0.14	2.80 ± 0.32
3	$(5.5 \pm 5.8) \times 10^{-2}$	2.10 ± 0.24	2.89 ± 0.39
4	$(3.57 \pm 8.9) \times 10^{-4}$	4.20 ± 0.79	3.03 ± 0.86

certain Z_{ck}, the cross section $\sigma_{Z,Z-k}$ $(k \geqq 2)$ might start to decrease with increasing Z, as is the case in Fig. 7.22 for $k = 2$.

Resonant three-electron capture has been experimentally observed in slow $Kr + Kr^{3+}$ collisions [7.179], showing a cross section of $(6-4) \times 10^{-16}$ cm^2 in the energy range 0.3–5 keV. The cross section has a typical $\sigma^{1/2} = a + b \ln(1/v)$ $(a, b = \text{const})$ behaviour, as for the single- and double-resonant charge-transfer processes. This suggests that the process is determined by the phase interference effects between the gerade and ungerade states of the A_2^{3+} system. The corresponding adiabatic energy splitting, $\Delta^{(3)}(R)$ for this system has been recently calculated [7.180], showing the asymptotic behaviour

$$\Delta^{(3)}(R) = P(R) \exp[-(2\alpha_1 + \alpha_2) R], \qquad (7.200)$$

where $P(R)$ is a power function of R, and $\alpha_i = (2 I_A^{(i)})^{1/2}$, $i = 1, 2$. It should be noted that the exponential behaviour of $\Delta^{(3)}(R)$ does not depend on $\alpha_3 = (2 I_A^{(3)})^{1/2}$. Analogously, the four-electron exchange interaction $\Delta^{(4)}(R)$ asymptotically behaves as

$$\Delta^{(4)}(R) \sim \exp[-(2\alpha_1 + 2\alpha_2) R], \qquad (7.201)$$

and does not depend on the $I_A^{(3)}$ and $I_A^{(4)}$ ionization potentials. These properties of the multi-electron exchange interactions $\Delta^{(k)}(R)$ $(k \geqq 2)$ explain the relatively large cross sections of the multi-electron resonant charge-transfer reactions at low energies. At medium and high collision velocities $(v \gtrsim 1)$, the multi-electron-capture processes are seldom investigated. In this region, multi-electron transition processes are characterized by a large number of reaction channels, of which the ionization one is dominant.

8. Collisions of Atoms (Ions) with Highly Charged Ions: Excitation and Ionization

Excitation and ionization processes in collisions of atoms with multiply charged ions may occur as a result of either one-electron or many-electron transitions in the system. Generally speaking, the single-electron excitation and ionization processes in the adiabatic energy region ($v \ll 1$) are characterized by cross sections which are much smaller than the cross section for electron capture. In the intermediate-energy region ($v \sim 1$), single-electron excitation and ionization become as effective as the charge-transfer process, and at high collision energies ($v \gg 1$) they become dominant inelastic processes in the ion-atom collision. For the multi-electron-transition processes such general relationships between the cross sections of the excitation, ionization, and charge-transfer processes cannot be ruled out. Moreover, combined processes, such as simultaneous capture and ionization (or excitation) become dominant and, with increasing ionic charge, their role in the collision dynamics becomes increasingly pronounced.

The excitation and ionization processes in collisions between atoms and highly charged ions have so far been much less investigated than the corresponding electron-capture processes. This holds both for theory and experiment. The information on multi-electron-transition processes is especially limited. Although most of the general theoretical methods, described in Chap. 6, (see also [8.1, 2]), can be applied to the study of single-electron excitation and ionization, such applications are still rather scarce and restricted to collisions of low-charged ions. Also, special methods which would reflect and make use of the specific aspects of the highly charged ion-atom collision dynamics have not been developed to such an extent as in the case of the charge-transfer processes. Despite this, the information gained is still sufficiently abundant and versatile to allow one to build a coherent picture of the general properties of the excitation and ionization processes in collisions of highly charged ions with neutral atomic species (and low charged ions, as well). Several review papers are now available containing detailed information about the recent theoretical and experimental investigations of these processes [8.3 – 7].

In this chapter, we shall confine our consideration to the direct excitation and ionization processes only, which usually involve electrons from the outer electronic shells. The Auger and related processes, which result from multi-electron transitions in the collision system, will be treated in the next chapter.

8.1 Basic Electron Transition Mechanisms

The physical mechanisms that govern the excitation and ionization processes in atom-highly charged ion collisions are numerous and differ in various regions of the collision energies and impact parameters. In the adiabatic energy region ($v \ll 1$), or for very distant collisions, the direct single- or multi-electron excitation and ionization processes are adiabatically improbable. The probability for excitation may become significant at higher energies only, when the collision system is brought high into the discrete quasi-molecular spectrum, and transitions between the states become possible due to the strong nonadiabatic (radial or rotational) coupling in certain regions of the internuclear separation. For still higher energies, a sequence of such nonadiabatic transitions may bring the system either directly into the continuum (ionization), or close to its edge, when decay can occur due to different adiabatic perturbations. A coupled-channel molecular-orbital formalism describing these processes should evidently include an adequately large basis which, in the case of ionization, has to contain states that, as $R \to \infty$, describe autoionizing or ionized atomic levels. With some simplifying assumptions, the single-electron excitation problem may also be described by the multichannel Landau-Zener model.

At intermediate collision energies, the excitation and ionization processes involve the strong coupling of a large number of atomic states (infinite, in the case of ionization) and a coupled-channel formalism, employing a finite basis set expansion, must include suitable chosen pseudostates. In this energy range, ionization may also occur at large distances by an electron tunnelling mechanism, analogous to the one in multiphoton ionization of atoms. A classical description of the ionization process in the intermediate-velocity region ($v \sim 1-2$) has also proved to be useful.

For highly charged ions, the collision velocities at which the ion-atom interaction may successfully be described by first-order methods lie in the region where relativistic effects have to be taken into account. Therefore, for a proper description of the ionization channel in the high-energy region, higher-order methods have to be used. This requirement is considerably relaxed for the excitation of low-lying target states.

8.2 Excitation Processes

Both experimental and theoretical investigations of the excitation processes in atom-highly charged ion collisions

$$A + B^{Z+} \rightarrow A^* + B^{Z+} \tag{8.1}$$

are very scarce. The production of $3\,^3P$ and $4\,^1D$ excited states of He in collisions with 100 keV Ne^{Z+} ($Z = 1-4$) ions has been observed by *Winter* et al. [8.8] through their radiative decay. It was found that with increasing projectile-ion charge, the emission cross section for the corresponding transition decreases. This decrease is related to the more pronounced role of the electron-capture process for higher Z. Other experimental data involve mainly singly charged ions and are discussed elsewhere [8.1, 2].

Below, we shall present in more detail some of the theoretical methods which have been applied to the excitation problem when multicharged ions are explicitly involved.

8.2.1 Close-Coupling Treatments

The close-coupling methods discussed in Sect. 6.2, based either on a molecular- or atomic-orbital expansion, may in principle be used to calculate the cross section of reaction (8.1) at low and intermediate collision velocities. Unfortunately, such calculations have so far been carried out only for $H + H^+$ collisions using both the standard MO-CC or AO-CC techniques (with ETF's included) [8.9, 10] and expansions which include pseudostates [8.11–13] or Sturmian functions [8.14, 15]. Similarly to the case of the charge-exchange process, the size of the basis dramatically increases with increasing ionic charge, particularly for transitions into higher excited states.

For the intermediate- to high-energy region, the single-electron excitation process can be considered as being caused by the ion-atom interaction

$$V(\boldsymbol{r}, \boldsymbol{R}) = -Z/|\boldsymbol{R}(t) - \boldsymbol{r}|\,, \tag{8.2}$$

where $\boldsymbol{r}$ is the radius vector of the active electron with respect to the centre of the atom, and $\boldsymbol{R}$ is the internuclear distance vector. Assuming, further, that only the initial and final electron states in reaction (8.1) are strongly coupled, and that the dipole term in the interaction $V(\boldsymbol{r}, \boldsymbol{R})$ dominates at the relevant distances, the excitation problem for an $n\,s \rightarrow n'\,p$ transition is reduced to the following system of coupled equations [8.16]:

$$\mathrm{i}\frac{da}{dt}=\frac{Z\lambda}{R^2}(b_0\cos\theta+b_1\sin\theta)\,\mathrm{e}^{-\mathrm{i}\omega t}, \tag{8.3a}$$

$$\mathrm{i}\frac{db_0}{dt}=\frac{Z\lambda}{R^2}a\cos\theta\,\mathrm{e}^{\mathrm{i}\omega t}, \tag{8.3b}$$

$$\mathrm{i}\frac{db_1}{dt}=\frac{Z\lambda}{R^2}a\sin\theta\,\mathrm{e}^{\mathrm{i}\omega t}, \tag{8.3c}$$

$$\cos\theta=v\,t/R\,,\quad \sin\theta=\varrho/R\,,\quad \lambda=(f_{01}/2\,\omega)^{1/2}, \tag{8.4}$$

where ω is the energy difference between the atomic states $|n\,s\rangle=|0\rangle$ and $|n'\,p\rangle=|1\rangle$, f_{01} is the oscillator strength for the $n\,s\rightarrow n'\,p$ transition and ϱ is the impact parameter. a, b_0 and b_1 are the amplitudes of the states $n\,s$, $n'\,p$ $(m=0)$ and $n'\,p$ $(|m|=1)$, respectively, satisfying the initial conditions

$$a(-\infty)=1\,,\quad b_{0,1}(-\infty)=0\,. \tag{8.5}$$

The integration of the system of coupled equations (8.3) with the initial conditions (8.5) can be performed by using a generalization of the VPS method [8.17]. The result for the transition probability is [8.18]

$$W(\varrho)=W_0(\varrho)+W_1(\varrho)\,, \tag{8.6}$$

$$W_0(\varrho)=|\,b_0(+\infty)\,|^2=\frac{P_0}{1+P_0+P_1}, \tag{8.7}$$

$$W_1(\varrho)=|\,b_1(+\infty)\,|^2=\frac{P_1}{1+P_0+P_1}, \tag{8.8}$$

$$P_0=\left|\int_{-\infty}^{\infty}dt\left(\frac{Z\lambda v\,t}{R^3}\right)\sin\left\{\int_0^t dt'\left[\omega^2+4\left(\frac{Z\lambda}{R^2}\right)^2\right]^{1/2}\right\}\right|^2, \tag{8.9}$$

$$P_1=\left|\int_{-\infty}^{\infty}dt\left(\frac{Z\lambda\varrho}{R^3}\right)\cos\left\{\int_0^t dt\left[\left(\frac{\omega}{2}\right)^2+\left(\frac{Z\lambda}{R^2}\right)^2\right]^{1/2}+\frac{\omega\,t}{2}\right\}\right|^2. \tag{8.10}$$

This dipole-approximation close-coupling (DACC) theory can, in principle, be extended to include the contribution from the transitions through intermediate states. Quadrupole transitions can also be treated within an analogous scheme. In fact, the asymptotic dipole potential $(Z\lambda/R^2)$ in (8.3a–c) and in the integrals (8.9, 10) can be replaced by the complete dipole radial matrix element $U_{1s-2p}(R)$. An important property of the DACC theory lies in the fact that the transition probability (8.6–10) is properly normalized and has a non-singular behaviour as $\varrho\rightarrow 0$. This property remains valid also in the perturbation limit $(\omega/2\gg Z\lambda/R^2)$, when

P_0 and P_1 go over into the well-known result [8.19]

$$P_0 = \left[\frac{2Z\lambda\omega}{v^2} K_0\left(\frac{\omega\varrho}{v}\right)\right]^2, \quad P_1 = \left[\frac{2Z\lambda\omega}{v^2} K_1\left(\frac{\omega\varrho}{v}\right)\right]^2, \tag{8.11}$$

where $K_0(x)$ and $K_1(x)$ are the MacDonald functions. Introducing in (8.9, 10) the scaling transformation $\varrho = (Z\lambda/v)\, y$, $v\,t = (Z\lambda/v)\, x$, defining a dimensionless parameter β by

$$\beta = Z\lambda\omega/v^2, \tag{8.12}$$

and integrating $W(\varrho)$ over the impact parameters, one obtains the following expression for the excitation cross section [8.18]:

$$\sigma_{\text{exc}} = 2\pi\left(\frac{Z\lambda}{\omega}\right) D(\beta), \tag{8.13}$$

$$D(\beta) = \beta \int_0^\infty y\,dy \frac{P_0(y,\beta) + P_1(y,\beta)}{1 + P_0(y,\beta) + P_1(y,\beta)}. \tag{8.14}$$

For $\beta > 2$ (adiabatic energy region) and $\beta \ll 0.01$ (validity region of the Born approximation), the function $D(\beta)$ can be approximated by

$$D(\beta) \simeq \tfrac{1}{2}\beta\left(1 - \tfrac{1}{8}\beta^{-3/2}\right) e^{-(2\beta)^{1/2}}, \quad \beta > 2, \tag{8.15}$$

$$D(\beta) = 4\beta \ln(1.4/\beta), \quad \beta \ll 0.01. \tag{8.16}$$

The values of the function $D(\beta)$ in the interval $0.04 \leq \beta \leq 2$, calculated with the full radial dipole matrix element $U_{1s-2p}(R)$ in the integrals P_0 and P_1, are given in Table 8.1. In the adiabatic and the Born limits, the excitation cross section has the form

$$\sigma_{\text{exc}}^{(\text{ad})} = \pi\left(\frac{Z\lambda}{v}\right)^2 \exp\left[-\frac{1}{v}(2Z\lambda\omega)^{1/2}\right], \tag{8.17}$$

$$\sigma_{\text{exc}}^{(\text{B})} = 8\pi\left(\frac{Z\lambda}{v}\right)^2 \ln\left(\frac{1.4\,v^2}{Z\lambda\omega}\right). \tag{8.18}$$

Table 8.1. Values of the function $D(\beta)$, (8.14) in the interval $0.04 \leq \beta \leq 2$

β	$D(\beta)$	β	$D(\beta)$
0.04	0.202	0.200	0.401
0.05	0.221	0.25	0.406
0.066	0.268	0.28	0.410
0.083	0.291	0.333	0.399
0.100	0.332	0.500	0.385
0.143	0.365	1.00	0.212
0.1666	0.381	2.00	0.132

The expression for $\sigma_{\mathrm{exc}}^{(\mathrm{B})}$ also follows from the general Bethe-Born theory [8.20].

The above results of the DACC theory reveal the following general features of the $n\,s \to n'\,p$ excitation cross section in atom-multicharged ion collisions:

I) The reduced excitation cross section

$$\tilde{\sigma}_{\mathrm{exc}}(\tilde{E}) = \frac{\omega}{Z\,\lambda}\,\sigma_{\mathrm{exc}} = 2\pi\,D\,(\beta) \tag{8.19}$$

is represented by the universal function $2\pi\,D(\beta)$. The reduced energy parameter $\tilde{E}$ for the excitation process is given by, see (8.12),

$$\tilde{E} = \frac{E}{Z\,\lambda\,\omega}, \tag{8.20}$$

where E is expressed in units of keV/amu.

II) The scaling of the cross section with respect to the ionic charge Z is different in various energy regions, see (8.14, 17, 18).

III) The relative velocity v_{m} at which the cross section maximum occurs is given by

$$v_{\mathrm{m}} = 1.89\,(Z\,\lambda\,\omega)^{1/2} \tag{8.21}$$

and scales with Z like $Z^{1/2}$. Relation (8.21) follows from the fact that $D\,(\beta)$ has a maximum at $\beta \approx 0.28$ (see Table 8.1). The value of the cross section maximum is

$$\sigma_{\mathrm{exc,\,m}} \simeq 0.82\,\pi \left(\frac{Z\,\lambda}{\omega}\right), \tag{8.22}$$

i.e. it scales linearly with Z.

Janev and *Presnyakov* [8.18] have performed DACC cross section calculations for the excitation transitions $\mathrm{H}\,(1\,s) + Z \to \mathrm{H}\,(n'\,p) + Z$, with $n' = 2, 3, 4, 5$ and $Z = 1, 2, 4, 6, 8, 10$. Some illustrative examples are shown in Fig. 8.1.

The validity of the DACC theory is restricted to relatively high values of Z and not too low values of the parameter $v/Z^{1/2}$. When these conditions are not satisfied, either the dipole or the two-state approximation is violated.

8.2.2 UDWA Treatment

The unitarized distorted-wave approximation (UDWA), which has been discussed in Sect. 7.3.2, also provides an easy way of estimating the excita-

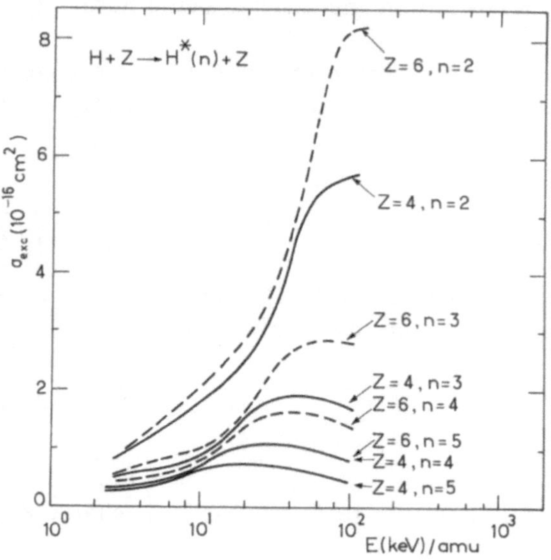

Fig. 8.1. Cross sections for the $H(1s)+Z \to H(np)+Z$ excitation ($Z=4, 6$; $n=2-5$) obtained from DACC theory [8.18]

tion cross section in atom-highly charged ion collisions at intermediate collision energies. The probability for the $0 \to n$ transition is given by [8.21]

$$P_{0n}^{\mathrm{exc}} = p^{-1} \mid t_{0n}^{\mathrm{ex}} \mid^2 \sin^2 p^{1/2} , \tag{8.23}$$

where t_{0n}^{exc} is the transition matrix element, calculated in the distorted-wave Born approximation, cf. (7.90),

$$p = p_{\mathrm{exc}} + p_{\mathrm{ion}} + p_{\mathrm{cx}} \tag{8.24}$$

with p_{exc}, p_{ion} and p_{cx} being the total un-normalized probabilities for excitation, ionization and charge exchange, respectively. *Ryufuku* [8.21] has carried out cross section calculations for the $0 \to n$ excitation cross sections in the $H(1s) + Z$ collisions for $Z = 1, 3, 5, 6$ and 14. The results for the total excitation cross sections,

$$\sigma_{\mathrm{exc}} = \sum_n \sigma_n^{\mathrm{exc}} = \sum_n \left[2\pi \int_0^\infty \varrho \, d\varrho \, P_{0n}^{\mathrm{exc}}(\varrho) \right] , \tag{8.25}$$

are shown in Fig. 8.2. They are compared with the experimental data for the $Z = 1$ case [8.22] and with the DACC results for $Z = 1, 2$ and 6. It should be noted that below ~ 10 keV/amu, the UDWA model for excitation is not expected to be valid, since transitions via intermediate states in this region are important.

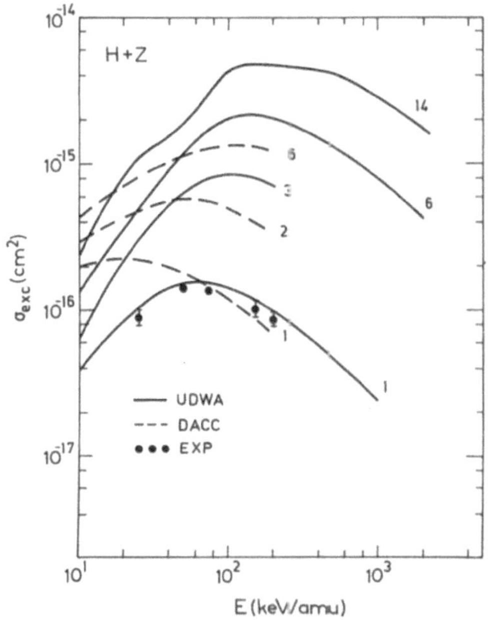

Fig. 8.2. Total excitation cross sections for the $H(1s)+Z$ collisions ($Z = 1, 2, 3, 6$ and 14). (—) UDWA [8.21], (- - -) DACC theory [8.18], (●) experimental data for $Z = 1$ [8.22]

8.2.3 Perturbation Theory and VPS Approximation

The dipole excitation of atoms by a charged particle in first-order perturbation theory (or the Born approximation) is presented in many standard books [8.20, 23] and we shall confine ourselves here to some remarks. For an $ns \to n'p$ transition, the un-normalized excitation probability is given by, cf. (8.6, 11),

$$\tilde{W}(\varrho) = P_0(\varrho) + P_1(\varrho) = \left(\frac{2Z\lambda\omega}{v^2}\right)^2 \left[K_0^2\left(\frac{\omega\varrho}{v}\right) + K_1^2\left(\frac{\omega\varrho}{v}\right)\right]. \tag{8.26}$$

Since $W(\varrho)$ is singular as $\varrho \to 0$, a normalization procedure is necessary. One possibility is to use (8.6–8). Procedures used earlier include introduction of a minimum impact parameter ϱ_0, defined by [8.24] $\tilde{W}(\varrho_0) = 1/2$, so that the normalized probability is

$$W^{\mathrm{N}}(\varrho) = \min\{1/2, \tilde{W}(\varrho)\}. \tag{8.27}$$

The excitation cross section is then represented as

$$\sigma_{\mathrm{exc}}^{\mathrm{N}} = \pi\varrho_0^2 + 2\pi\int_{\varrho_0}^{\infty} \tilde{W}(\varrho)\,\varrho\,d\varrho. \tag{8.28}$$

The integration in (8.28) can be carried out to give

$$\sigma_{\rm exc}^{\rm N} = \pi \varrho_0^2 + 2\pi \left(\frac{2Z\lambda\omega}{v}\right)^2 \left(\frac{\omega\varrho}{v}\right) K_0\left(\frac{\omega\varrho_0}{v}\right) K_0'\left(\frac{\omega\varrho_0}{v}\right), \tag{8.29}$$

where it is assumed that $(\omega\varrho_0/v) > 1$. For high values of the ionic charge Z, ϱ_0 is large and the constant-value approximation of $\tilde{W}(\varrho)$ in the region $\varrho \leq \varrho_0$ may lead to a significant uncertainty in the calculated cross section. In this respect, normalization of the excitation probability $W(\varrho)$ using (8.6–8) would be much more appropriate.

Another possibility for "normalizing" $\sigma_{\rm exc}$ is to smoothly join the probability $\tilde{W}(\varrho)$ with some more correct expression for $W(\varrho)$ in the region $\varrho < \varrho_0$. $W(\varrho)$ by itself may be incorrect in the region $\varrho > \varrho_0$. Such an expression for $W(\varrho)$ is provided by the sudden approximation and will be discussed in Sect. 8.3, in the context of single-electron high-energy ionization.

The excitation cross section can also be calculated if the transition probability between the initial and final states is calculated in the VPS approximation (6.150). If the interaction $V(\boldsymbol{r}, \boldsymbol{R})$ between the atom and the ion is represented by a multipole expansion, then for each term $V_n(R)$ of the expansion, taken in the form

$$V_n(R) = \xi_n R^{-n}, \tag{8.30}$$

the VPS excitation cross section can be obtained in the form [8.25]

$$\sigma_{\rm VPS}^{(n)} = 2\pi \left(\frac{\xi_n}{v}\right)^{2/(n-1)} \exp\left[-2\,(2^{2/n}\alpha_n)^{1/2} \sin\left(\frac{\pi}{2n}\right)\right] I_n(\alpha_n), \tag{8.31}$$

$$I_n(\alpha_n) = \int_0^\infty x\,dx \sin^2\left(\frac{c_n}{x^{n-1}}\right)$$

$$\times \exp\left\{2\,(2^{2/n}\alpha_n)^{1/2} \sin\left(\frac{\pi}{2n}\right) - 2\left[2^{2/n} \sin^2\left(\frac{\pi}{2n}\right) + \alpha_n^{n/(n-1)}\right]^{1/2}\right\}, \tag{8.32}$$

$$\alpha_n = \xi_n^{2/n}\, \omega^{2(n-1)/n}\, v^{-2}, \quad c_n = \pi^{1/2}\,\Gamma[(n-1)/2]/\Gamma(n/2). \tag{8.33}$$

When $\alpha_n \to 0$ $(v \to \infty)$, $I_n(\alpha_n)$ behaves as

$$I_2(\alpha_2) \sim \ln(1/\alpha_2), \quad I_{n>2}(0) = \text{const}. \tag{8.34}$$

In the opposite case $(\alpha_n \gg 1$, v small), one has

$$I_n(\alpha_n) \simeq (1/8)\,\alpha_n^{-n/(n-1)} \left[2\,(2^{2/n}\alpha_n)^{1/2} \sin^2\left(\frac{\pi}{2n}\right) + 1\right]. \tag{8.35}$$

Table 8.2. Values of the integrals $I_2(\alpha_2)$ and $I_3(\alpha_3)$ (8.32), for different values of α_2 and α_3

α_2 or α_3	$I_2(\alpha_2)$	$I_3(\alpha_3)$
0.00	$\to \ln(1/\alpha_2)$	$\pi/2$
0.02	21.6	1.42
0.04	15.6	1.32
0.08	10.7	1.16
0.16	6.25	0.954
0.32	2.94	0.699
0.64	1.01	0.436
1.28	0.268	0.224
2.56	8.29×10^{-2}	9.80×10^{-2}
5.12	2.43×10^{-2}	4.24×10^{-2}

The result (8.34) which gives a constant limit for $I_n(\alpha_n)$ $(n > 2)$ as $\alpha_n \to 0$, indicates that with the multipole interaction in the form (8.30), the high-velocity behaviour of $\sigma^{(n)}$ would be $\sigma^{(n)} \sim v^{-2/(n-1)}$. This behaviour is the result of the fact that $V_n(R)$ has a pole at $R = 0$. The real multipole potential is only valid for $R \geq r_0$, where r_0 is the dimension of the atom. If this restriction on $V_n(R)$ is taken into account [by setting $V_n(R) = \xi_n/(R^2 + r_0^2)^{n/2}$, for instance], the cross section $\sigma^{(n)}$ behaves for $v \gg 1$ as $\sigma^{(n)} \sim v^{-2}$ for all values of n. At low velocities, the pole of $V_n(R)$ is unimportant.

The values of the integrals $I_n(\alpha_n)$ for the dipole $(n = 2)$ and the quadrupole $(n = 3)$ interactions are given in Table 8.2 [8.25]. Note that for the dipole interaction, $\xi_2 = Z\lambda$ and $\alpha_2 = \beta$, where λ and β are defined by (8.4) and (8.12), respectively.

8.2.4 Excitation in Ion-Ion Collisions

The methods discussed in the preceding sub-sections can also be applied to the excitation problem in ion-ion collisions if the Coulomb character of the trajectory is explicitly taken into account. We shall give some results obtained with first-order perturbation theory and the VPS approximation. If Z_0 and Z are the charges of the target and the projectile ion, respectively, the trajectory of relative motion is defined parametrically by

$$R = a(\varepsilon \cosh \zeta + 1), \quad t = (a/v)(\varepsilon \sinh \zeta + \zeta), \tag{8.36}$$

$$\varepsilon = [1 + (\varrho/a)^2]^{1/2}, \quad a = Z_0 Z / M v^2, \tag{8.37}$$

where M is the reduced mass of the ions. The probability of a transition being induced by the dipole part of the ion-ion interaction is (using first-order perturbation theory) [8.26]

$$\tilde{W}^{\mathrm{ii}}_{\mathrm{exc}}(\varrho) = \left(\frac{2\xi_2\,\omega}{v^2}\right) \mathrm{e}^{-\pi\nu}\left[K^2_{i\nu}(\nu\,\varepsilon) + \tfrac{1}{2}K^2_{i\nu+1}(\nu\,\varepsilon) + \tfrac{1}{2}K^2_{i\nu-1}(\nu\,\varepsilon)\right], \tag{8.38}$$

where $\nu = a\,\omega/v$.

For a hydrogen-like target ion, and for transitions with no change in the principal quantum number n ($n, l \to n, l+1$), ξ_2 is given by

$$\xi_2 = \left[\frac{3\,(l+1)\,n^2\,[n^2-(l+1)^2]}{4\,Z_0^2\,(2l+1)}\right]^{1/2}. \tag{8.39}$$

Using (8.38) for $\tilde{W}(\varrho)$, the cross section is obtained in closed form [8.26]:

$$\sigma^{\mathrm{ii,BI}}_{\mathrm{exc}} = 2\pi\left(\frac{2\xi_2}{v}\right)^2 \mathrm{e}^{-\pi\nu}\left[-\nu\,K_{i\nu}(\nu)\,K'_{i\nu}(\nu)\right] \tag{8.40}$$

$$= 2\pi\left(\frac{2\xi_2}{v}\right)^2 \Phi(\nu), \tag{8.41}$$

where the function $\Phi(\nu)$ is tabulated [8.26]. Introducing a cut-off impact parameter ϱ_0 by the equation $\tilde{W}(\varrho_0) = 1/2$, one obtains the normalized cross section in the form

$$\sigma^{\mathrm{ii,N}}_{\mathrm{exc}} = \pi\,\varrho_0^2 + 2\pi\left(\frac{2\xi_2}{v}\right)^2 \mathrm{e}^{-\pi\nu}\left[-\nu\,\varepsilon_0\,K_{i\nu}(\nu\,\varepsilon_0)\,K'_{i\nu}(\nu\,\varepsilon_0)\right], \tag{8.42}$$

where $\varepsilon_0 = \varepsilon(\varrho_0)$.

The presence of the exponential factor $\exp(-\pi\nu)$ in the expressions (8.40, 42) for σ_{exc} indicates that the excitation cross section in ion-ion collisions is exponentially small unless ω is very small.

Within the VPS approximation, the excitation cross section in ion-ion collisions can be obtained in closed form for the general multipole interaction (8.30). The result is [8.25]

$$\sigma^{\mathrm{ii}(n)}_{\mathrm{VPS}} = 2\pi\left(\frac{\xi_n}{v}\right)^{2/(n-1)} \exp\left[-2\,(2^{2/n}\alpha_n)^{1/2}\sin\left(\frac{\pi}{2n}\right)\right] I_n(\alpha_n, \gamma_n) \tag{8.43}$$

where

$$\begin{aligned} I_n(\alpha_n,\gamma_n) &= \int_0^\infty x\,dx\,\sin^2\left(\frac{c_n}{(x^2+\gamma_n^2)^{(n-1)/2}}\right) \\ &\times \exp\left\{2\,(2^{2/n}\alpha_n)^{1/2}\sin\left(\frac{\pi}{2n}\right) - 2\left[2^{2/n}\sin^2\left(\frac{\pi}{2n}\right) + \alpha_n^{n/(n-1)}\right]^{1/2}\right. \\ &\left. - \alpha_n^{n/2(n-1)}\,\gamma_n\left(\pi - 2\arctan\frac{\gamma_n}{x}\right)\right\}, \end{aligned} \tag{8.44}$$

$$\gamma_n = a\,(\xi_n/v)^{n-1}, \tag{8.45}$$

Table 8.3. Values of the integral $I_2(\alpha_2, \gamma_2)$

α_2 \ γ_2	0.064	0.128	0.256	0.512	1.024	2.048
0.02	21.6	21.5	21.5	21.5	21.2	19.3
0.04	15.9	15.9	15.9	15.8	15.5	13.9
0.08	10.7	10.7	10.7	10.5	10.1	8.43
0.16	6.24	6.23	6.20	6.06	5.60	3.96
0.32	2.93	2.92	2.88	2.73	2.29	1.14
0.64	1.01	0.996	0.956	0.826	0.546	0.126
1.28	0.264	0.253	0.217	0.137	4.73×10^{-2}	2.03×10^{-3}
2.56	7.88×10^{-2}	6.82×10^{-2}	4.28×10^{-2}	1.09×10^{-2}	5.76×10^{-4}	6.69×10^{-7}
5.12	2.29×10^{-2}	1.52×10^{-2}	4.92×10^{-3}	2.42×10^{-4}	1.44×10^{-7}	8.5×10^{-14}

Table 8.4. Values of the integral $I_3(\alpha_3, \gamma_3)$

α_3 \ γ_3	0.064	0.128	0.256	0.512	1.024	2.048
0.02	1.42	1.41	1.40	1.34	1.111	0.324
0.04	1.31	1.31	1.30	1.22	0.984	0.255
0.08	1.16	1.15	1.14	1.06	0.803	0.170
0.16	0.951	0.943	0.923	0.833	0.574	8.76×10^{-2}
0.32	0.695	0.686	0.662	0.567	0.330	2.92×10^{-2}
0.64	0.433	0.422	0.396	0.306	0.132	4.72×10^{-3}
1.28	0.220	0.210	0.184	0.115	2.92×10^{-2}	2.23×10^{-4}
2.56	9.41×10^{-2}	8.45×10^{-2}	6.42×10^{-2}	2.49×10^{-2}	2.40×10^{-3}	1.44×10^{-6}
5.12	3.84×10^{-2}	3.01×10^{-2}	1.48×10^{-2}	2.37×10^{-3}	3.71×10^{-5}	2.94×10^{-10}

and α_n and c_n have been defined earlier. For $\gamma_n > 1$, the integral $I_n(\alpha_n, \gamma_n)$ behaves as

$$I_n(\alpha_n, \gamma_n) \sim I_n(\alpha_n)\, e^{-\pi\nu}, \quad \gamma_n > 1 \tag{8.46}$$

and the cross section is exponentially small. Note that for $\gamma_n = 0$, the cross section $\sigma_{\mathrm{VPS}}^{\mathrm{ii},(n)}$ goes over into $\sigma_{\mathrm{VPS}}^{(n)}$, defined by (8.31–33). The values of the integrals $I_n(\alpha_n, \gamma_n)$ for the dipole and quadrupole interactions for certain values of α_n and γ_n are given in Tables 8.3, 4, respectively.

8.3 Single-Electron Ionization

Theoretical treatments of the ionization process

$$A + B^{Z+} \rightarrow A^+ + B^{Z+} + e \tag{8.47}$$

are always considerably more complicated in comparison with the processes involving transitions within the discrete spectrum, due to the nature of continuum states. A coupled-channel formalism for the ionization process cannot be practically constructed without introducing pseudostates to represent the continuum. The positive energy pseudostates should adequately span the energy spectrum of ejected electrons, which implies a large basis set. Attempts at such an elaborate coupled-channel treatment of the ionization process have been made, but only for the $H + H^+$ collisions [8.13] and inner-shell ionization [8.27, 28]. When highly charged ions are involved in the collision, the coupled-channel problem for ionization becomes extremely difficult, and it is necessary to introduce drastic simplifications. Even the perturbation approach to the ionization problem with highly charged ions is inappropriate, since with increasing Z, the convergence of the perturbation series becomes increasingly slower. In this situation, classical methods have been widely applied to describe the collision dynamics of the ionization process. We shall now present some of the methods which have been used or which could potentially be useful in the treatment of single-electron ionization.

8.3.1 Perturbation Treatments

The amplitude of the transition from a bound state $|0\rangle$ into a continuum state $|k\rangle$ is given in first-order time-dependent perturbation theory ($t \to \infty$) as

$$a_{0k}^{\mathrm{Bl}}(\infty) = -\mathrm{i} \int_{-\infty}^{\infty} dt\, \mathrm{e}^{\mathrm{i}\omega_k t} \langle k | V | 0 \rangle\,, \tag{8.48}$$

where $\omega_k = I_0 + k^2/2$ is the transition energy, $\boldsymbol{k}$ is the momentum of the ejected electron, I_0 is the ionization potential of the target atom and V is the ion-atom interaction. In the dipole approximation for $V(\boldsymbol{r}, \boldsymbol{R})$ and in the straight-line-trajectory approximation, the calculation of $a_{0k}^{\mathrm{Bl}}(\infty)$ with hydrogenic states can be carried out analytically [8.23]. The transition probability summed over the projections of the final-state angular momentum at a given impact parameter ϱ is given by [8.23]

$$W_{\mathrm{i}}^{\mathrm{Bl}}(\varrho) = \frac{4Z^2}{3v^2} \sum_k \omega_k^2 r_{0k}^2 \left[K_0^2\left(\frac{\omega_k \varrho}{v}\right) + K_1^2\left(\frac{\omega_k \varrho}{v}\right) \right], \tag{8.49}$$

where the sum runs over the states of the continuum, and r_{0n}^2 is the square of the dipole matrix element. Introducing an impact parameter $\varrho_{\min}$ such that the total excitation [see (8.26) for W_{exc}] and ionization probability is equal to unity ($\varrho_{\min} \approx 2Z/v$), one can represent the normalized ionization cross section in the form

$$\sigma_i^N = \pi \varrho_{min}^2 S + 2\pi \int_{\varrho_{min}}^{\infty} W_i^{Bl}(\varrho)\, \varrho\, d\varrho\,, \tag{8.50}$$

where

$$S = \sum_k \frac{r_{0k}^2}{\langle r \rangle^2}\,. \tag{8.51}$$

For the hydrogen atom S has the value $S = 0.283$. Carrying out the integration in (8.50) in the limit $(Z/v^2) \ll 1$, one obtains

$$\sigma_i^N = \pi A \frac{Z^2}{v^2} \ln \frac{a v^2}{Z}\,, \tag{8.52}$$

where for the hydrogen atom target the constants A and a have the values [8.29]

$$A_H = 2.29\,, \quad a_H = 1.51\,. \tag{8.53}$$

This first-order time-dependent (or semi-classical) perturbation theory for ionization has also been elaborated in detail for complex atoms [8.30, 31].

The normalization procedure leading to (8.50) presupposes that ϱ_{min} is sufficiently small so that the inaccuracy introduced in σ_i^N from the region $\varrho < \varrho_{min}$ is not large. Since $\varrho_{min} \approx 2Z/v$, this condition is fulfilled only for $v \gtrsim Z$, which is consistent with the more general condition $v^2 \gg Z$ for the validity of the perturbation treatment.

Instead of performing the normalization of the ionization cross section in the above-mentioned way, one can calculate $W(\varrho)$ in the strong interaction region $\varrho < \varrho_c \sim Z/v$ by using the first-order Magnus (or sudden) approximation [8.32, 33]. The transition amplitude in this approximation is given by

$$a_{0k}^M = \left\langle k \left| \exp\left[-\mathrm{i} \int_{-\infty}^{\infty} V^{int}(t)\, dt \right] \right| 0 \right\rangle, \tag{8.54}$$

where $V^{int}(t)$ is the interaction representation of the ion-atom potential $V(t) = V(\boldsymbol{r}, \boldsymbol{R})$,

$$V^{int}(t) = e^{\mathrm{i}H_0 t}\, V(t)\, e^{-\mathrm{i}H_0 t}\,, \tag{8.55}$$

and H_0 is the Hamiltonian of the atom. The sudden approximation (8.54) of the full Magnus amplitude is justified if the collision energy is such that the collision time is small compared with the characteristic atomic time. It is seen that the first Born amplitude (8.48) is obtained as the first non-vanishing term in the expansion of a_{0k}^M in terms of V^{int}. By matching the probabilities $W^{M}(\varrho) = |a_{0k}^{M}|^2$ and $W^{B1}(\varrho)$ at some $\varrho = \varrho_c$ where they are equal, one obtains a representation of the ionization probability valid in the whole region of ϱ. Such a procedure has been applied by *Salop* and *Eichler* [8.34] for calculation of the ionization cross section in $H + He^{2+}$ and $H + C^{6+}$ collisions in the energy range from a few hundreds of keV to a few MeV.

The wave-mechanical version of first-order perturbation theory for ionization (plane-wave Born approximation) is discussed in detail elsewhere [8.35].

8.3.2 DACC Theory for Ionization

The formalism of the two-state dipole-approximation close-coupling (DACC) theory for single-electron excitation (Sect. 8.2.1) can also be extended to the ionization process [8.18]. For the $n\,s \to k\,p$ ionization transition, the continuum may be represented by two pseudostates $|\widetilde{k\,p_0}\rangle$ and $|\widetilde{k\,p_1}\rangle$, which can be chosen so that they have the same oscillator strength as the atomic (single-electron ionization) continuum. The corresponding coupled equations for the state amplitudes then coincide with (8.3) for the excitation problem, in which the coupling constant λ is now defined by

$$\lambda_{\text{eff}} = (f_{\text{eff}}/2\,\omega_i)^{1/2}\,, \tag{8.56}$$

where ω_i is the ionization potential of the atom and f_{eff} is the effective oscillator strength [8.36]. To allow for multipole transitions other than the dipole transition, f_{eff} can be written as

$$f_{\text{eff}} = s\, f_{\text{cont}}\,, \tag{8.57}$$

where f_{cont} is the usual continuum oscillator strength. For the hydrogen atom $s = 1.5$, $f_{\text{cont}} = 0.4350$ and $\omega_i = 1/2$. Defining, in analogy with (8.12), a dimensionless parameter β_i by

$$\beta_i = Z\,\lambda_{\text{eff}}\,\omega_i/v^2\,, \tag{8.58}$$

and following the procedure leading to (8.13), one obtains the DACC ionization ($n\,s \to k\,p$) cross section in the form [8.18]

$$\sigma_i^{s-p} = 2\pi\,(Z\,\lambda_{\text{eff}}/\omega_i)\,D(\beta_i) \tag{8.59}$$

where $D(\beta_i)$ is given by the integral (8.14). The contribution of the most important two-step ionization process to the total ionization, namely, the one going via the resonant atomic level, can also be estimated. If ω_{0r} is the resonance transition frequency and λ_{0r} and β_{0r} are the parameters λ and β for the ground-to-resonant-state transition, then the ionization cross section summed over the $n\,s \to k\,p$ and $n\,s \to n_r\,p \to (k\,s + k\,d)$ transitions is given by [8.18]

$$\sigma_i = 2\pi Z \left[\frac{\lambda_{eff}}{\omega_i} D(\beta_i) + \frac{1}{8}\frac{\lambda_{0r}}{\omega_{0r}} D(\beta_{0r})\right]. \tag{8.60}$$

The second term in (8.60) contributes (by 10–20%) to σ_i only in the energy region near the cross section maximum. In the asymptotic regions $\beta \gg 1$ ($v \ll 1$) and $\beta \ll 1$ ($v \gg 1$), σ_i is determined only by the direct transition and reads

$$\sigma_i^{ad} \simeq \pi \left(\frac{Z\lambda_{eff}}{v}\right)^2 \exp\left[-(1/v)(2Z\lambda_{eff}\omega_i)^{1/2}\right], \quad v \ll (Z\lambda_{eff}\omega_i)^{1/2}, \tag{8.61 a}$$

$$\sigma_i^{B1} \simeq 8\pi \left(\frac{Z\lambda_{eff}}{v}\right)^2 \ln\left(\frac{1.4\,v^2}{Z\lambda_{eff}\omega_i}\right), \quad v \gg (Z\lambda_{eff}\omega_i)^{1/2}. \tag{8.61 b}$$

The velocity v_m at which the ionization cross section has its maximum is approximately given by

$$v_m \simeq 1.89\,(Z\lambda_{eff}\omega_i)^{1/2}. \tag{8.62}$$

The cross section value at $v = v_m$ is

$$\sigma_{i,m} \simeq 0.82\,\pi Z \left(\frac{\lambda_{eff}}{\omega_i} + \frac{1}{8}\frac{\lambda_{0r}}{\omega_{0r}}\right). \tag{8.63}$$

Thus, the maximum of σ_i scales linearly with Z, while the velocity at which it appears scales as $Z^{1/2}$. Since the contribution to σ_i of the second term in (8.60) is small, it follows that the ionization cross section can be represented in the reduced form

$$\tilde{\sigma}_i(\tilde{E}) = \frac{\omega_i}{Z\lambda_{eff}}\sigma_i = 2\pi\, f_\zeta(\tilde{E}), \quad \tilde{E} = \frac{E}{Z\lambda_{eff}\omega_i}, \tag{8.64}$$

where the collision energy E is expressed in keV/amu, and

$$f_\zeta(\tilde{E}) = D\left(\frac{1}{\tilde{E}}\right) + \left(\frac{\zeta}{8}\right) D\left(\frac{1}{\zeta\tilde{E}}\right), \quad \zeta = \frac{\lambda_{0r}\,\omega_i}{\lambda_{eff}\,\omega_{0r}}. \tag{8.65}$$

The maximum of the reduced cross section $\tilde{\sigma}_i(\tilde{E})$ occurs at $\tilde{E}_m \simeq 35.7$. For the case of $H(1s) + Z$ collisions, the ionization cross section can be parametrized in a simpler form:

$$\tilde{\sigma}_{i,H}(\tilde{E}) = \sigma_i/Z = 2\pi f_H(\tilde{E}), \quad \tilde{E} = E/Z, \tag{8.66}$$

$$f_H(\tilde{E}) = 1.62\,D(0.404/\tilde{E}) + 0.246\,D(0.279/\tilde{E}). \tag{8.67}$$

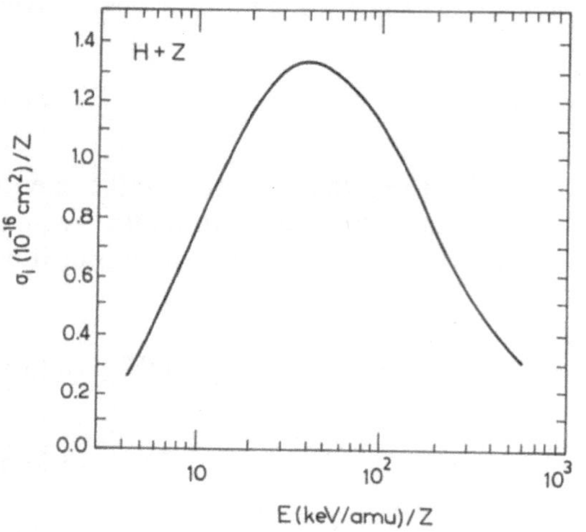

Fig. 8.3. Scaled ionization cross section for the H+Z system vs. scaled energy, from DACC theory [8.18]

The scaled cross section $\sigma_{i,H}(\tilde{E})$ is presented in Fig. 8.3. For the He+Z collisions, the scaled ionization cross section $\tilde{\sigma}_i = \sigma_i/Z$ can be represented as [8.37]

$$\tilde{\sigma}_{i,He}(\tilde{E}) = 2.726 \times 10^{-16} F(\tilde{E}) \quad [\mathrm{cm}^2], \tag{8.68}$$

$$F(\tilde{E}) = D(\beta_i) + 0.096\, D(\beta_i/2), \quad \beta_i = \frac{15.788}{\tilde{E}\,[\mathrm{keV/amu}]}. \tag{8.69}$$

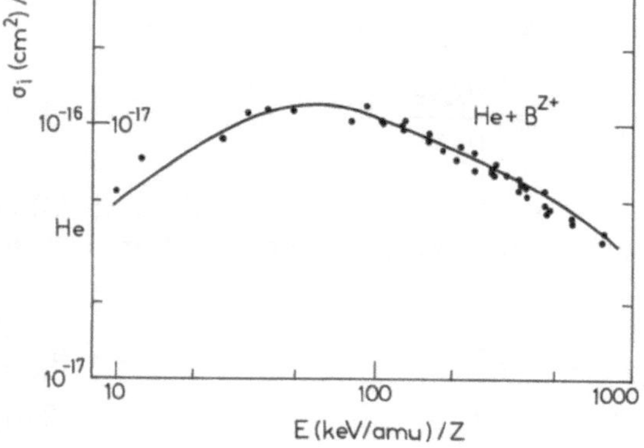

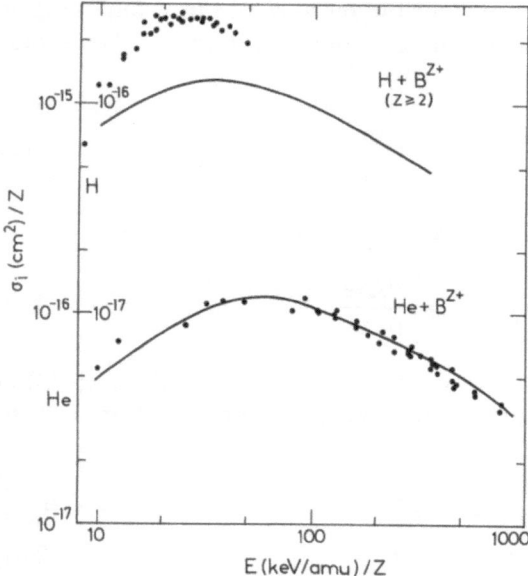

Fig. 8.4. Scaled experimental and DACC ionization cross sections for the H+B^{Z+}($Z \geqq 2$) and He+B^{Z+} collisions [8.41]

Using three-state DACC theory, ionization cross sections have been calculated for the H $(1s) + Z$ system $(1 \leq Z \leq 32)$ [8.18] and the He $(1s^2) + Z$ system $(2 \leq Z \leq 32)$ [8.37]. Experimental ionization cross sections for the $H + B^{Z+}$ collisions, with $B^{Z+} = H^+$, He^{2+}, $Li^{(1-3)+}$, $C^{(2-4)+}$ and $O^{(2-5)+}$ have been measured by *Shah* and *Gilbody* [8.38], and for $He + B^{Z+}$ ($B^{Z+} = He^{2+}$, O^{Z+}, Au^{Z+}) by *Hvelplund* and associates [8.39, 40]. In Fig. 8.4, the three-state DACC cross section results for the $H + Z$ and $He + Z$ system are compared with the experimental data on $H + B^{Z+}$ $(Z \geq 2)$ and $He + B^{Z+}$ $(Z \geq 2)$ collision systems in a reduced representation [8.41]. It should be noted that in the region $E < 60$ keV/amu, the capture into continuum states, not accounted for in the DACC theory, may significantly contribute to the ionization of hydrogen atoms [8.42]. The experimental ionization cross sections, however, do include this contribution.

8.3.3 UDWA Calculations

Within the unitarized distorted-wave approximation (UDWA), the probability for the ionization $|0\rangle \to |k\rangle$ transition is given by [8.21]

$$P_{0k}^{\rm ion} = p^{-1} \, |t_{0k}^{\rm ion}|^2 \sin^2 p^{1/2} \, , \tag{8.70}$$

where $t_{0k}^{\rm ion}$ is the transition matrix element in the distorted-wave Born approximation (DWBA), and p is defined by (8.24). In obtaining the ionization cross section σ_i, $P_{0k}^{\rm ion}$ is integrated both over the states $|k\rangle$ of the continuum and the impact parameters. In the calculation of the DWBA matrix element $t_{0k}^{\rm ion}$, see (7.92), the wave function ψ_k of the ejected electron

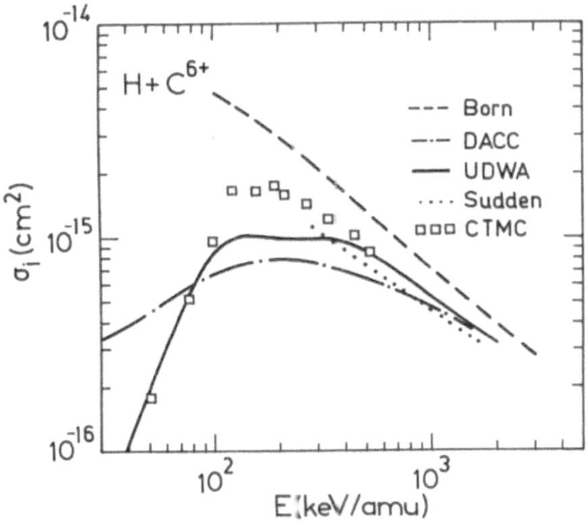

Fig. 8.5. Ionization cross sections for the $H + C^{6+}$ collisions in different theoretical models: Born approximation [8.34], DACC theory [8.18], UDWA [8.21], sudden approximation [8.34]. CTMC method [8.43]

is centred on the projectile ion, which automatically includes the effect of the "capture into continuum". This choice of the coordinate origin for ψ_k is natural for high values of Z; however for low Z it can lead to a significant (~50%) overestimation of the DWBA cross section. Therefore, the calculated t_{0k}^{ion} (DWBA) has to be accordingly reduced before being used in the calculations of the UDWA cross section. Using such a correction procedure, UDWA ionization cross sections have been calculated for the $H + H^+$, Li^{3+}, B^{5+}, C^{6+} and Si^{14+} collisions [8.21]. The correction procedure, however, introduces some uncertainty into the results. The UDWA cross section for the reaction $H + C^{6+} \rightarrow H^+ + C^{6+} + e$ is presented in Fig. 8.5, and compared with the other theoretical predictions.

8.3.4 Classical Models

The classical trajectory Monte Carlo (CTMC) method, discussed in Sect. 6.4.2, has been extensively used for calculation of the ionization cross section in the $H + Z$ $(Z = 1-32)$ [8.43] and $He + Z$ $(Z = 2-8)$ [8.44] collisions. (In the latter case an effective charge of $Z_{\text{eff}} = 1.69$ has been assigned to the ionic core He^+.) The CTMC method has also been applied to collisions of hydrogen atoms with the incompletely stripped ions B^{Z+}, C^{Z+}, N^{Z+} and O^{Z+} by using a suitably chosen effective charge for the ion [8.43]. The results for $H + C^{Z+}$ $(Z = 3-6)$ are shown in Fig. 8.6, together with the corresponding charge-exchange cross sections.

The classical impulse or binary-encounter approximation (BEA, see Sect. 6.4.1) also provides a means for approximate estimation of the ionization cross section at large collision energies. The integration of the differential energy-transfer cross section $\Delta\sigma_i$ with a delta-function velocity distribution of the bound electron leads to the following form of the BEA cross section for ionization [8.45]:

$$\sigma_i^{\text{BEA}} = 0, \quad \alpha < (\sqrt{2} - 1)/2, \quad \alpha = v/v_0 \tag{8.71 a}$$

$$= \frac{\pi}{3}\left(\frac{Z_0 Z}{I_0 \alpha}\right)^2 \left(2\alpha^3 - \frac{1 + 2\sqrt{2}}{2(3 + 2\sqrt{2})} + \frac{3}{8(\alpha + 1)}\right), \quad (\sqrt{2} - 1)/2 < \alpha < (\sqrt{2} + 1)/2 \tag{8.71 b}$$

$$= \frac{\pi}{3}\left(\frac{Z_0 Z}{I_0 \alpha}\right)^2 \left(5 - \frac{3}{4(\alpha^2 - 1)}\right), \quad \alpha > (\sqrt{2} + 1)/2, \tag{8.71 c}$$

where Z_0 and I_0 are, respectively, the charge of the core and the electron binding energy of the target atom, v and Z are, respectively, the projectile

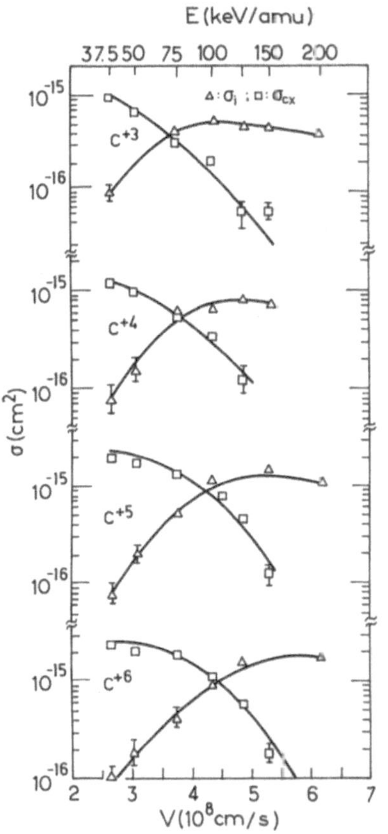

Fig. 8.6. The CTMC ionization and charge-exchange cross sections for the $H+C^{Z+}$ system for $Z=3-6$ [6.43]

ion velocity and its charge, and v_0 is the classical velocity of the bound electron. The above expressions for σ_i^{BEA} are valid within an $O(1/M)$ accuracy, where M is the mass of the projectile ion. In the high-velocity limit, $\alpha \gg 1$, the asymptotic form of σ_i^{BEA} is

$$\sigma_i^{BEA} \simeq \frac{5\pi}{3}\left(\frac{Z_0 Z}{I_0\, v/v_0}\right)^2 . \tag{8.72}$$

It can be shown that this asymptotic behaviour of σ_i^{BEA} does not depend on the form of the electron velocity distribution and follows directly from the virial theorem [8.46]. The high-velocity asymptotic value of σ_i^{BEA} lies below the first Born cross section σ_i^{B1}, and for increasingly high velocities, σ_i^{BEA} diverges from σ_i^{B1}. In this limit the CTMC ionization cross section should converge towards σ_i^{BEA}.

The binary-encounter approximation has been extended to treat complex atomic targets also [8.47], its applications being, however, restricted to low-charged ions. Close in nature to BEA is the classical treatment of ionization by *Thomson* [8.48], which gives

$$\sigma_i^{Th} = 2\pi \frac{Z^2}{I_0 v^2}. \tag{8.73}$$

In the low-velocity region ($v \lesssim 1$), *Bohr* [8.49] treated the ionization of atoms by heavy charged particles by using purely classical arguments and going beyond the adiabatic concept. An energy-independent ionization cross section follows from his treatment [8.49, 39] (H-atom case),

$$\sigma_i^{Bohr} = 4\pi \frac{Z}{(2I_0)^{3/2}}, \quad v < (2I_0 Z^2)^{1/4}. \tag{8.74}$$

The linear Z dependence of σ_i^{Bohr} is consistent with the predictions of the DACC theory for the energy region in which the maximum of σ_i lies.

8.3.5 Higher-Order High-Energy Approximations

Of the higher-order approximations for high-energy collisions discussed in Sect. 6.3.5, only the eikonal (Glauber) and the continuum distorted-wave (CDW) approximations have been used in the ionization problem involving multiply charged ions. Neglecting the longitudinal momentum transfer in the eikonal amplitude (6.203), one obtains the scattering amplitude in the Glauber approximation [8.50, 51],

$$a_{0k}^G(\boldsymbol{q}, \boldsymbol{k}) = -i(2\pi/v)\langle\phi_k|\, T^G(\boldsymbol{q})\,|\phi_0\rangle, \tag{8.75}$$

$$T^G(\boldsymbol{q}) = \int d^2\varrho \exp(i\,\boldsymbol{q}\,\boldsymbol{\varrho})\,\Gamma(\varrho, v), \tag{8.76a}$$

$$\Gamma(\varrho, v) = 1 - \exp\left[\frac{i}{v}\int_{-\infty}^{+\infty} V(\boldsymbol{r}, \boldsymbol{R})\, dz\right], \tag{8.76b}$$

where $V(\boldsymbol{r}, \boldsymbol{R})$ is the interaction potential, $\boldsymbol{R} = \boldsymbol{\varrho} + \boldsymbol{v}\,t = \boldsymbol{\varrho} + \boldsymbol{z}$, $\boldsymbol{q}$ is the transferred momentum and $\phi_0(\boldsymbol{r})$, $\phi_k(\boldsymbol{r})$ are the initial and final state electron wave functions, respectively. By expanding the exponential function in $\Gamma(\varrho, v)$ in a power series, one obtains the first Born term exactly, as well as the higher Born terms approximately. The Glauber cross section for ionization is given by

$$\sigma_i^G = \int |a_{0k}^G(\boldsymbol{q}, \boldsymbol{k})|^2\, d\boldsymbol{q}\, d\boldsymbol{k}. \tag{8.77}$$

Ionization cross section calculations within the Glauber approximation have been performed recently by *McGuire* [8.52] for the H+Z ($Z = 1-6$) system

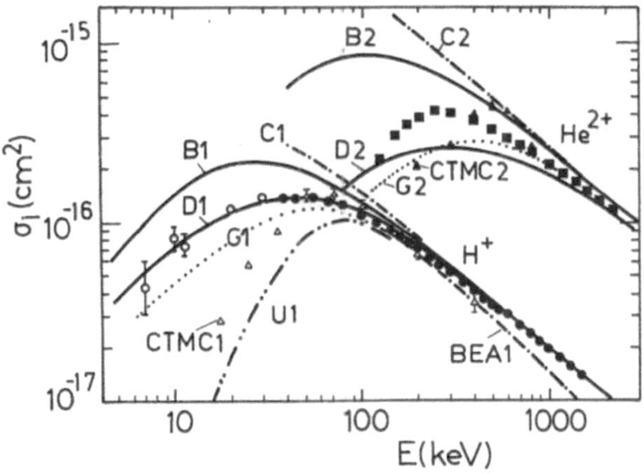

Fig. 8.7. Comparison of experimental and theoretical cross sections for ionization of H by H^+ and He^{2+} ions. Theoretical calculations: Born approximation (B1, B2) [8.55], DACC theory (D1, D2) [8.18], UDWA (U1) [8.21], Glauber approximation (G1, G2) [8.51, 52], CDW approximation (C1, C2) [8.53, 54], CTMC method (△, ▲) [8.43], classical impulse approximation (BEA) [8.47]. Experimental data: (○) (H^+) [8.56], (●) (H^+) and (■) (He^{2+}) [8.38]

in the region 20–2000 keV/amu, in addition to some earlier calculations for low-charged ions.

The CDW approximation has also been extended to the ionization problem in one-electron collision systems [8.53, 54] and applied to the $H+H^+$ and $H+He^{2+}$ collisions. It is worthwhile noting that while the Glauber approximation approaches the first Born cross section from below, the CDW approximation converges (much faster) towards σ_i^{B1} from above. The CDW and Glauber results for the $H+H^+$ and $H+He^{2+}$ ionizing collisions are shown in Fig. 8.7. They are compared with the results of other theories (B1, DACC, CTMC, UDWA, BEA) as well as with experimental data [8.38, 56].

8.3.6 Ion-Ion and Ion-Molecule Ionization

a) Ion-Ion Collisions

The theoretical methods for treating ion-atom ionizing collisions can be directly applied to the ion-ion ionization process if the Coulomb repulsion is taken into account in the description of the relative motion of the ions. It should , however, be noted that the Coulomb repulsion has a non-negligible effect on the cross section only in the energy region of the cross section maximum and below. The exponential decrease of the ionization cross section in the adiabatic region with decreasing collision energy is significantly enhanced due to Coulomb repulsion (also in an exponential fashion) only for $Z_0 Z \gtrsim E$, where Z_0, Z are, respectively, the target and projectile charges and E is the collision energy. In the simplest descriptions of the ionization process, the ion-ion collision cross section can be obtained from the corresponding ion-atom cross section by applying certain scaling relationships. Thus, in the Born approximation, the ionization cross section for the reaction

$$A^{(Z_0-1)+} + B^{Z+} \rightarrow A^{Z_0+} + B^{Z+} + e, \tag{8.78}$$

where A^{Z_0+} and B^{Z+} are fully stripped ions, can be written as [8.55]

$$\sigma_i^{(A)}(E) = \frac{Z^2}{Z_0^4}\,\sigma_i^{(H)}(\tilde{E}), \quad \tilde{E} = E/M_B Z_0^2 \tag{8.79}$$

where $\sigma_i^{(H)}$ is the Born ionization cross section for the $H + H^+$ collisions. In the classical impulse approximation, the ion-ion cross section is given by (8.71).

The experimental investigations of the ionization process in ion-ion collisions have so far been restricted to low-charged ions and have recently been reviewed elsewhere [8.57]. It has been shown that within a factor of two, the experimental data allow a reduced BEA representation,

$$\sigma_i^{BEA}(\tilde{E}) = \frac{1}{N}\left(\frac{Z_0 I_0}{Z I_H}\right)^2 \sigma_i(E), \tag{8.80}$$

where N is the number of equivalent outer-shell electrons and $\tilde{E}$ is given by (8.79).

b) Ion-Molecule Collisions

The ionization process in collisions of highly charged ions with molecules has only been treated experimentally [8.58–60]. The target used most often in these studies was the H_2 molecule, whereas the projectiles included ions of Li^{Z+}, C^{Z+}, N^{Z+}, O^{Z+}, Fe^{Z+}, Nb^{Z+} and Pb^{Z+} ranging from $Z = 2$ (for the lighter ions) up to $Z = 59$ (for the heaviest ones). For the heavier ions, the data cover only the high-energy region, beyond the cross section maximum. The attempts at comparing the ionization cross sections $\sigma_i(H_2)$ with the theoretical cross sections $\sigma_i(H)$ are based on the assumption that the ratio $r = \sigma_i(H)/\sigma_i(H_2)$ is close to $r = 0.5$. Recent experimental investigations of *Shah* and *Gilbody* [8.60] have demonstrated that this ratio may vary significantly ($\sim \pm 20\%$) around the value $r = 0.5$, depending on the collision energy and the ionic charge. At high energies, r tends to become smaller than 0.5. Apart from direct ionization, the collision of a multiply charged ion with a molecule may also lead to dissociative ionization. Experimental investigations [8.60] have shown that at lower energies the cross section of this ionization channel attains its maximum at an energy somewhat smaller than that for the maximum of the direct process, and for energies beyond the cross section maximum it decreases faster with increasing energy than does the direct ionization cross section.

8.3.7 Scaling Laws for Single-Electron Ionization

The dipole-approximation close-coupling (DACC) theory provides a suitable basis for discussion of the scaling laws for the single-electron ionization cross section in different collision energy regions. The natural scaled quantities, which come out of the DACC theroy, are (Sect. 8.3.2)

$$\tilde{\sigma}_i = \sigma_i/Z, \quad \tilde{E} = E/Z \quad (\text{or } \tilde{v} = v/Z^{1/2}). \tag{8.81}$$

In terms of the scaled velocity $\tilde{v}$, the reduced cross section $\tilde{\sigma}_i$ in the DACC approximation has the behaviour

– adiabatic region ($\tilde{v} \ll 1$):

$$\tilde{\sigma}_i \approx \frac{C_1}{\tilde{v}^2}\, e^{-C_2/v^2}, \quad C_1, C_2 = \text{const}, \tag{8.82}$$

– region of the maximum ($\tilde{v} \sim 1$):

$$\tilde{\sigma}_{i,\,max} \approx 0.82\,\pi\left(\frac{\lambda_{eff}}{\omega_i} + \frac{1}{8}\frac{\lambda_{0r}}{\omega_{0r}}\right), \quad \tilde{v}_m \simeq 1.89\,(\lambda_{eff}\,\omega_i)^{1/2}, \tag{8.83}$$

– high-velocity region ($\tilde{v} \gg 1$):

$$\tilde{\sigma}_i \simeq 8\,\pi\frac{\lambda_{eff}^2}{\tilde{v}^2}\ln\left(\frac{1.4\,\tilde{v}^2}{\lambda_{eff}\,\omega_i}\right). \tag{8.84}$$

The constant value of $\tilde{\sigma}_i$ in the region $\tilde{v} \lesssim 1$ is also predicted by Bohr's classical model, see (8.74). For collision velocities $v \gtrsim v_m$, the Z-dependence of the ionization cross section can, thus, be represented as

$$\sigma_i(Z) = \sigma_{0i}\, Z^{\beta(v)}, \tag{8.85}$$

where σ_{0i} is independent of $\tilde{v}$ in the region $v \sim v_m$, and for $\tilde{v} \gg 1$ it depends logarithmically on $\tilde{v}$. The function $\beta(\tilde{v})$ is

$$\beta(\tilde{v}) = 1, \quad \tilde{v} \approx \tilde{v}_m, \tag{8.86a}$$

$$\beta(\tilde{v}) = 2, \quad \tilde{v} \geqq \tilde{v}_B \gg \tilde{v}_m. \tag{8.86b}$$

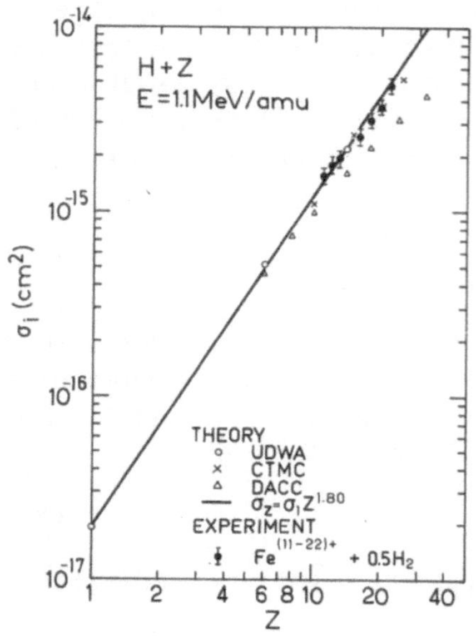

Fig. 8.8. Z-dependence of the ionization cross section predicted by UDWA [8.21], CTMC method [8.61] and DACC theory [8.18]. Experimental points are the cross sections for $H_2 + Fe^{Z+}$, divided by two

The region $\tilde{v}_m \lesssim \tilde{v} \lesssim \tilde{v}_B$, in which $1 \lesssim \beta \lesssim 2$, may be rather large for highly charged ions. In the system $H_2 + B^{Z+}$ B = C, N, O; $2 \leqq Z \leqq 5$) at a collision energy of $E = 145$ keV/amu ($\tilde{v} \approx 1-2$), the value of $\beta = 1.18 \pm 0.02$ has been experimentally obtained [8.60], while in other experiments at $E = 277$ and 1100 keV/amu with Fe^{Z+} ($Z \leqq 25$) [8.61], the values $\beta = 1.40$ and $\beta = 1.43$ have been obtained.

The Z-dependence of σ_i for the collision system $H+Z$ at an energy of 1.1 MeV/amu, as predicted by different theories, is shown in Fig. 8.8.

8.3.8 Direct Multiple-Ionization

Multi-electron ionization may occur either by a simultaneous (direct) transition of two or more electrons into the continuum or as a result of the reorganization of the electronic structure of the target atom in the field of the highly charged ion (Auger-type processes). This type of process will be considered in more detail in the next chapter.

The two- and more-electron ionization processes take place predominantly at relatively high energies when the collision time is sufficiently short. Under these conditions the electrons can be considered in the independent-particle approximation, while the dynamics can be described in terms of the sudden approximation. The probability for removal of n electrons from a shell containing N electrons is then given by [8.62]

$$P_n(\varrho) = \binom{N}{n} P_s^n(\varrho)\,[1 - P_s(\varrho)]^{N-n}, \tag{8.87}$$

where $P_s(\varrho)$ is the ionization probability in the one-electron model. The integration of (8.87) over the impact parameters gives the cross section $\sigma_i(n)$ for simultaneous removal of n electrons from a given shell (or subshell). One can also define a "gross ionization cross section"

$$\sigma_i\,(\text{gross}) = \sum_{n=1}^{N} n\,\sigma_i\,(n)\,, \tag{8.88}$$

which gives the total number of ejected electrons from a shell (subshell) with N electrons.

The calculation of the direct multiple-ionization cross section $\sigma_i(n)$ requires knowledge of the single-electron transition probability $P_s(\varrho)$. This probability can be calculated by using either of the high-energy methods for ionization. Calculations of $\sigma_i(n)$ and σ_i (gross) for the He$-Z$ and Ar$+Z$ ionizing collisions have been performed by using the CTMC method [8.63]. The results for He$+Z$ at $E = 1.1$ MeV/amu are shown in Fig. 8.9 and are compared with experimental data on σ_i (gross) [8.64]. The Z dependence of $\sigma_i(2)$ and σ_i (gross) in Fig. 8.9 is quadratic, except at the higher values of Z. This indicates that the scaled energy parameter for multiple ionization is again $\tilde{E} = E/Z$. A constant, and even decreasing, Z-dependence of $\sigma_i(2)$ has been observed for the He$+$Au^{Z+} collisions at $E = 100$ keV/amu, when Z was increased from $Z = 5$ to $Z = 21$ [8.65].

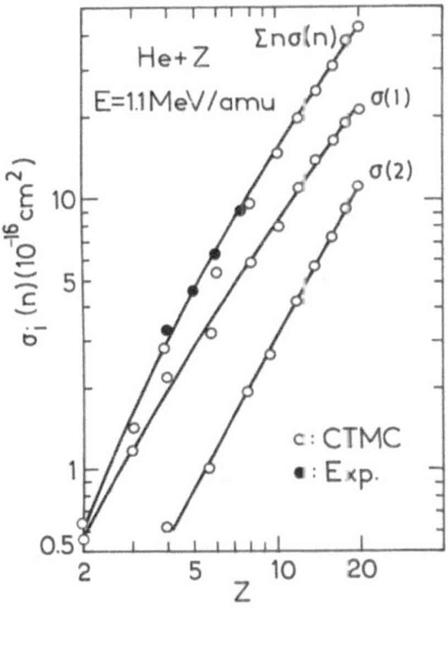

Fig. 8.9. One- and two-electron ionization cross sections, $\sigma_i(1)$ and $\sigma_i(2)$, for He$+Z$ collisions at $E = 1.1$ MeV/amu, obtained by the CTMC method [8.63]. The "gross ionization" cross section $\sum_n n\,\sigma(n)$ is compared with experimental data (●) [8.64]

8.4 Electron Loss and Stripping Processes

The removal of an electron (or more electrons) from the target by either electron capture or ionization is termed electron loss. The similar process, when an electron (or electrons) is removed from the projectile, is termed stripping. The stripping process takes place at high collision energies and for neutral targets it is determined exclusively by the ionization channel. The typical relative magnitude of the cross sections for single-electron capture (σ_c), ionization (σ_i), loss ($\sigma_{loss} = \sigma_c + \sigma_i$) and stripping ($\sigma_s$) are shown in Fig. 8.10 for He+O^{7+} collisions [8.66]. Since σ_s is usually much smaller than σ_{loss}, attention in this section will be mainly devoted to the electron loss processes. Although this process is simply a combination of the electron-capture and ionization processes, it deserves special consideration in view of its direct importance in some applications and because it is very often experimentally investigated as an independent process. Since the electron-capture and ionization channels dominate the electron loss process in different velocity regions, there exists a relatively narrow velocity range in which σ_c and σ_i are of the same order of magnitude where the two channels are strongly coupled. In this region ($Z^{1/4} \lesssim v < Z^{1/2}$), the electron loss process is governed by a unique mechanism, connected with the extraction of the electron from the target potential well by the field of the highly charged ion. In our consideration of the electron loss process we shall concentrate on this region.

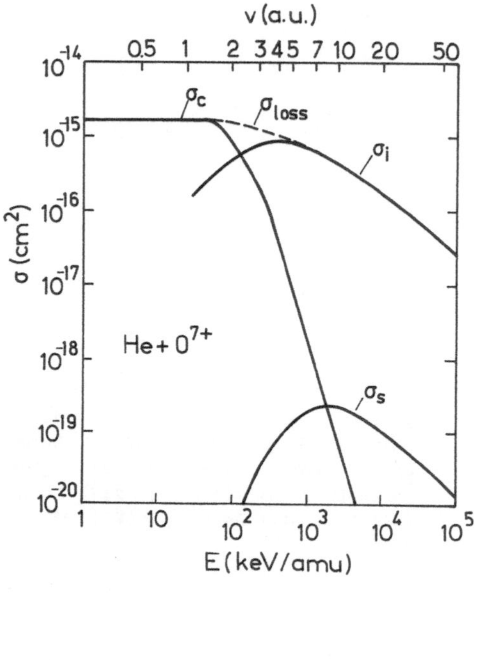

Fig. 8.10. Comparison of the electron capture (σ_c), ionization (σ_i), electron-loss (σ_{loss}) and stripping (σ_s) cross sections for the He+O^{7+} collision [8.66]

8.4.1 The Keldysh Quasi-Classical Method

The decay of an atomic state due to the time-dependent external electric field $\mathscr{E} = Z/(\varrho^2 + v^2 t^2)$ of a multiply charged ion is analogous to the problem of dynamic field ionization of atoms. Therefore, the electron loss process can be treated by the quasi-classical method developed by *Keldysh* [8.67] for the multi-photon ionization of atoms in a strong electromagnetic field. This analogy has been noticed by *Duman* et al. [8.68], who calculated the decay probability in the form

$$W_{\text{loss}}(\varrho) \sim \exp\left[-\frac{2\varrho^2}{3Z} g\left(\frac{\varrho v}{Z}\right)\right], \quad \text{where} \tag{8.89}$$

$$g(x) = \frac{3}{2x}(1 - p^2 + x^2)\sin\varphi \tag{8.90}$$

and $p(x)$ and $\varphi(x)$ are solutions of the system of equations

$$p \sin\varphi = \varphi, \quad (x^2 + p^2 + 1)\cos\varphi = 2p. \tag{8.91}$$

As is seen from (8.89), the Keldysh method does not enable one to obtain the pre-exponential factor in the expression for the electron transition probability. However, since the formula (8.89) for $W_{\text{loss}}(\varrho)$ has the correct exponential behaviour in a fairly large region around $v_0 \sim Z^{1/4}$, one can use it to connect the corresponding probabilities from the regions $v \ll v_0$ (where charge exchange is the dominant process) and $v \gg v_0$ (where ionization is dominant). Using such a procedure one can obtain $W_{\text{loss}}(\varrho)$ for the H$-Z$ collisions in the form [8.68]

$$W_{\text{loss}}(\varrho, v) = C(\varrho, v)\exp\left[-\frac{2\varrho^2}{3Z} g\left(\frac{\varrho v}{Z}\right)\right], \tag{8.92}$$

$$C(\varrho, v) = 8.7\frac{\varrho^2}{v Z^{1/2}}, \quad \varrho v/Z \leqq 1 \tag{8.93a}$$

$$= 1.14\frac{Z^2}{\varrho^2 v^2}, \quad \varrho v/Z \geqq 1. \tag{8.93b}$$

The electron loss cross section is now given by

$$\sigma_{\text{loss}} = \int_0^\infty \{1 - \exp[-W_{\text{loss}}(\varrho, v)]\} 2\pi\varrho\, d\varrho = ZF(v/Z^{1/2}), \tag{8.94}$$

where the values of the universal function $F(y)$ are given in Table 8.5 for a sufficiently wide range of its argument.

Table 8.5. Values of the function $F(y)$ in (8.94)

y	$F(y)$	y	$F(y)$
0.002	55.7	0.2	35.2
0.004	52.1	0.4	19.4
0.008	48.5	0.8	6.6
0.01	47.3	1.0	4.8
0.02	43.7	2.0	2.1
0.04	40.2	4.0	1.0
0.08	37.0	8.0	0.5
0.10	36.2	10	0.36

From (8.94) it follows that the electron loss cross section for H+Z collisions can be represented in the reduced form

$$\tilde{\sigma}_{\text{loss}}(\tilde{E}) = F(\tilde{E}^{1/2}) = f(\tilde{E}), \tag{8.95}$$

$$\tilde{\sigma}_{\text{loss}} = \sigma_{\text{loss}}/Z, \quad \tilde{E} = E/Z, \tag{8.96}$$

where E is given in units of keV/amu.

8.4.2 Classical Treatments

In the intermediate-velocity region ($v \sim Z^{1/4}$), the electron loss process in the H+Z system can be successfully described by the methods of classical mechanics. Two approaches have been developed along this line: an analytical treatment, based on the overbarrier electron transitions [8.69] and the CTMC method [8.43].

The classical analytical model starts with the same assumptions as in the case of electron-capture overbarrier transitions (Sect. 7.3.3). When particles approach each other, the potential barrier separating the proton and ion wells becomes equal to the binding energy of the electron in its initial state. Representing the initial electron state as a classical ensemble of non-interacting particles (with a given distribution of the states in configuration space), one can show [8.69] that, after the barrier disappears, the only particles of the classical ensemble which can leave the proton well are those moving within a cone of aperture $\theta_m(R)$, defined by

$$\cos\theta_m(R) = \frac{R+2Z}{4(1+\sqrt{Z})} + \frac{1-Z}{R+2Z}(1+\sqrt{Z}) \tag{8.97}$$

where R is the internuclear distance. Setting $\theta_m = 0$, one obtains from (8.97) the maximum distance at which the removal of the electron from the proton well becomes possible,

$$R_m = 2(2\sqrt{Z} + 1). \tag{8.98}$$

The electron loss probability is given by the ratio of the particles moving within the above-defined cone to the total number of particles at the distance of closest approach $R = R_{min} = \varrho$, i.e.

$$W_{cl}(\varrho) = \tfrac{1}{2}[1 - \cos\theta_m(\varrho)]. \tag{8.99}$$

Integrating $W_{cl}(\varrho)$ over the impact parameters one obtains the classical cross section for electron loss [8.69]

$$\sigma_{loss}^{cl} = \pi \left\{ \frac{(2 + 2\sqrt{Z} - Z) R_m^2}{4(1+\sqrt{Z})} - \frac{R_m^3}{12(1+\sqrt{Z})} + (Z-1)(1+\sqrt{Z}) \left[R_m - 4Z \ln\left(1 + \frac{1}{\sqrt{Z}}\right)\right]\right\}. \tag{8.100}$$

In the limit of $Z^{1/2} \gg 1$, σ_{loss}^{cl} reduces to

$$\sigma_{loss}^{cl} = 4\pi(\tfrac{4}{3}Z - \tfrac{1}{5}). \tag{8.101}$$

With an increased collision velocity, not all of the particles can leave the Coulomb well of the proton. The velocity dependence of σ_{loss}^{cl} can be accounted for by using for W_{cl}, not its value at $R_{min} = \varrho$, but rather an expression of the form [8.69]

$$W_{cl}(\varrho) = \frac{1}{4\pi} \int_{-T_0/2}^{+T_0/2} [1 - \cos\theta_m(R)]\, dt \tag{8.102}$$

where $R = R(t)$ is the classical trajectory of the nuclei and

$$T_0 = \frac{2}{v}(R_m^2 - \varrho^2)^{1/2} \tag{8.103}$$

is the time during which the nuclei traverse the electron loss region R_m. By using (8.102) for $W_{cl}(\varrho)$, cross section calculations have been performed for the H+Z ($Z = 1, 2, 18$) systems, which agree well with the CTMC results [8.43].

Numerous CTMC electron loss cross section calculations have been performed for the H+Z ($1 \leqq Z \leqq 36$) system [8.43, 70], as well as for the He+Z ($1 \leqq Z \leqq 8$) system [8.71]. An effective charge for the He^+ ion core has been used in the latter case. Figure 8.11 shows some examples of CTMC electron loss cross sections.

The classical concept for the electron loss process allows one to obtain a simple scaling law for σ_{loss} with respect to both the ionic charge Z and the binding energy I_0 of the bound electron. By linking the low-energy expres-

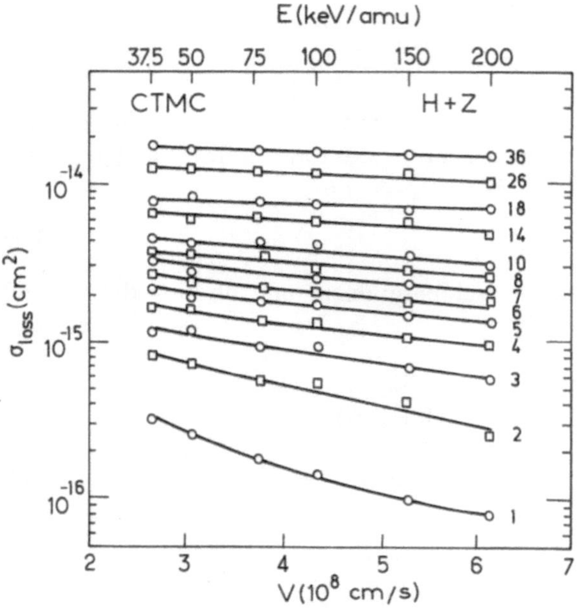

Fig. 8.11. Electron-loss cross sections for the H + Z collisions obtained by CTMC method [8.43]

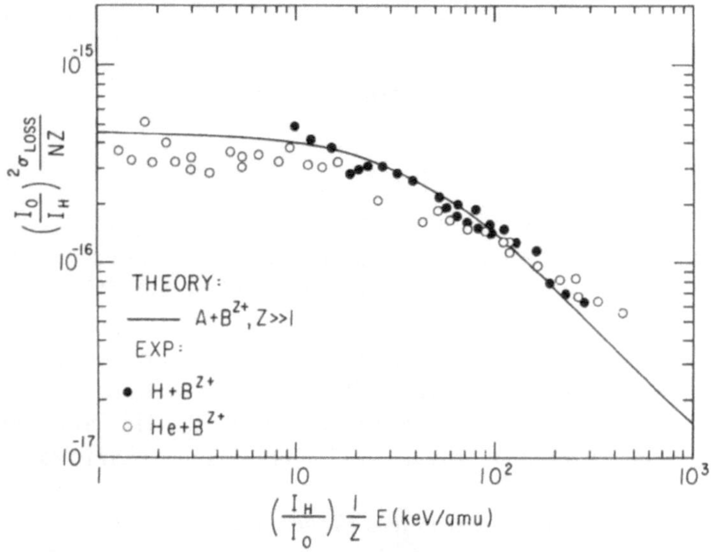

Fig. 8.12. Scaled electron-loss cross section $\tilde{\sigma}^{cl}_{loss}$ vs. scaled energy $\tilde{E}$ (——). Experimental data are for H + B^{Z+} (●) [8.6, 38]) and He + B^{Z+} (○) [8.39, 74]

sion (8.101) (which can easily be generalized to a non-hydrogenic case) with the high-energy asymptote of the BEA ionization cross section (8.72) (for $Z_0 = 1$), one obtains [8.72]

$$\tilde{\sigma}_{\text{loss}}^{\text{cl}} = \left(\frac{I_0}{I_{\text{H}}}\right)^2 \frac{\sigma_{\text{loss}}^{\text{cl}}}{NZ} = \frac{16\,\pi}{3} \left\{\frac{31.25}{\tilde{E}} \left(1 - \mathrm{e}^{-\tilde{E}/31.25}\right)\right\}, \tag{8.104}$$

$$\tilde{E} = \left(\frac{I_{\text{H}}}{I_0}\right) \frac{E\,[\text{keV/amu}]}{Z}, \tag{8.105}$$

where I_{H} is the hydrogen-atom ionization potential and N is the number of equivalent electrons in the shell from which the electron is removed. Relation (8.104) is valid for $\tilde{E} \gtreqless 1$ and $Z^{1/2} \gg 1$. It is intersting to note that (8.104), derived on purely analytical grounds, reproduces the CTMC electron loss results for H+Z and He+Z systems [8.43, 71, 73]. Figure 8.12 shows the dependence of $\tilde{\sigma}_{\text{loss}}^{\text{cl}}$ on the reduced energy $\tilde{E}$. Comparison is made with the experimental data on $\text{H} + \text{B}^{Z+}$ [8.6, 38] and $\text{He} + \text{B}^{Z+}$ [8.39, 74].

8.4.3 Stripping Reactions

The stripping reactions

$$\text{A} + \text{B}^{Z+} \rightarrow \text{A} + \text{B}^{(Z+1)+} + \text{e} \tag{8.106}$$

have been extensively studied in the past in connection with the stopping-power problem [8.49]. Reaction (8.106) is also frequently termed electron loss. On the basis of adiabatic considerations, it follows that the maximum of the stripping cross section lies at a collision velocity $v_{\text{m}} = \gamma Z$ ($\gamma \sim 1$), which for high Z belongs to the MeV energy region.

Theoretical treatments of stripping reactions are based mostly on the classical models [8.49] and Born approximation [8.75–77].

In the classical model, the electron removal from a one-electron projectile ion results from the energy transfer during its scattering on the target atom, considered as a system of a charge Z_0 and Z_0 electrons. The stripping cross section is [8.49]

– for light targets:

$$\sigma_{\text{s}}^{\text{Bohr}} = 4\,\pi \frac{1}{v^2} \left(\frac{Z_0^2 + Z_0}{Z^2}\right), \quad v \gg Z, Z > Z_0, \tag{8.107}$$

– for heavy targets:

$$\sigma_{\text{s}}^{\text{Bohr}} = \pi\, Z_0^{2/3} \frac{1}{v\, Z}, \quad v \gg Z, Z \lesssim Z_0^{1/3}. \tag{8.108}$$

The first term in (8.107) ($\sim Z_0^2$) comes from the elastic scattering on the target nucleus, whereas the second one ($\sim Z_0$) originates from collisions with target electrons. In the free-electron approximation for the target, the Born approximation gives [8.75]

$$\sigma_s = 4\pi \frac{1}{v^2 Z^2}\left[Z_0^{*2} + Z_0^* - \frac{Z_0^{*2} Z^2}{v^2}\left(\frac{1}{4} + \frac{1}{Z_0^*}\right)\right], \quad v \gg Z, Z > 2 Z_0^*, \tag{8.109}$$

where Z_0^* is the effective target nuclear charge, accounting for the screening effects. In the same limit, with a correct description of the target electrons, the Born approximation predicts [8.76]

$$\sigma_s = 4\pi \frac{1}{v^2 Z^2}\{Z_0^2[1 + 0.56 \ln(Z/2 Z_0^*)] + Z_0[1 + 0.56 \ln(Z/Z_0^*)]\}. \tag{8.110}$$

The experimental cross section data on the stripping reactions are rather numerous (for references, see [8.78, 79]). In Fig. 8.13, we give some results for the He + one-electron-ion systems to demonstrate the energy and charge dependence of σ_s [8.78–80].

Multiple stripping reactions have also been investigated experimentally [8.81, 82]. We note that σ_s, as a function of Z_0, shows a non-monotonic behaviour, reflecting the shell structure of the target [8.81, 82].

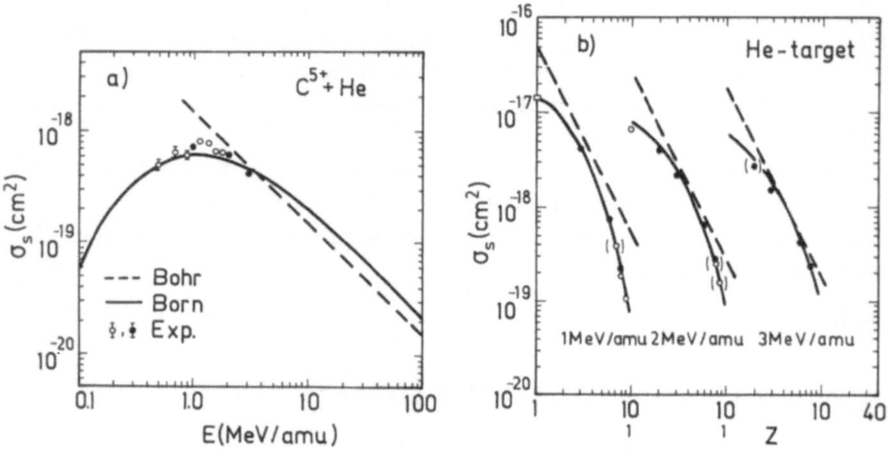

Fig. 8.13 a, b. Cross sections for stripping reactions. (**a**) Energy dependence of σ_s for the C^{5+}+He collisions. (**b**) The Z-dependence of σ_s for same He + one-electron-ion collision systems at three energies. Full curves-Born approximation [8.78], dashed curves-Bohr classical model, (8.107); experimental data are from [8.78] (●), [8.79] (○) and [8.80] (□)

9. Auger, Inner-Shell and Related Processes

The collision system consisting of a highly charged ion and an atom represents a super-excited quasi-molecular complex. The large amount of potential energy stored in such a system can be disseminated at low collision energies through a variety of electron rearrangement processes. Most of these processes involve correlated transitions of two or more electrons. For sufficiently high collision energies, the collision partners may come close to each other and electrons from the inner shells become involved in the collision process. Creation of inner-shell vacancies, with subsequent emission of x-rays and/or Auger electrons, is a prominent feature of such collisions. At still higher energies, non-characteristic (continuous) x-ray emission may also be observed. In addition, if the sum of projectile and target nuclear charges exceeds some critical value Z_{cr}, appearance of specific quantum-electrodynamical phenomena may be expected.

In this chapter we shall consider in more detail the above-mentioned processes occurring in atom-highly charged ion collisions. Particular attention will be paid to correlated electron processes involving electron transitions from both the outer and inner shells.

Theoretical descriptions of these processes are much less developed in comparison with the processes discussed in the preceding two chapters. This circumstance will necessarily restrict some of our discussions to a qualitative level. The typical inner-shell processes (such as inner-shell vacancy creation, K x-ray or Auger electron emission, etc.) will be considered only briefly, since they are already well represented in the monograph literature [9.1, 2]. Since they are not specific to the current atomic physics of highly charged ions, questions connected with the processes in super-critical electric fields [$(Z \gtrsim Z_{cr})$ will also be discussed only qualitatively.

9.1 Collisional Auger Processes Involving Outer Shells

Processes in which correlated transitions of two or more electrons take place and which lead to a non-radiative relaxation of the "super-excited" collision complex are called Auger processes. In this section we restrict ourselves to collisional Auger processes in which only electrons from the outer target shells are involved. The processes considered always include capture of one or more target electrons by the ion, with simultaneous or subsequent transitions of other target electrons (or captured electrons) in the discrete or continuous spectrum (with conservation of total electronic energy). For this reason, these Auger processes are frequently referred to as capture-excitation and capture- (or transfer-)ionization [9.3]. The capture-ionization may proceed via several different channels (see Sect. 9.2) and may involve transitions of more than two target electrons (Sect. 9.3). The Auger processes have recently been briefly reviewed [9.4]. Below, we consider these processes in more detail.

9.1.1 Capture-Excitation

The process of electron capture with excitation of a target electron may proceed via two channels

– simultaneous capture and excitation:

$$A + B^{Z+} \rightarrow A^{+*} + B^{(Z-1)+}, \qquad (9.1\,a)$$

– capture into an excited state followed by resonant excitation transfer:

$$A + B^{Z+} \rightarrow A^{+} + B^{(Z-1)+*} \rightarrow A^{+*} + B^{(Z-1)+}. \qquad (9.1\,b)$$

The one-step process (9.1 a) involves a two-electron coupling which combines a one-electron exchange interaction and an excitation matrix element. The two-step process (9.1 b) involves a one-electron exchange interaction in the first step and a two-electron exchange interaction in the second one. The two-electron coupling in the process (9.1 a) and the exchange interactions in the process (9.1 b) usually operate in well localized regions of internuclear separations and are associated either with a Landau-Zener pseudocrossing or with a region where the exchange interaction equals the energy difference of two parallel (or nearly parallel) diabatic potentials (Demkov's nonadiabatic coupling region). The one-electron, $\Delta_{1c}(R)$, and two-electron, $\Delta_{ex,tr}(R)$, exchange interactions involved in the process (9.1 b) are well known: expressions for $\Delta_{1c}(R)$ for the $A+B^{Z+}$ system are presented in Chap. 6, and for $\Delta_{ex,tr}(R)$ in the monograph by *Smirnov* [9.5]. The two-electron coupling interaction $\Delta_{c,ex}(R)$ involved in the reaction (9.1 a) can be calculated in an analogous manner to the two-electron exchange interaction responsible for the simultaneous two-electron-capture process (Sect. 7.7.1). The result for the interaction $\Delta_{c,ex}(R)$ has the form [9.6]

$$\Delta_{\mathrm{c,ex}}(R) = DR^{(Z+1)/\alpha_1-1-m_1}\,\mathrm{e}^{-\alpha_1 R}, \quad \alpha_1 R \gg 1, \tag{9.2}$$

$$D = 2\langle \varphi_{1a}(\boldsymbol{r}_{1a})\,\varphi_{ab}(\mu, \nu, \phi_{1b})\,|\,V_{\mathrm{d}}\,|\,\varphi_{2a}(\boldsymbol{r}_{1a})\,\varphi_{\mathrm{b}}(\boldsymbol{r}_{1b})\rangle, \tag{9.3}$$

$$V_{\mathrm{d}} = \boldsymbol{r}_{1b}\,\boldsymbol{r}_{2a} - 3(\boldsymbol{n}\cdot\boldsymbol{r}_{1b})(\boldsymbol{n}\cdot\boldsymbol{r}_{2a}), \quad \boldsymbol{n} = \boldsymbol{R}/R. \tag{9.4}$$

In (9.2–4), the indices "1" and "2" are associated with the transferred and excited electron, respectively, "a" and "b" refer to the centres A and B^{Z+}, φ_{1a}, φ_{2a} and φ_b are the unperturbed wave functions of the corresponding one-electron atomic (ionic) states, and $\varphi_{ab}(\mu, \nu, \phi_{1b})$ is the perturbed wave function of the transferred electron in the vicinity of ion B^{Z+} [9.7], μ, ν, ϕ_{1b} being the parabolic coordinates with respect to this ion.

A characteristic feature of the interaction $\Delta_{\mathrm{c,ex}}(R)$ is that its R-dependence is determined only by the first ionization potential $\alpha_1^2/2$ of atom A, the ionic charge Z and the magnetic quantum number of the initial electron state. The coefficient D in (9.2) contains matrix elements of the dipole type and for any $A+B^{Z+}$ pair can easily be calculated. The factor D is different from zero only for dipole-allowed transitions of the "second" electron, "2". The expression (9.2) for $\Delta_{\mathrm{c,ex}}(R)$ is valid for $\alpha_1 R \gg 1$.

Evidence for the occurrence of capture-excitation processes (9.1) has been given by *Winter* et al. [9.8] in the collisions of Ne^{Z+} $(Z = 1-4)$ with a He target, at an energy of 100 keV. Emission from the He^{+*} $(4\,^2F^o)$ state has been observed at 4686 Å $(4\,^2F^o \rightarrow 3\,^2D$ transition). It was found that the emission cross section increases monotonically when Z increases from $Z = 2$ to $Z = 4$. The inspection of diabatic potential energy diagrams indicates that both (9.1 a) and (9.1 b) contribute to the population of the $4\,^2F^o$ state of He^{+*}. In the case of $Ne^{4+}+He$ collisions, both a Landau-Zener and a Demkov coupling mechanism control the formation of He^{+*}, whereas in the case of $Ne^{3+}+He$ collision, the production of He^{+*} is dominated by the two-step exchange reaction (9.1 b).

If the captured electron in the capture-excitation process originates from a non-valence electronic shell (or subshell), the excited target ion is left in an autoionizing state, A^{+**}. Such a process can be observed by the detection of Auger electrons emitted during the decay of the doubly excited A^{+**} ion. This type of Auger excitation has been observed in the $Ne^{4+}+Ne$ collisions at an energy of 100 keV [9.9]:

$$Ne + Ne^{4+} \rightarrow Ne^{+**}(2s\,2p^5\,nl) + Ne^{3+}. \tag{9.5a}$$

Formation of autoionizing product target ion states may be achieved also by the two-step capture-excitation process (9.1 b), as in the reaction

$$\begin{aligned} Ne + Ne^{3+} &\rightarrow Ne^{+}(2s\,2p^6) + Ne^{2+*} \\ &\rightarrow Ne^{+**}(2s\,2p^5\,nl) + Ne^{2+}. \end{aligned} \tag{9.5b}$$

Auger electron spectra have been observed from decay of doubly excited Ne^{+**} ($2s\,2p^5\,n\,l$) ions, created in the 100 keV collisions of Ne^{3+} with Ne [9.9].

9.1.2 Capture-Ionization

The process of capture- (or transfer-)ionization in atom-highly charged ion collisions may occur via the following channels

– proper Auger ionization:

$$A + B^{Z+} \rightarrow A^{2+} + B^{(Z-1)+} + e\,, \tag{9.6}$$

– electron capture followed by Penning ionization:

$$A + B^{Z+} \rightarrow A^{+} + B^{(Z-1)+*} \rightarrow A^{2+} + B^{(Z-1)+} + e\,, \tag{9.7}$$

– two-electron capture into an autoionizing projectile ion state, followed by non-radiative relaxation:

$$A + B^{Z+} \rightarrow A^{2+} + B^{(Z-2)+**} \rightarrow A^{2+} + B^{(Z-1)+} + e\,, \tag{9.8}$$

– electron capture followed by electron promotion into an autoionizing projectile state:

$$A + B^{Z+} \rightarrow A^{+} + B^{(Z-1)+*} \rightarrow A^{2+} + B^{(Z-2)+**} \rightarrow A^{2+} + B^{(Z-1)+} + e\,. \tag{9.9}$$

The processes (9.6–9) are possible only if the recombination energy of the projectile ion exceeds the threshold for double ionization of the target; they can be called "exothermic ionization". However, they do not require, in principle, kinetic energy from the relative motion and, therefore, can occur at very low energies.

Kishinevskii and *Parilis* [9.10] were the first to consider the capture-ionization processes (9.6–8) as decay channels of the atom-multicharged ion autoionizing system. Within such a concept they were able to explain the only ionization experimental data available at that time for atom-multi-charged ion systems, namely $Xe + Ne^{Z+}$ ($Z \leq 4$) [9.11]. The possibility of the decay of an autoionizing state of the $A + B^{Z+}$ system was introduced into the treatment by a factor

$$w_k(\varrho) = \exp\left[-\int_{-t_1}^{t_1} W_k(R(t)\,dt\right], \tag{9.10}$$

where k indicates the decay channel in question, i.e. one of the reactions (9.6–8), $W_k(R)$ is the corresponding decay probability per unit time and

$[-t_i, t_i]$ is the time interval in which the decay channel is operative. Ascribing indices I, II and III to the decay channels (9.6–8), respectively, and assuming that the diabatic crossing points R_1 and R_2 for single- and two-electron transfer are ordered as in Fig. 7.19a (i.e., $R_2 > R_1$), one obtains the following expressions for the ionization probabilities P_{I}, P_{II} and P_{III} [9.10]:

$$P_{\mathrm{I}}(\varrho) = p_1 p_2 (1 - w_{\mathrm{I}}), \tag{9.11}$$

$$P_{\mathrm{II}}(\varrho) = p_2 (1 - p_1)(1 - w_{\mathrm{II}}), \tag{9.12}$$

$$P_{\mathrm{III}}(\varrho) = (1 - p_2)[1 - w_{\mathrm{III}}(1 - p_2)] + p_2 (1 - p_2)[p_1^2 w_{\mathrm{I}} + (1 - p_1)^2 w_{\mathrm{II}}], \tag{9.13}$$

where p_i $(i = 1, 2)$ is the corresponding single-crossing Landau-Zener probability at the crossing point R_i.

For the calculation of probabilities p_1 and p_2, the respective one- and two-electron exchange interactions are required. The determination of these quantities has been discussed in Sects. 6.1.3 and 7.7.1, respectively. For the calculation of the decay factors $w_{\mathrm{I-III}}$ one needs the Auger transition probabilities $W_{\mathrm{I-III}}(R)$. Since the two-electron transition is a result of the electron-electron interaction r_{12}^{-1}, $W_{\mathrm{I-III}}(R)$ are given by

$$W_{\mathrm{I}}(R) = 2\pi |\langle \psi_{\mathrm{B}}(1) \psi_{\mathrm{E}}(2) | r_{12}^{-1} | \psi_{\mathrm{A}}(1) \psi_{\mathrm{A}}(2) \rangle|^2 g_{\mathrm{f}}, \tag{9.14}$$

$$W_{\mathrm{II}}(R) = 2\pi | V_{\mathrm{II}}^{(1)} + V_{\mathrm{II}}^{(2)} |^2 g_{\mathrm{f}}, \tag{9.15}$$

$$V_{\mathrm{II}}^{(1)} = \langle \psi_{\mathrm{B}}(1) \psi_{\mathrm{E}}(2) | r_{12}^{-1} | \psi_{\mathrm{B}^*}(1) \psi_{\mathrm{A}}(2) \rangle, \tag{9.16a}$$

$$V_{\mathrm{II}}^{(2)} = \langle \psi_{\mathrm{B}}(1) \psi_{\mathrm{E}}(2) | r_{12}^{-1} | \psi_{\mathrm{A}}(1) \psi_{\mathrm{B}^*}(2) \rangle, \tag{9.16b}$$

$$W_{\mathrm{III}}(R) = 2\pi |\langle \psi_{\mathrm{B}}(1) \psi_{\mathrm{E}}(2) | r_{12}^{-1} | \psi_{\mathrm{B}^{**}}(1) \psi_{\mathrm{B}^{**}}(2) \rangle|^2 g_{\mathrm{f}}, \tag{9.17}$$

where ψ_{A} and ψ_{B} are the quasi-molecular wave functions of the electrons, which go over into the atomic wave functions as $R \to \infty$, and ψ_{E} is the wave function of the electron in the continuum. For brevity, the two-electron functions are written in non-symmetrized form. The quantity g_{f} in (9.14, 15, 17) denotes the density of final states. The matrix elements $V_{\mathrm{II}}^{(1)}$ and $V_{\mathrm{II}}^{(2)}$ are connected with the following two possibilities for process (9.7): (1) a direct emission into the continuum of an atomic electron, and (2) an "exchange" process, with the emission of an electron from the ion.

For large internuclear distances the product $\psi_{\mathrm{A}}(1) \cdot \psi_{\mathrm{B}}(1)$ in (9.14) is localized around the ion B^{Z+}, since ψ_{B} describes an energetically deeper state, whereas the product $\psi_{\mathrm{A}}(2) \cdot \psi_{\mathrm{E}}(2)$ is localized in the vicinity of the atom A. Therefore, the region in the six-dimensional configuration space of electronic coordinates which predominantly determines the matrix element in (9.14) is the one where r_{1b} and r_{2a} are small. For such an electronic configuration, the interaction $1/r_{12}$ can be expanded in powers of $1/R$, the largest term being given by (dipole approximation)

$$\frac{1}{r_{12}} = \frac{1}{R^3} [\boldsymbol{r}_{1b} \cdot \boldsymbol{r}_{2a} - 3 (\boldsymbol{r}_{1b} \cdot \boldsymbol{n}) (\boldsymbol{r}_{2a} \cdot \boldsymbol{n})] + \dots, \quad \boldsymbol{n} = \boldsymbol{R}/R\,. \tag{9.18}$$

Applying analogous arguments for the matrix elements contained in $W_{\rm II}$, and calculating the perturbed one-electron wave function $\psi_{\rm A}(1)$ in the vicinity of the ion $\rm B^{Z+}$, one can obtain the following expressions for $W_{\rm I-II}$ [9.10]:

$$W_{\rm I} = A_{\rm I} R^{-6} \mathscr{D}^2 (R)\,, \tag{9.19}$$

$$A_{\rm I} = 2^9 \omega^{-1} c\, \sigma_{\rm ph}(\omega)\, Z^5 (Z+\gamma)^{-2Z/\gamma-6} (Z-\gamma)^{2Z/\gamma-4}\,, \tag{9.20 a}$$

$$\frac{\gamma^2}{2} = I_{\rm A}^{(2)} + \frac{Z}{R}\,, \quad \omega = \tfrac{1}{2}(Z^2 - \gamma^2)\,, \tag{9.20 b}$$

$$V_{\rm II}^{(1)} = A_{\rm II}^{(1)} (\pi \omega R^3)^{-1} [c\, \sigma_{\rm ph}(\omega) f_{\rm B}]^{1/2}\,, \tag{9.21}$$

$$V_{\rm II}^{(2)} = A_{\rm II}^{(2)} \mathscr{D}(R)\,, \tag{9.22}$$

where $A_{\rm II}^{(1)}$ is a numerical factor of the order of unity, $A_{\rm II}^{(2)} \simeq 0.1-0.01$, $\sigma_{\rm ph}$ is the photoionization cross section, $f_{\rm B}$ is the oscillator strength for the transition of the ion $\rm B^{(Z-1)+*}$ to the ground state, and c is the velocity of light. The function $\mathscr{D}(R)$ is connected with the one-electron exchange effects and for $s-s$ transitions is given by [9.10]

$$\mathscr{D}(R) = A_1 \Gamma(1 - Z/\gamma)\, (2\gamma)^{Z/\gamma} R^{(Z-2)/\gamma-1} \mathrm{e}^{-\gamma R}\,, \tag{9.23}$$

where A_1 is the "normalization" constant in the asymptotic form of the initial-state one-electron wave function. In view of the discussions in Sect. 6.1.3, some amendments to the form of $\mathscr{D}(R)$ are necessary for highly charged ions.

The decay probability $W_{\rm III}$ of the autoionizing state $B^{(Z-2)+**}$ can be estimated by well-known methods (see Sect. 2.5), and is of the order $10^{-3}-10^{-4}$.

We note that the estimates of $W_k(R)$ at distances of the order of a few atomic units have a magnitude of about $10^{15}-10^{16}\,\rm s^{-1}$. These values are higher than the Auger transition probabilities in isolated atoms. With increasing R, the probabilities $W_{\rm I}$ and $W_{\rm II}$ decrease sharply due to both the exponential dependence of $\mathscr{D}(R)$ and the ω^{-3} dependence of $\sigma_{\rm ph}$. For $R \gg 1$, $V_{\rm II}^{(1)} \gg V_{\rm II}^{(2)}$; however, at small internuclear distances, where $\mathscr{D}(R) \sim 1$, $V_{\rm II}^{(2)}$ may exceed $V_{\rm II}^{(1)}$ by several times. It should be added that $V_{\rm II}^{(2)}$ differs from zero not only for the resonant but also for the metastable excited states of the ion $B^{(Z-1)+*}$. The dependence of W_k on the charge Z of the multi-charged ion is essentially connected with the energy dependence of V_k, as is seen from (9.20b).

The capture-ionization cross sections for the channels I–III can be obtained by integrating $P_{\mathrm{I-III}}$ over the impact parameters. Bearing in mind that the single-crossing probabilities p_i are given by

$$p_i = \exp\left(-\frac{s_i}{v_{r_i}}\right), \quad s_i = \frac{\pi}{2}\frac{\Delta_i^2(R_i)}{|\Delta F_i|}, \quad i = 1, 2, \tag{9.24}$$

where v_{r_i} is the radial velocity at the crossing point R_i, and representing W_{I} and W_{II} in the form

$$W_k = B_k \, \mathrm{e}^{-\beta_k R}, \quad k = \mathrm{I, II}, \tag{9.25}$$

one obtains [9.10] for the cross sections $\sigma_{\mathrm{I-III}}$,

$$\sigma_{\mathrm{I}} \simeq \frac{\pi}{\beta_{\mathrm{I}}^2}\exp\left(-\frac{s_1+s_2}{v}\right)\left[\ln\left(\frac{2.5\,B_{\mathrm{I}}}{\beta_{\mathrm{I}}\,v}\ln\frac{5\,B_{\mathrm{I}}}{\beta_{\mathrm{I}}\,v}\right)\right]^2, \tag{9.26}$$

$$\sigma_{\mathrm{II}} \simeq \frac{\pi}{\beta_{\mathrm{II}}^2}\exp\left(-\frac{s_2}{v}\right)\left[1-\exp\left(-\frac{s_1}{v}\right)\right]\left[\ln\left(\frac{2.5\,B_{\mathrm{II}}}{\beta_{\mathrm{II}}\,v}\ln\frac{5\,B_{\mathrm{II}}}{\beta_{\mathrm{II}}\,v}\right)\right]^2, \tag{9.27}$$

$$\sigma_3 \simeq 2\pi R_2^2\left\{\bar{w}_{\mathrm{III}}\,F_3\left(\frac{s_2}{v}\right)+\left(1-\frac{R_1^2}{R_2^2}\right)F_3\left(\frac{R_2\,s_2}{v\,(R_2^2-R_1^2)^{1/2}}\right)\right.$$
$$\left.+\left[\frac{1}{2}-E_3\left(\frac{s_2}{v}\right)\right](1-\bar{w}_{\mathrm{III}})\right\}, \tag{9.28}$$

where

$$E_3(\eta) = \int_1^\infty \exp(-\eta x)\,x^{-3}\,dx, \quad F_3(\eta) = E_3(\eta) - E_3(2\eta), \tag{9.29}$$

and $\bar{w}_{\mathrm{III}}$ is a certain mean value of W_{III}. In order to illustrate the velocity dependence and the typical magnitudes of $\sigma_{\mathrm{I-III}}$, an example is presented in Fig. 9.1, obtained with the following values of the parameters of the problem: $s_1 = 3\times10^7\ \mathrm{cm\,s^{-1}}$, $s_2 = 10^6\ \mathrm{cm\,s^{-1}}$, $R_1 = 2\ \text{Å}$, $R_2 = 2.5\ \text{Å}$, $B_{\mathrm{I}} = 6\times10^{17}\ \mathrm{s^{-1}}$, $B_{\mathrm{II}} = 2\times10^{17}\ \mathrm{s^{-1}}$, $\beta_{\mathrm{I}} = 5\ \text{Å}^{-1}$, $\beta_{\mathrm{II}} = 4\ \text{Å}^{-1}$, $\bar{w}_{\mathrm{III}} = 5\times10^{13}\ \mathrm{s^{-1}}$. The cross sections σ_{I} and σ_{II} reach their maximum values at velocities several times greater than s_1 and s_2. The cross section σ_{III} does not decrease at small velocities. It is seen from the figure that the typical values of capture-ionization cross sections are of the order of $10^{-15}-10^{-16}\ \mathrm{cm^2}$. An analysis of (9.26–28) shows that with increasing Z, the cross sections $\sigma_{\mathrm{I-III}}$ also increase.

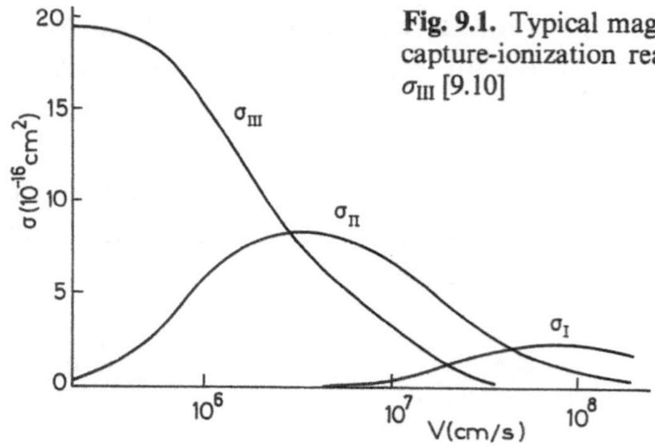

Fig. 9.1. Typical magnitudes and velocity behaviour of the capture-ionization reactions (9.6): σ_{I}, (9.7): σ_{II}, and (9.8): σ_{III} [9.10]

An important question in connection with Auger ionization processes is the energy spectrum of the ejected electrons. The number of Auger transitions in a spherical layer dR with radius R for a unit flux in the kth channel is

$$dN_k(R) = 2\pi \int_0^R w_k(\varrho, R)\, \varrho\, d\varrho\, W_k(R)\, dR = 2\pi\, I_k(R)\, W_k(R)\, dR\,. \tag{9.30}$$

The finite lifetime of the initial state causes a spectral broadening represented by the Breit-Wigner function

$$f(E_k', E_k) = \frac{1}{2\pi}\frac{W_k}{(E_k' - E_k)^2 + (W_k/2)^2}\,, \tag{9.31}$$

where $E_k = E_k(R)$ is the energy difference between the potential energy of the initial state and the state corresponding to the channel k. The energy distribution of the ejected electrons is now given by

$$\frac{dN_k}{dE_k} = 2\pi \int I_k(R)\, W_k(R) \left|\frac{dR}{dE_k'}\right| f(E_k', E_k)\, dE_k'\,. \tag{9.32}$$

If $E_k(R)$ possesses a minimum at a certain $R = R_{\mathrm{m}}$ (which is usually the case), then $W_k(R)$ and dR/dE_k are maximal in its vicinity. Therefore for $E_k = E_k(R_{\mathrm{m}})$ the spectrum of the Auger electrons has a maximum, given by

$$\left(\frac{dN_k}{dE_k}\right)_{\max} = 2\pi\, I_k(R_{\mathrm{m}})\, W_k^{1/2}(R_{\mathrm{m}}) \left(\frac{d^2E_k}{dR^2}\right)_{R=R_{\mathrm{m}}}^{-1/2}. \tag{9.33}$$

Usually the energies $E_k(R)$ differ from each other as well as their minima. Therefore, in the total energy spectrum, the maxima corresponding to different decay channels should be distinguishable. *Kishinevskii* and *Parilis* [9.10] pointed out that beside the maximum described above, an additional peak in the energy spectrum for each channel may appear on the right side if the Auger transition has time to occur at large distances where $I_k(R)$ decreases rapidly at low velocities. Unlike the first maximum, which does not change its position with varying collision energy, this second maximum shifts to the left when the velocity increases.

The transfer-ionization processes may also be described in terms of the Fano theory [9.12] for the interaction of a bound state with a continuum (or continua) of states. Such an approach, in which the decaying character of one or all of the interacting states is explicitly taken into account, has been followed by a number of authors [9.13–17].

In an orthonormalized diabatic basis $\{\varphi_j, \omega\}$, where $|\varphi_j\rangle$ $(j = 1, 2, \ldots, N)$ are discrete states with energies E_{0j} and $|\omega\rangle$ are continuum states with energies ω, the Hamiltonian $H(t)$ of the non-stationary system is given by

$$H(t) = H_0(t) + \hat{V}(t) = \sum_{j=1}^{N} E_{0j}(t)\,|\varphi_j\rangle\langle\varphi_j| + \int d\omega\,\omega\,|\omega\rangle\langle\omega| + V(t)\,, \tag{9.34}$$

$$\langle\varphi_j|\varphi_k\rangle = \delta_{jk}\,,\quad \langle\varphi_j|\omega\rangle = 0\,,\quad \langle\omega|\omega'\rangle = \delta(\omega-\omega') \tag{9.35a}$$

and $V(t)$ is the interaction operator with

$$\langle\omega|V(t)|\omega'\rangle = 0\,. \tag{9.35b}$$

The solution of the time-dependent Schrödinger equation can be represented as

$$|\Psi\rangle = \sum_{j=1}^{N} a_j(t)\,|\varphi_j\rangle + \int d\omega\, b(\omega, t)\,\mathrm{e}^{-\mathrm{i}\omega t}\,|\omega\rangle\,. \tag{9.36}$$

Assuming that the continuum states of interest are not too close to the continuum edge [i.e., the corresponding integrals over $d\omega$ can be extended to the interval $(-\infty, \infty)$], and that matrix elements $\langle\varphi_j|V|\omega\rangle$ weakly depend on ω, the following coupled equations for the coefficients $a_j(t)$ can be derived [9.15]:

$$\mathrm{i}\frac{da_j}{dt} = \left[E_{0j}(t) - \frac{\mathrm{i}}{2}\Gamma_j(t)\right]a_j + \sum_{\substack{k=1\\(k\neq j)}}^{N}\left[V_{kj}(t) - \frac{\mathrm{i}}{2}(\Gamma_k\Gamma_j)^{1/2}\right]a_k\,, \tag{9.37}$$

$$V_{kj}(t) = \langle\varphi_k|\hat{V}(t)|\varphi_j\rangle\,,\quad \Gamma_j(t) = 2\pi\,|\langle\varphi_j|\hat{V}(t)|\omega\rangle|^2\,. \tag{9.38}$$

The amplitudes $b(\omega, t)$ are expressed in terms of $a_j(t)$ through the relation

$$b(\omega, t) = -\mathrm{i}\sum_{j=1}^{N}\int_{t_0}^{t}\left(\frac{\Gamma_j(t')}{2\pi}\right)^{1/2} a_j(t')\,\mathrm{e}^{\mathrm{i}\omega t'}\,dt'\,. \tag{9.39}$$

The lower limit t_0 in the integral in (9.39) is the moment at which the initial conditions are specified; for $t \leq t_0$ only the discrete states are populated.

In such a formulation of the problem, the interaction of the state $|j\rangle$ with the continuum is represented by the width $\Gamma_j(t)$, while the quantity $(\Gamma_k \Gamma_j)^{1/2}$ represents the interaction of the states $|k\rangle$ and $|j\rangle$ through the continuum. It should be noted that the energies $E_{0j}(t)$ in (9.37) are calculated from the lowest continuum. To the lowest order of adiabatic theory, the decay of the state $|\alpha\rangle$ takes place due to the interaction with its own continuum. Equation (9.37) then immediately gives

$$a_\alpha(t) = a_\alpha(t_0) \exp\left[-\mathrm{i}\int_{t_0}^{t} E_\alpha(t')\,dt'\right], \tag{9.40}$$

where

$$E_\alpha(t) = E_{0\alpha}(t) - \frac{\mathrm{i}}{2}\Gamma_\alpha(t). \tag{9.41}$$

The amplitude of the continuum state $|\omega\rangle$ is now given by

$$b_\alpha(\omega) = -\mathrm{i}\,a_\alpha(t_0)\int_{t_0}^{t}\left(\frac{\Gamma_\alpha(t')}{2\pi}\right)^{1/2} \exp\left[\mathrm{i}\,\omega\, t' - \mathrm{i}\int_{t_0}^{t'} E_\alpha(t'')\,dt''\right] dt'. \tag{9.42}$$

The above integral can be calculated by the stationary-phase method, and the result is

$$b_\alpha(\omega) = -\mathrm{i}\,a_\alpha(t_0)\sum_n \left(\frac{1}{2}\right)\left[\frac{-\Gamma_\alpha(t)}{dE_\alpha/dt}\right]^{1/2}_{t=t_n} \exp\left[\mathrm{i}\,\omega\, t_n(\omega) - \mathrm{i}\int_{t_0}^{t_n(\omega)} E_\alpha(t')\,dt'\right] \tag{9.43}$$

where $t_n(\omega)$ are the stationary-phase points, determined from the equation

$$E_\alpha(t_n) = E_{0\alpha}(t_n) - \frac{\mathrm{i}}{2}\Gamma_\alpha(t_n) = \omega. \tag{9.44}$$

Summation in (9.43) is carried out over those points t_n through which the integration contour passes after its deformation according to the stationary-phase method. The coherent electron emission at different t_n (with the same energy ω) gives rise to interference effects in $b_\alpha(\omega)$. Since the difference of the phases in (9.43) depends on ω, oscillations appear in the energy spectrum $|b_\alpha(\omega)/a_\alpha(t_0)|^2$ of the emitted electrons.

Usually several decay channels $\alpha = 1, 2, \ldots, M$, are open for a given autoionizing state $|j\rangle$ in the course of the collision. Moreover, two quasi-discrete states $|j\rangle$ and $|k\rangle$ may be coupled via a common continuum α.

In such a situation the energy spectrum $W(\omega)$ of emitted Auger electrons is found by summing over all decay channels [9.17],

$$W(\omega)=\sum_{\alpha=1}^{M}\sum_{j,k=1}^{N} S_{jk}^{(\alpha)}\, b_{j\alpha}(\omega)\, b_{k\alpha}^{*}(\omega) \quad \text{where} \tag{9.45}$$

$$b_{j\alpha}(\omega)=-\mathrm{i}\int_{t_0}^{\infty}\left[\frac{\Gamma_{j\alpha}}{2\pi}\right]^{1/2} a_j(t)\exp\left[\mathrm{i}\,\omega\, t-\mathrm{i}\int^{t} E_\alpha(t')\,dt'\right]dt \tag{9.46}$$

and $S_{jk}^{(\alpha)}$ is the overlap of the proper continua of states $|j\rangle$ and $|k\rangle$. It is defined by the relation

$$\langle j\,\omega\,\alpha\,|\,k\,\omega'\,\beta\rangle = S_{jk}^{(\alpha)}\,\delta_{\alpha\beta}\,\delta(\omega-\omega')\,. \tag{9.47}$$

If $S_{jk}^{(\alpha)}=\delta_{kj}$, i.e. each of the states decays into its own continuum, $W(\omega)$ has the form

$$W(\omega)=\sum_{j=1}^{N}|\,b_{j\alpha}(\omega)\,|^2\,. \tag{9.48}$$

If the proper continua overlap, then interference of decay channels occurs.

The decaying character of the states involved in capture-ionization processes is also reflected in the form of two-state transition probabilities in the Landau-Zener or Demkov nonadiabatic coupling regions. The system of coupled equations (9.37) now reduces to

$$\mathrm{i}\frac{da_1}{dt}=\left[E_{01}(t)-\frac{\mathrm{i}}{2}\Gamma_1(t)\right]a_1+\left[V_{12}(t)-\frac{\mathrm{i}}{2}\Gamma_{12}\right]a_2\,, \tag{9.49a}$$

$$\mathrm{i}\frac{da_2}{dt}=\left[E_{02}(t)-\frac{\mathrm{i}}{2}\Gamma_2(t)\right]a_2+\left[V_{21}(t)-\frac{\mathrm{i}}{2}\Gamma_{12}\right]a_1\,, \tag{9.49b}$$

$$\Gamma_{12}=(\Gamma_1\Gamma_2)^{1/2}\,. \tag{9.49c}$$

The system (9.49) is not Hermitian; the quantity $d(|a_1|^2+|a_2|^2)/dt$ gives the total decay probability for both of the states.

The system (9.49) has been considered using different assumptions for the time variation of the functions E_{0i}, Γ_i $(i=1,2)$ and V_{12} [9.13–17]. The most extensive analysis is given in [9.14] where the collision situations corresponding to the Landau-Zener, Demkov and Nikitin models are considered. For the interaction model,

$$\left(E_{02}-\frac{\mathrm{i}}{2}\Gamma_2\right)-\left(E_{01}-\frac{\mathrm{i}}{2}\Gamma_1\right)\equiv\Delta E\,(1-\beta\,\mathrm{e}^{2\alpha t})-\frac{\mathrm{i}}{2}\Delta\Gamma\,, \tag{9.50a}$$

$$V_{12}=V_{21}=V_0\,\mathrm{e}^{\alpha t}\,,\quad t\le 0\,, \tag{9.50b}$$

$$\Gamma_{12}=0\,, \tag{9.50c}$$

where ΔE is the "resonance defect" and β and V_0 are constants ($\alpha \to -\alpha$ for $t > 0$). With the initial conditions

$$|a_1(t_0)| = 1\,, \quad t_0 \to -\infty\,, \tag{9.51 a}$$

$$|a_2(t_0)| = 0\,, \quad t_0 \to -\infty\,, \tag{9.51 b}$$

the probability $w_{12}(t)$ for the $1 \to 2$ transition after one pass through the strong coupling region ($t \sim 0$) is given by

$$w_{12}(t) = e^{\Gamma_1 t_0} \left(\frac{\Delta E\, \beta}{2\alpha}\right)^{-\Delta\Gamma/2\alpha} |\Gamma(\tfrac{1}{2} + q)|^2$$

$$\times \frac{1 - \exp(-2\,\mathrm{i}\,\pi p \,\mathrm{sgn}\,\{\Delta E\, \beta\})}{2\pi \exp(-\pi |\Delta E| \,\mathrm{sgn}\,\{\beta/2\alpha\})} e^{-\Gamma_2 t}\,, \tag{9.52}$$

$$p = -\mathrm{i}\, V_0^2/2\alpha\,\beta\,\Delta E\,, \quad q = \left(\frac{\Delta E}{2\alpha}\right)\left(\mathrm{i} + \frac{\Delta\Gamma}{2\Delta E}\right), \tag{9.53}$$

where $\mathrm{sgn}\,\{x\} = x/|x|$ and $\Gamma_{1,2}$ are considered to be independent of time. If the pseudocrossing region gives the main contribution to the transition probability, i.e. if the condition

$$|\Delta E/\alpha| \gg \max\,\{1, |V_0^2/\alpha\,\beta\,\Delta E|\} \tag{9.54}$$

holds, then expanding the gamma function in (9.52) for large values of the parameter q, one obtains

$$w_{12}(t) = \exp[-\Gamma_1 (t_c - t_0)] \left[1 - \exp\left(-\frac{\pi V_0^2}{|\Delta E|\,\alpha\,\beta}\right)\right]$$

$$\times \exp\left[\frac{\Delta E}{\alpha} \arctan\left(\frac{\Delta\Gamma}{2\Delta E}\right) - \frac{\Delta\Gamma}{2\alpha}\right] \exp[-\Gamma_2 (t - t_c)]\,, \tag{9.55}$$

where

$$t_c = -\left(\frac{1}{2\alpha}\right) \ln \left\{\beta \left[1 + \left(\frac{\Delta\Gamma}{2\Delta E}\right)^2\right]\right\} \tag{9.56}$$

is the moment of "pseudocrossing" of the complex energies. In the treatment of *Kishinevski* and *Parilis* [9.10], the decay factors $\exp(-\Gamma_k \tau_k)$ are introduced phenomenologically. For $\Gamma_{1,2} \to 0$, the expression (9.55) for w_{12} goes over into the Landau-Zener transition probability p_{LZ} for one passage of the curve-crossing region. The ratio w_{12}/p_{LZ} increases with increasing values of the parameters $\delta = \Delta\Gamma/|\Delta E|$ and $x = |\Delta E|/2\alpha$.

In the case when the main contribution to the transition is given by a region where the potential energy curves are parallel and when the resonance defect is small, i.e.

$$|V_0^2/\Delta E\,\alpha\,\beta| \gg \max\{1, |\Delta E|/\alpha\}, \tag{9.57}$$

the transition probability $\tilde{w}_{12}(t)$ after one passage through the strong coupling region is given by

$$\tilde{w}_{12} = \exp[-\Gamma_1(t_1 - t_0)]\frac{|\Gamma(1/2+q)|^2}{2\pi\exp(-\pi|\Delta E|/2\alpha)} \times \exp[-(\Gamma_1+\Gamma_2)(t_2-t_1)]\exp[-\Gamma_2(t-t_2)], \tag{9.58}$$

$$t_1 = \frac{1}{\alpha}\ln(2\alpha/V_0), \quad t_2 = \frac{1}{\alpha}\ln(V_0/|\Delta E|\beta). \tag{9.59}$$

In the time interval (t_1, t_2) the electron belongs to a molecular state which decays with a width $(\Gamma_1+\Gamma_2)$. If one sets $\beta = 0$ in (9.50a), the case considered above corresponds to the Demkov coupling of two decaying states. The transition probability after two passages of the transition zone $(t \sim \pm t_1)$, has the form

$$\tilde{W}_{12} = \exp[-\Gamma_1(t_1-t_0)]|\cos\pi q|^{-2}\sin^2(2V_0/\alpha) \times \exp[2(\Gamma_1+\Gamma_2)t_1]\exp[-\Gamma_2(t+t_1)], \quad t_1 = -(1/\alpha)\ln(V_0/2\alpha). \tag{9.60}$$

For $\Gamma_1 = \Gamma_2 = 0$, $\tilde{W}_{12}$ reduces to the Demkov formula

$$P_D = \operatorname{sech}^2(\pi\Delta E/2\alpha)\sin^2(2V_0/\alpha). \tag{9.61}$$

The ratio $\tilde{W}_{12}/P_D$ increases with increasing values of the parameters $\delta = \Delta\Gamma/|\Delta E|$ and $x = |\Delta E|/2\alpha$.

Another Nikitin-like model, considered by *Bazylev* et al. [9.14], is defined by

$$\left[E_{02}(t) - \frac{i}{2}\Gamma_2(t)\right] - \left[E_{01}(t) - \frac{i}{2}\Gamma_1(t)\right] \equiv \Delta E(1 - b\,e^{\alpha t}), \tag{9.62a}$$

$$V_{12} = V_{21} = V_0\,e^{\alpha t}, \quad t \le 0, \tag{9.62b}$$

$$\Gamma_{12} = U_0\,e^{\alpha t}, \qquad t \le 0, \tag{9.62c}$$

with $\alpha \to -\alpha$ for $t > 0$ and $b = \beta(1 + i\Delta\Gamma/2\Delta E)$. As can be seen from (9.62c), this model also takes into account the interaction of the states $|1\rangle$ and $|2\rangle$ through their common continuum. Exact solutions of (9.49) may be obtained within the model (9.62), allowing one also to consider the Landau-Zener and Demkov coupling cases as special limit situations. For example, the two-pass probability $\tilde{\tilde{W}}_{12}$ for the Demkov coupling case is given by

$$\tilde{\tilde{W}}_{12} = \exp[-(\Gamma_1 + \Gamma_2)/\alpha] \operatorname{sech}^2(\pi \Delta E/2\alpha) |\sin(2V_0/\alpha - \mathrm{i}\, U_0/\alpha)|^2 \quad (9.63)$$

which, for $\Gamma_1 = \Gamma_2 = U_0 = 0$, reduces to (9.61). The interaction of the states $|1\rangle$ and $|2\rangle$ through the continuum gives a non-vanishing charge-exchange probability even in the case when the direct electron exchange is zero ($V_0 = 0$); in this case one has

$$\tilde{\tilde{W}}_{12} = \exp[-(\Gamma_1 + \Gamma_2)/\alpha] \operatorname{sech}^2(\pi \Delta E/2\alpha) \sinh^2(U_0/\alpha)\,. \quad (9.64)$$

It is interesting to note that when Γ_1 and Γ_2 are close to each other ($\Gamma_1 = \Gamma_2$) and have very large values ($\Gamma_{1,2} \rightarrow \infty$), the capture probability is still finite and is given by

$$\tilde{\tilde{W}}_{12} = \tfrac{1}{4} \operatorname{sech}^2(\pi \Delta E/2\alpha)\,. \quad (9.65)$$

The above analysis of the nonadiabatic interaction of two quasi-stationary molecular states demonstrates that the theory of *Kishinevskii* and *Parilis* [9.10] for capture-ionization processes is incomplete due to the neglect of decay in the course of the electronic transition. This decay usually leads to an increase of the effective transition zone and thereby to an increase of the elementary transition probability. On the other hand, a theory which takes into account both the decaying character of the states (throughout the collision) and the specific dynamic mechanisms for each of the capture-ionization channels, has not been formulated as yet. A particular problem along this line would also be the determination of complex energies of interacting quasi-stationary states.

In experimental studies of transfer-ionization processes coincidence measurements of the projectile and target product charge states must be made as well as an analysis of the energy spectra of Auger electrons. The first measurements [9.8, 9, 18] have been performed without charge-state analysis of product ions, and information about the ionization channels has been extracted on the basis of either purely qualitative arguments [9.8, 18], or by analysis of the recorded Auger electron energy spectra and using rough diabatic potential energy diagrams [9.9]. For example, the total electron-production cross sections σ_e in the $Ne^{Z+} + He$ and $Ne^{Z+} + Ar$ ($Z = 1-4$) collisions for energies from 25 to 800 keV are shown in Fig. 9.2 [9.8]. The decrease of σ_e from $Z = 1$ to $Z = 3$ in the $Ne^{Z+} + He$ case can be attributed to the decrease of single-target-electron ionization, while the jump of σ_e for

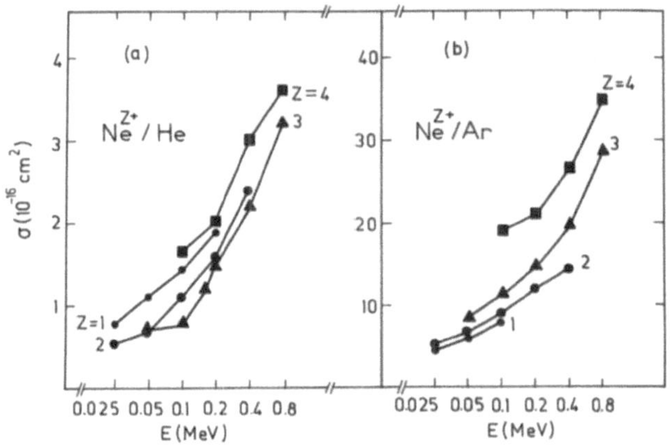

Fig. 9.2 a, b. Cross sections for production of slow electrons in collisions of Ne^{Z+} ($Z = 1-4$) ions with (**a**) He and (**b**) Ar [9.8]

the $Z = 4$ case results from the opening of the Auger ionization channels (9.6) and (9.7). The behaviour of the σ_e cross sections for $Ne^{Z+} + Ar$ collisions can also be explained on the basis of the increased role of capture-ionization processes for this system. The transfer-ionization channel (9.8) in the $Ne^{4+} + Ar$, Kr collisions at an energy of 100 keV [9.9] has been clearly identified from the Auger electron spectra and the corresponding diabatic potential energy curves.

Since the product states of both single- and double-electron-capture processes involved in the transfer-ionization channels (9.7) and (9.8) may also decay radiatively, it is of interest to investigate the ratio of radiative and non-radiative channels in these reactions. Figure 9.3 shows the experimental ratio

$$S = \sigma_A / \sigma_{CT} \qquad (9.66)$$

of the total cross section σ_A for all Auger processes to the total single-electron-capture cross section σ_{CT}, for $Ar^{Z+} + X$ (X = inert gas atom) collisions at 200 keV [9.18]. Since σ_{CT} also includes the radiative decay channels (as intermediate steps), Fig. 9.3 shows that with increasing Z, the role of non-radiative decay channels becomes increasingly important.

A more precise distinction between the transfer-ionization channels and the electron-capture and direct ionization processes can be achieved by coincidence charge-state analysis of the product ions in atom-highly charged ion collisions. Such measurements have recently been performed by *Cocke* and associates [9.19, 20] *Damsgaard* et al. [9.21] and by *Salzborn* and collaborators [9.22–24]. Figure 9.4 shows the transfer-ionization (TI) cross sections

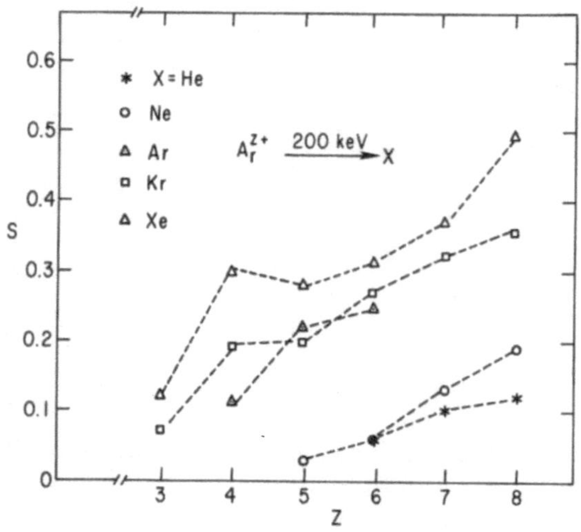

Fig. 9.3. Ratio S of the total cross section for all Auger processes to the total single-electron-capture cross section vs. projectile ionic charge in $Ar^{Z+} + X$ collisions ($Z = 1-8$, X = inert gas atom) [9.18]

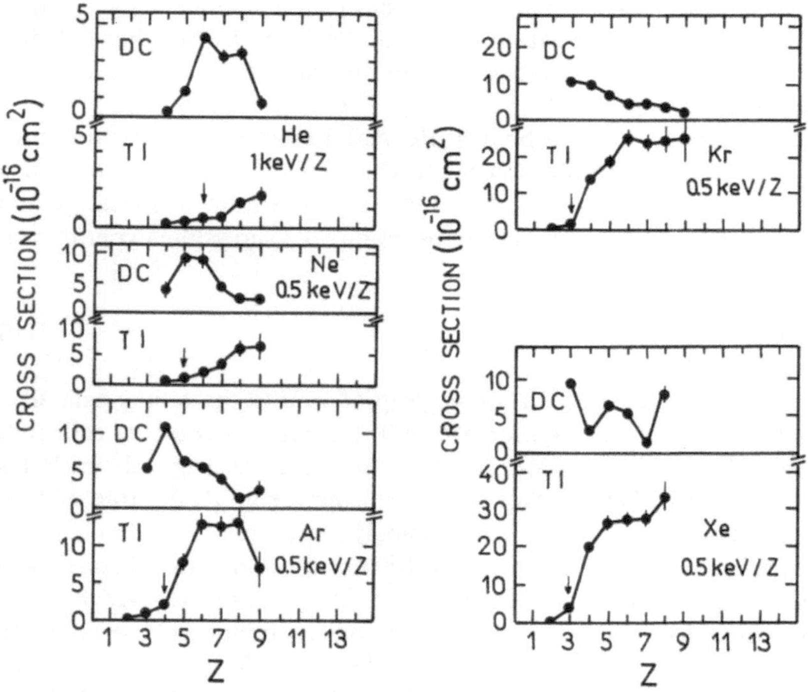

Fig. 9.4. Comparison of transfer-ionization (TI) and double-capture (DC) cross sections in $Ar^{Z+} + X$ collisions (X = inert gas atom) [9.19]

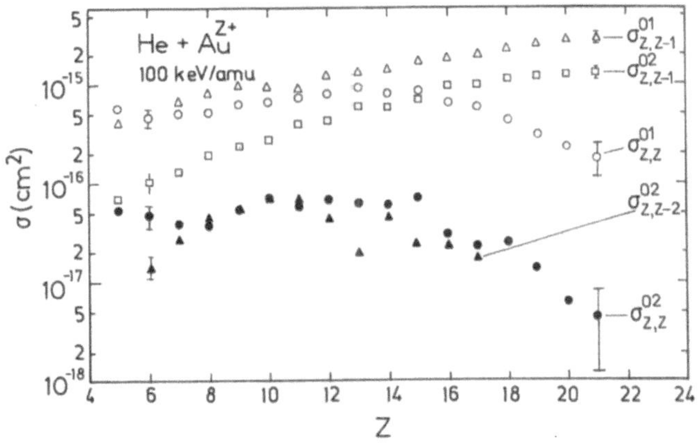

Fig. 9.5. Comparison of the transfer-ionization cross section $\sigma^{02}_{Z,Z-1}$ with the single- and double-electron-capture ($\sigma^{01}_{Z,Z-1}$ and $\sigma^{02}_{Z,Z-2}$), and single and double direct ionization ($\sigma^{01}_{Z,Z}$ and $\sigma^{02}_{Z,Z}$) cross sections for He + Au^{Z+} collisions at an energy of 100 keV/amu [9.21]

for Ar^{Z+} + X (X = inert gas atom) collisions as a function of Z, compared with the corresponding double-capture (DC) cross sections. There is an evident correlation between the values of the cross sections for these two processes: when the transfer-ionization channel (9.8) becomes exoergic (indicated by an arrow in the figures), its cross section begins to rise, while the corresponding double-capture cross section begins to fall. This behaviour indicates that the most effective transfer-ionization channel for these reactions (in the particular range of Z and collision energy) is the two-electron capture into an autoionizing projectile state, followed by a subsequent decay. Unfortunately, these experiments have not been suplemented by measurements of the Auger electron energy spectra to obtain more precise information on the mechanism of the transfer-ionization process.

A more complete product charge-state analysis of the ionization and electron-capture processes occurring in He+Au^{Z+} ($Z = 5-21$) collisions at an energy of 100 keV/amu has been performed by *Damsgaard* et al. [9.21]. Cross sections for single- and double-electron capture ($\sigma^{01}_{Z,Z-1}$ and $\sigma^{02}_{Z,Z-2}$), single and double direct ionization ($\sigma^{01}_{Z,Z}$, $\sigma^{02}_{Z,Z}$) and transfer-ionization ($\sigma^{02}_{Z,Z-1}$) have been measured in coincidence as a function of ion charge Z. The cross sections are shown in Fig. 9.5. Due to the wide range of variation of Z in this experiment, it is likely that more than one of the mechanisms (9.6–8) is responsible for producing the observed transfer-ionization cross section. For high Z values, Auger electrons originating from the decay of autoionizing projectile-product-ion states, have been recorded, giving evidence for the transfer-ionization channel (9.8).

9.1.3 Multiple Transfer Ionization

When the target atom contains many valence electrons, multiple transfer ionization

$$A^{Z+} + B \rightarrow A^{(Z-k)+} + B^{i+} + (i-k)\,e + \Delta E \tag{9.67}$$

becomes an important process at low and intermediate collision energies. With increasing projectile-ion charge, the potential energy brought into the system increases and the degree of internal excitation of the whole collisional complex is larger. The relaxation of such a highly excited system involves several types of multi-electron transition processes many of which lead to multiple transfer ionization (9.67). Some of the three-electron-transition Auger-type processes are listed below:

$$A^{Z+} + B \rightarrow A^{(Z-1)+} + B^{3+} + 2\,e, \tag{9.68a}$$

$$A^{Z+} + B \rightarrow A^{(Z-3)+**} + B^{3+} \rightarrow A^{(Z-1)+} + B^{3+} + 2\,e, \tag{9.68b}$$

$$A^{Z+} + B \rightarrow A^{(Z-2)+**} + B^{2+} \rightarrow A^{(Z-1)+} + B^{3+} + 2\,e, \tag{9.68c}$$

$$A^{Z+} + B \rightarrow A^{Z+} + B^{2+**} + 2\,e \rightarrow A^{Z+} + B^{3+} + 3\,e. \tag{9.68d}$$

Analogous processes involving four- or five-electron transitions can also occur. In analyzing the electron production data from $Ar^{Z+} + X$ ($X =$ inert gas atom) collisions at $E = 200$ keV, *Winter* et al. [9.18] have come to the conclusion that for $Z \geqq 6$ the multi-electron Auger-type processes contri-

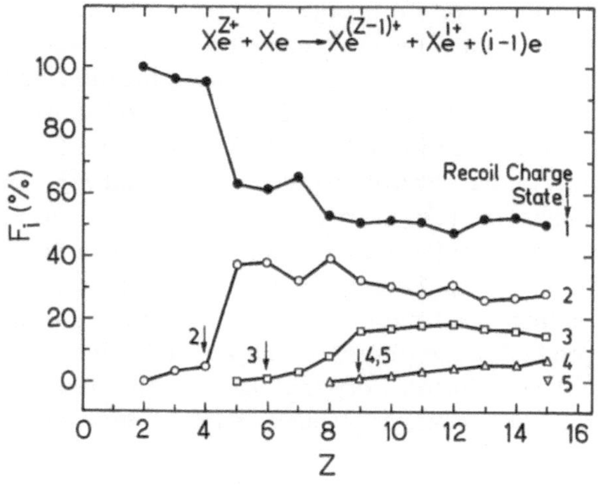

Fig. 9.6. Z-dependence of the recoil-ion fractions in multiple transfer-ionization $Xe^{Z+} + Xe$ collisions with one-electron capture [9.24]

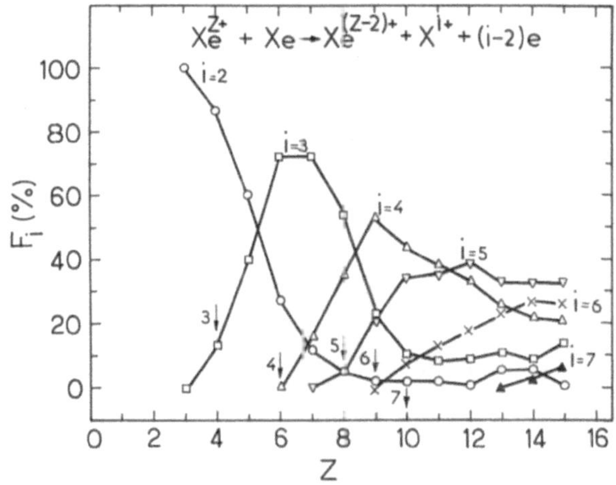

Fig. 9.7. Z-dependence of the recoil-ion fractions in multiple transfer-ionization $Xe^{Z+} + Xe$ collisions with two-electron capture [9.24]

bute significantly to the total electron production cross section. For Kr and Xe targets, the contribution of three- and more-electron-transition processes is larger than that of two-electron Auger processes. Recent charge-state coincidence measurements [9.21, 24] have shed more light on the dynamics of multiple transfer-ionization processes. Figures 9.6 and 7 show the Z-dependence of various recoil-ion fractions in the $Xe^{Z+} + Xe$ collisions of type (9.67), when, respectively, one and two electrons are captured by the projectile ($k = 1$ and $k = 2$) [9.24]. The values of the fractions do not depend appreciably on the collision energy in the interval from 25 to 150 keV. In the case of single-electron capture the fractions Xe^{i+} tend to flatten when Z increases (Fig. 9.6), and the pure electron capture always dominates, whereas in the case of two-electron capture flattening of the curves does not take place, and for $Z \geqq 6$ the pure two-electron capture becomes negligible. Similar curves to those presented in Figs. 9.6 and 7 have been obtained for other inert gas collision partners [9.24]. The pronounced similarity of the recoil-ion fractions indicates that a common mechanism governs the multiple transfer ionization process in these systems, which is not too sensitive to the details of their electronic structure. The only physical parameter which has been found to correlate with this experimental fact is the reaction exothermicity ($\Delta E \geqq 0$), which gives the excess of potential energy available in the system for ionization of target electrons. The values of the parameter ΔE for each of the transfer-ionization reactions is shown by an arrow in Figs. 9.6 and 7. The observed energy and species independence of the recoil-ion fractions suggests that a potential electron emission mechanism might be responsible for the multiple transfer ionization process.

9.2 Inner-Shell Processes

The inner-shell processes in ion-atom collisions have been an extremely broad and active field of research for more than two decades. The accumulated information on these processes is very abundant and represented in many review articles [9.1, 2, 25–30]. We do not intend here to go into a detailed description of inner-shell processes. Rather, we shall give a brief account of the basic concepts from the point of view of atom-highly charged ion collisions. A specific aspect of these collisions is that a highly stripped ion brings into the colliding system a large number of vacant, deeply lying electronic levels, so that the variety of processes associated with inner-shell electron transitions may be very large. In particular, correlated multi-electron transitions are very likely in such a situation. Further, if the collision energy is not too high, one can expect a pronounced coupling of the inner-shell processes with those involving outer-shell electrons. All these questions have not been much investigated and we shall therefore restrict our discussions to single-electron transition processes.

9.2.1 Basic Electron Transition Mechanisms

The occurrence of inner-shell processes presumes a violent collision in which the nuclei of the colliding particles approach close to each other (close collisions). However, the dynamical mechanism governing the inner-shell processes depends on the value of the parameter v/v_0, where v is the relative nuclear velocity and v_0 is the classical velocity of the electron which participates in the process. For $(v/v_0) \gg 1$, the dominant transition mechanism is the Coulomb interaction of the bound electron with the Coulomb field of the nucleus of the incident ion, which leads to direct transitions either into the unoccupied states in the target atom spectrum (direct excitation or ionization) or into unoccupied states of the projectile ion (electron capture). Direct processes within the target spectrum are often referred to as Coulomb processes. The conventional high-energy methods (quantum or classical) may be applied to describe direct electron transition processes. Coulomb ionization is the dominant direct process in this energy region. The simplest description of inner-shell ionization is provided by the classical impulse approximation (BEA) (see Sects. 6.4.1 and 8.3.6). General expressions for K- and L-shell direct ionization are also available in the first Born approximation both in its quantum [9.31] and semiclassical [9.32, 33] form, which include the electron screening effects. The electron capture from inner-shells in this energy region has been widely described by the Brinkman-Kramers approximation and some results are given in Sect. 7.4.1.

The mechanisms of inelastic inner-shell processes in the low-velocity region, $(v/v_0) \ll 1$, are much different from the ones described above. In this region, the collision may be considered as slow and a quasi-molecular picture of the collision dynamics is appropriate. Being energetically far from the outer-shell electrons and influenced strongly by the Coulomb field of the colliding nuclei, the inner-shell electrons may be successfully described within the independent-particle model. The one-electron molecular orbital (MO) concept (see Sect. 6.1.5) can be fully exploited in the formulation of collision dynamics. Within the MO model, inner-shell processes predominantly take place through strong interactions in the regions of internuclear distance R where two MO's are nearly degenerate. Since for inner-shell MO's these regions are usually well separated, the whole collision dynamics may be treated as a series of separable two-state close-coupling problems. The situation is illustrated in Fig. 9.8, where the MO energies for a charge asymmetric system are shown. The symbols "h" and "l" refer to the heavier and lighter collision partner, respectively. Only some of the strongly interacting states are shown. It can be seen that the principal mechanisms responsible for inelastic transitions of inner-shell electrons are the radial and rotational coupling between the degenerate, or nearly degenerate MO's, having appropriate symmetries.

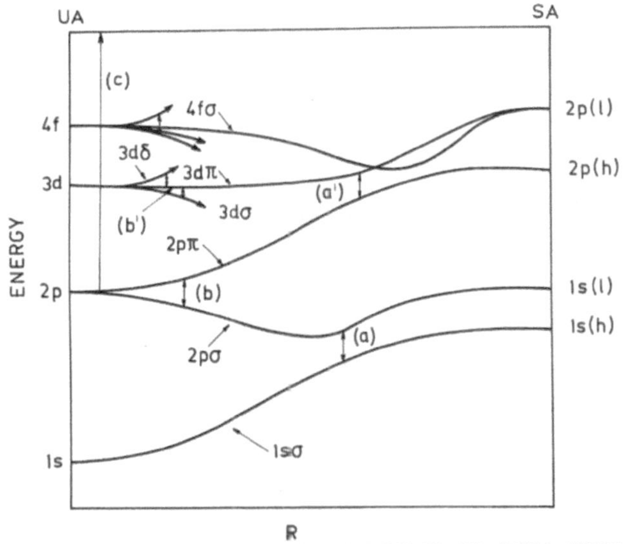

Fig. 9.8. Schematic MO energy diagram of the lower states of an asymmetric system ($Z_h > Z_l$). The regions where different transition mechanisms operate are shown: (*a*), (*a'*): radial coupling, (*b*), (*b'*): rotational coupling, (*c*): direct transitions to higher discrete states or into continuum

The one-electron MO picture, coupled with the localized character of electron transition zones and the phenomenon of state promotion (see Fig. 9.8 and Sect. 6.1.5) constitute the physical frame of the Fano-Lichten model [9.34, 35] for inelastic inner-shell processes. This model was able to explain the puzzling features of the energy loss experiments of *Afrosimov* and *Fedorenko* [9.36] and *Morgan* and *Everhart* [9.37], and continues to be the most useful key for the interpretation of modern experiments on inner-shell processes. Further, quantitative development of the electron promotion model for inner-shell processes has been made by *Briggs* and *Macek* [9.38, 39] and others [9.40].

9.2.2 K-, and L-Shell Excitations

Let us consider first the *K*-shell excitation of an atom by an ion which is stripped up to the *L* shell and has N_{2p} vacancies in the *L* shell. Assuming that the ion is heavier than the atom, the low-lying adiabatic energies of this system are qualitatively shown in Fig. 9.8. As is seen from the figure, excitation of one of the *K*-shell atomic electrons is possible by a rotational $2p\sigma - 2p\pi_x$ transition in the region close to the united atom limit. Let the probability of this rotational transition be $P_{\rm rot}$. If p_1 and p_2 are the probabilities that among the N_{2p} vacancies there are one and, respectively, two, $2p\pi_x$ vacancies available for the rotational transition, then the probabilities P_1 and P_2 for single and double *K*-shell excitation are, respectively, given by [9.38]

$$P_1 = p_1 P_{\rm rot} + 2 p_2 P_{\rm rot}(1 - P_{\rm rot}), \tag{9.69}$$

$$P_2 = p_2 P_{\rm rot}^2 . \tag{9.70}$$

The total cross section for *K*-shell excitation is, thus, given by

$$\sigma_k = 2\pi \int_0^\infty (P_1 + 2P_2)\, \varrho\, d\varrho = (p_1 + 2p_2)\, \sigma_{\rm rot} \equiv N_0\, \sigma_{\rm rot}, \tag{9.71}$$

where

$$\sigma_{\rm rot} = 2\pi \int_0^\infty P_{\rm rot}\, \varrho\, d\varrho \tag{9.72}$$

is the cross section for the $2p\sigma - 2p\pi_x$ transition, and $N_0 = p_1 + 2p_2$ is the $2p\pi_x$ vacancy number. For asymmetric systems $N_0 = (1/3)\, N_{2p}$. *Fastrup* et al. [9.40], argued that, in addition to the static vacancy number N_0, other vacancies $N(v)$ may be dynamically induced in the *L* shell in the course of the collision due to possible radial coupling processes involving the $2p\pi$ MO state [e.g., the transitions of type (a') in Fig. 9.8]. $N(v)$ depends on the collision velocity and its maximum value is two. Including $N(v)$ in (9.71) leads to the replacement $N_0 \to N_0 + N(v)$.

As follows from (9.71), the K-shell excitation problem is reduced to the calculation of $\sigma_{\rm rot}$. The proglem of $\sigma-\pi$ rotational transitions is investigated in detail both for symmetric [9.35, 38, 39, 41, 42] and asymmetric systems [9.43–46]. Assigning indices "1" and "2" to the $2p\sigma$ and $2p\pi_x$ states, the system of coupled equations for the problem is

$$\frac{da_1}{d\theta}=-f(R)\,a_2\,e^{-i\omega(\theta)}, \tag{9.73a}$$

$$\frac{da_2}{d\theta}=f(R)\,a_1\,e^{i\omega(\theta)}, \tag{9.73b}$$

where θ is the angle of rotation of the internuclear axis,

$$f(R)=\langle 2p\pi_x|\,i\,L_{y'}\,|2p\sigma\rangle$$

is the matrix element of the rotational operator and

$$\omega(\theta)=\int_0^\theta \frac{(E_2-E_1)}{\dot\theta}\,d\theta. \tag{9.74}$$

In (9.74), E_1 and E_2 are the adiabatic energies of the interacting states and $\dot\theta=\varrho v/R^2$ is the angular velocity. The equations (9.73a, b) are integrated between the limits $\theta=0$ and $\theta=\pi$ with the initial conditions

$$a_1(0)=1,\quad \alpha_2(0)=0. \tag{9.75}$$

The main contribution to the rotational coupling transitions comes from the region of R close to the united atom limit. The functions $f(R)$ and $\Delta_{21}(R)\equiv E_2-E_1$ in that region can be represented as [9.27]

$$f(R)\approx\frac{\varrho v}{R^2},\quad E_2-E_1\approx\alpha R^2, \tag{9.76}$$

where

$$\alpha=Z_1Z_2(Z_1+Z_2)^2/4C. \tag{9.77}$$

Taking into account that for any trajectory $\dot\theta=\varrho v/R^2$ and using θ as a new variable, (9.73), using (9.76) gets the form

$$\frac{da_1}{d\theta}=-\exp\left(-i\,\xi^3\int_0^\theta d\theta\,\sin^4\theta\right)a_2, \tag{9.78a}$$

$$\frac{da_2}{d\theta}=\exp\left(i\,\xi^3\int_0^\theta d\theta\,\sin^4\theta\right)a_1, \tag{9.78b}$$

where

$$\xi^3 = \alpha\,\varrho^3/v\,. \tag{9.79}$$

The transition probability $P_{rot} = |a_2(\theta_{max})|^2$ depends only on the parameter ξ. The cross section for rotational transitions can be represented as [9.43]

$$\sigma_{rot} = C(v/\alpha)^{2/3}, \quad \text{where} \tag{9.80}$$

$$C = 2\pi \int_0^\infty P_{rot}(\xi)\,\xi\,d\xi \tag{9.81}$$

is a universal constant which has to be obtained by numerical solution of the system of coupled equations (9.78). Its value is $C = 1.67$ [9.43, 46].

The system of coupled equations (9.78) may be solved in the adiabatic ($\xi \to \infty$) and sudden-perturbation ($\xi \to 0$) limits [9.46]. The results for the transition probability in the straight-line-trajectory approximation are

$$P_{rot}^{ad} = 2\exp\left(-\frac{8}{3}\frac{\alpha\,\varrho^3}{v}\right), \tag{9.82}$$

$$P_{rot}^{sa} = \frac{\pi^2}{\Gamma^2(4/3)}\left(\frac{2}{3}\right)^{2/3}\left(\frac{\alpha\,\varrho^3}{v}\right)^2. \tag{9.83}$$

In these two limits, the transition probability can also be obtained in closed form for the Coulomb-trajectory approximation. In this case, the result of the sudden-perturbation approximation is drastically changed in the region of small impact parameters, giving rise to a sharp "kinematic peak" at an angle of $\theta \approx 90°$ [9.38].

The above considerations show, see (9.79–83), that for the two-state rotational coupling problem one can introduce suitable reduced representations both for the variables ϱ and v, and for the quantities P_{rot} and σ_{rot}. The reduced probability and cross section, $\tilde{P}_{rot}$ and $\tilde{\sigma}_{rot}$, are

$$\tilde{P}_{rot}(\tilde{\varrho}, \tilde{v}) = P_{rot}(\lambda\,\varrho, v/\nu)\,, \tag{9.84}$$

$$\tilde{\sigma}_{rot}(\tilde{v}) = \sigma_{rot}(v/\nu)\,, \tag{9.85}$$

where λ and ν are given by [9.43]

$$\lambda = \{Z_1^2 Z_2^2 [(Z_1 + Z_2)/2]^4 M/(Z_A Z_B)\}^{1/7}, \tag{9.86}$$

$$\nu = \{Z_1 Z_2 [(Z_1 + Z_2)/2]^2 (Z_A Z_B/M)^3\}^{1/7}, \tag{9.87}$$

Z_A, Z_B being the nuclear charges of the colliding particles, $Z_1 \simeq (Z_A - 1)$, $Z_2 \simeq (Z_B - 1)$ are the screened charges, and M is the reduced mass of the

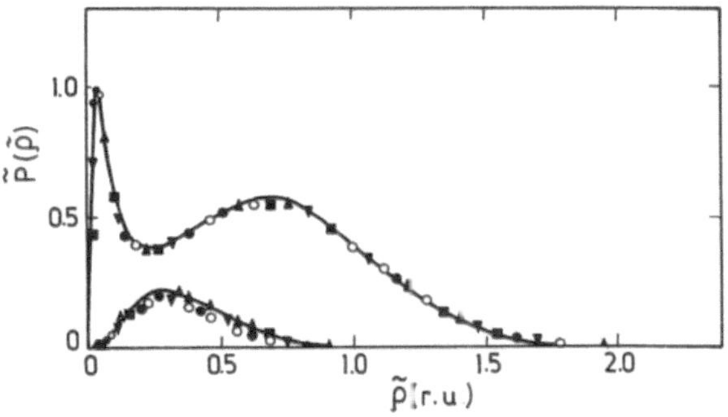

Fig. 9.9

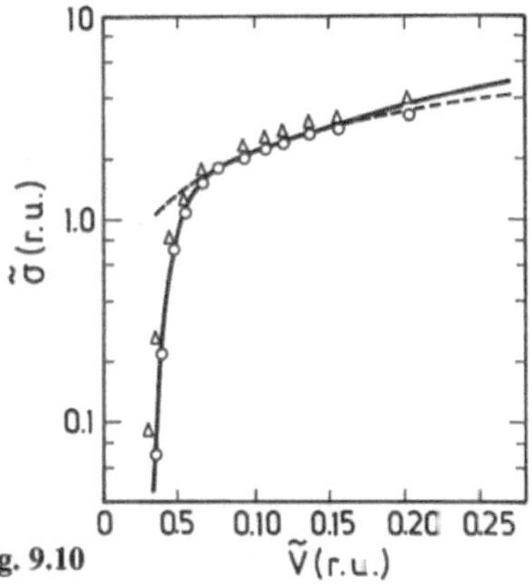

Fig. 9.10

Fig. 9.9. Reduced probability for $2p\sigma - 2p\pi$ rotational transitions vs. reduced impact parameter for two reduced velocities: $\tilde{v} = 0.1191$ r.u. (*upper curve*) and $\tilde{v} = 0.0377$ r.u. (*lower curve*). The full curves are the results of Hartree-Fock MO calculations for NeO. The symbols are the results of one-electron model calculations with $Q = Z_2/Z_1 = 1$ (○), 0.9 (●), 0.8 (▼), 0.7 (■), and 0.6 (▲), [9.45]

Fig. 9.10. Reduced cross section for rotational $2p\sigma - 2p\pi$ transitions vs. reduced collision velocity (——). Dashed curve is (9.80) in reduced form ($C = 1.67$). Open circles and triangles are coupled-state calculations for N-O and N-N systems, with Hartree-Fock MOs [9.45]

nuclei. *Taulbjerg* and *Briggs* [9.45] have determined that in the region of internuclear distances which contribute predominantly to rotational transitions, $R \lesssim 1$, the energy difference $\Delta_{12}(R)$ can also be represented in a reduced form. Since all quantities $[\omega, f(R), R(t)]$ in (9.73) scale to universal functions in the reduced units, the solution $a_2(\theta)$ will scale also. Figure 9.9 shows the reduced $2p\sigma - 2p\pi_x$ transition probability, $\tilde{P}_{\text{rot}}(\tilde{\varrho})$, for two values of the reduced velocity $\tilde{v}$ and different values of the ratio Z_2/Z_1. The scaled probabilities coincide within 10%. The "kinematic" peak is seen on the curve for the higher reduced velocity. The reduced cross section $\tilde{\sigma}_{\text{rot}}(\tilde{v})$ is shown in Fig. 9.10 (the full line) and within 10% coincides with the calculations for the N−N and N−O collisions performed on the basis of Hartree-Fock MO's. The universal functions $\tilde{P}_{\text{rot}}(\tilde{\varrho}, \tilde{v})$ and $\tilde{\sigma}_{\text{rot}}(\tilde{v})$ are tabulated in [9.45]. In the symmetric case ($Z_1 = Z_2 = Z$), the scaling of P_{rot} and σ_{rot} takes the particularly simple form

$$P_{\text{rot}}^{(Z)}(\varrho, v) \simeq P^{(1)}(Z\varrho, v/Z)\,, \tag{9.88}$$

$$\sigma_{\text{rot}}^{(Z)}(v) \simeq Z^{-2}\sigma_{\text{rot}}^{(1)}(v/Z)\,, \tag{9.89}$$

where the superscript "1" denotes the system with $Z = 1$.

The process of K-shell excitation through a $2p\sigma - 2p\pi$ rotational coupling in the united atom region leads to the creation of a vacancy (or two vacancies in the case of double excitation) in the $2\pi\sigma$ state and populates the $2p\pi$ state. In the further evolution of the colliding system, these two states may interact with the other states of the system (by some radial coupling mechanism, for instance; see regions a and a' in Fig. 9.8). In determining the final states of the reaction products, one has also to take into account these additional processes occurring on the outgoing part of the trajectory. At higher energies, there exist other mechanisms competing with the rotational coupling in the K-shell vacancy production (the direct process c in Fig. 9.8, for instance).

Despite certain unclear aspects, K-shell excitation is one of the best understood inner-shell processes, at least for not too heavy collision partners ($Z_A, Z_B \lesssim 15$). The excitations of L- and M-shells (when they properly are inner shells), are far less investigated processes within the electron promotion model. The excitation of the L-shell is associated with the promotion of the $4f\sigma$ MO. This orbital is very strongly promoted and crosses a large number of empty MO's over a short range of the internuclear distance, $R \lesssim R_c \sim 0.5$ a.u. For distances of closest approach smaller than R_c, one can assume a unit excitation probability and the L-shell vacancy production cross section may be written as [9.26]

$$\sigma_L = 4\pi R_c^2 (1 - E_c/E), \tag{9.90}$$

where E_c is the collision energy at which the distance of closest approach is equal to R_c at zero impact parameter. However, L-shell excitation may occur through many other excitation mechanisms [9.27].

The experimental studies of inner-shell excitation processes are numerous and are reviewed in [9.25, 26, 28, 30].

9.2.3 Vacancy Sharing

Referring to Fig. 9.8, it is obvious that in the absence of any additional electron transition mechanisms, the $2p\sigma$ vacancy created by the $2p\sigma - 2p\pi$ rotational coupling will end on the lighter partner after the collision. However, if there exists some other dynamical mechanism which couples the $2p\sigma$ and $1s\sigma$ states at large distances, then the $2p\sigma$ vacancy may be transferred to the K-shell of the heavier partner with a certain probability w. Thus, after the collision, the K-vacancy is "shared" by both collision partners. The cross section $\sigma_K(h)$ for producing a K-shell vacancy in the heavier partner by this vacancy-sharing mechanism is

$$\sigma_K(h) = w\,\sigma_K = w\,[\sigma_K(h) + \sigma_K(l)], \tag{9.91}$$

where σ_K is the total K-vacancy production cross section and $\sigma_K(l)$ is the cross section for vacancy production in the light partner. *Meyerhof* [9.47] assumed that the states $2p\sigma$ and $1s\sigma$ interact via a Demkov radial coupling. Solving the two-state coupled equations on the outgoing part of the trajectory, Meyerhof obtained

$$w_D = (1+e^{2x})^{-1}, \tag{9.92}$$

$$x = \frac{\pi\,\Delta\varepsilon}{2\,\gamma\,v_{R_c}}, \quad \gamma = \tfrac{1}{2}(\sqrt{2I_h} + \sqrt{2I_l}), \tag{9.93}$$

where $\Delta\varepsilon = |I_h - I_l|$ is the energy splitting of the $2p\sigma$ and $1s\sigma$ states at infinity, I_h, I_l are the ionization potentials of the K-shell electron in the heavier and lighter partner and v_{R_c} is the radial velocity at the distance R_c where the radial coupling is chiefly localized. The Rosen-Zener parameter x may be expressed in terms of the effective charges $Z^*_{h,l}$, so that for $2x$ one has

$$2x = \pi(Z^*_h - Z^*_l)/v_{R_c} = \pi\,\delta(Z_h - Z_l)/v_{R_c}, \tag{9.94}$$

where δ is some common average screening factor. For $3 \leq Z_{h,l} \leq 10$, $\delta = 0.87$ within 2%. Combining (9.91, 92), one can derive the following expression for the ratio R_K of the cross sections $\sigma_K(h)$ and $\sigma_K(l)$:

$$R_K = \frac{\sigma_K(h)}{\sigma_K(l)} = e^{-2x}. \tag{9.95}$$

The K-vacancy-sharing ratio R_K can be experimentally determined by measuring the K-LL Auger electron or x-ray emission yields in either of the collision partners. Most of the experimental findings are in accord with the

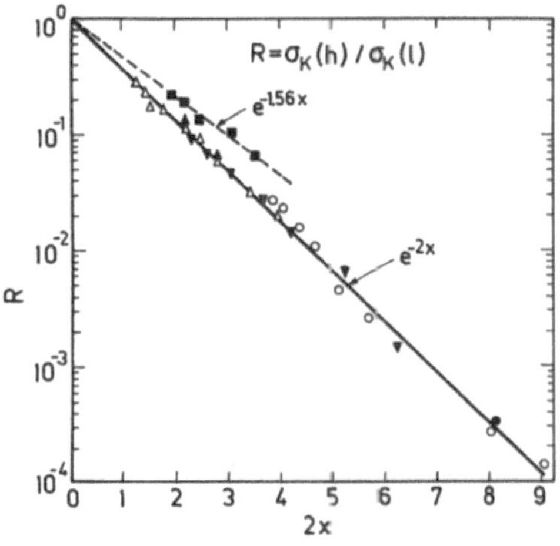

Fig. 9.11. K-vacancy-sharing ratio R. The full curve is the Meyerhof sharing; dashed curve is the best fit for $B + CH_4$ data; other symbols are various experimental data [9.28]

Meyerhof prediction (9.95), as illustrated in Fig. 9.11 [9.28]. For light collision partners, however, significant discrepancies between the experimental data and (9.95) have been found [9.30]. One example (for the $B+CH_4$ collision system) is given in Fig. 9.11 [the $\exp(-1.56\,x)$ curve]. This indicates that in such cases a mechanism different from Demkov coupling is responsible for the $2p\sigma-1s\sigma$ long-distance transitions. Another canditate for the description of the $2p\sigma-1s\sigma$ radial coupling transitions is the Nikitin model (see Sect. 6.2.4), as proposed recently [9.48]. The parameters of the model may be determined from MO calculations. The $2p\sigma-1s\sigma$ transition probability w_N in this case is given by [9.17]

$$w_N = \frac{\sinh[2\,x\sin^2(\theta/2)]}{\sinh(2\,x)}\exp[-2\,x\cos^2(\theta/2)], \tag{9.96}$$

where θ is a parameter in the Nikitin model. The K-shell vacancy-sharing ratio is now given by

$$R_K^{(N)}(\theta) = \frac{w_N}{1-w_N} = \frac{\exp\{2x[\sin^2(\theta/2)-\cos^2(\theta/2)]\}-e^{-2x}}{e^{2x}-\exp\{2x[\sin^2(\theta/2)-\cos^2(\theta/2)]\}}. \tag{9.97}$$

For $\theta=\pi/2$, $R_K^{(N)}(\theta)$ becomes equal to R_K, given by (9.95). A relation between the Nikitin parameter θ and the characteristics of the MO state energies has been established in [9.30].

In strongly asymmetric systems the K-vacancy may be shared not with the K-shell of the other collision partner, but rather with its L-shell. For this KL-vacancy sharing, the Nikitin model is frequently used to describe the long-distance radial coupling. Examples of KL-vacancy sharing, interpreted in terms of the Nikitin model, are found in the B-Ar and C-Ar systems [9.30]. Analogously, the L-vacancy, created rotationally in the $4f\sigma$ MO in heavier systems, may be shared with the $3d\sigma$ MO on the outgoing part of the trajectory. This LL-vacancy-sharing process has been considered in detail by *Meyerhof* et al. [9.49] in terms of the Demkov radial coupling.

The form of Auger-electron energy spectra after the vacancy-sharing process has been discussed by *Devdariani* et al. [9.17]. The spectra were calculated for both the Demkov and the Nikitin coupling mechanism.

9.2.4 Correlated Electron Transitions in Vacancy Production and Decay

Collisions of highly charged ions with atoms provide an efficient way of producing two or more vacancies in the inner electronic shells. Under favourable conditions, the nonadiabatic coupling mechanisms may provide cross sections for many-electron transition processes of the order of 10^{-16}–

10^{-18} cm^2. The double K-shell excitation probability via the $2p\sigma-2p\pi_x$ rotational coupling mechanism is given by (9.70). In symmetric systems, when the projectile is a completely stripped ion, a double K-shell vacancy may be created by a resonant two-electron-capture mechanism at distances far from the united atom limit. Other more complex mechanisms for double K-vacancy formation are also possible [9.50]. Double and multiple vacancies may also be created in the higher inner shells. When a double vacancy is created by a two-electron nonadiabatic transition mechanism, the correlation effects play a significant role and the independent-electron MO picture ceases to be valid.

The problem of double K-vacancy sharing has been considered by *Greenberg* et al. [9.50] in the case of Ni-Ni collisions.

If, at the end of the collision, a double K-vacancy is left in the many-electron target atom, its decay may be associated with either of the following processes: consecutive capture of two L electrons with emission of two Auger electrons or two $K-L$ x-ray photons, simultaneous capture of two L electrons with emission of one L Auger electron ($KK-LLL$ three-electron transition process), or with emission of one $K-L$ x-ray photon (two-electron-one-photon transition process, $KK-LL$, x process). The last two processes include strongly correlated electron transitions. They have recently been extensively studied both experimentally [9.51 – 54] and theoretically [9.55 – 60]. For the theoretical description of correlated two-electron-one-photon (or one Auger electron) processes, two different methods have been used: the shake-off model [9.55, 59], and the many-body (Feynman diagram) techniques [9.57, 58, 60]. While the shake-off model provides an easy way of estimating the transition probability, a comprehensive description of inter-electron correlations may be achieved only by using the quantum many-body methods.

9.3 Some Aspects of Quasi-Molecule and Quasi-Atom Formation in Heavy Ion Collisions

The recent development of heavy ion accelerators has made it possible to study ion-atom collision processes under physical conditions where the nuclei approach each other very closely. Under such conditions new physical phenomena are expected, the most remarkable of which is the transient formation of superheavy quasi-atoms.

Since the binding energy of the most strongly bound electrons increases rapidly with increasing nuclear charge, in close collisions of very high Z collision partners, the innermost electrons can adjust to the time-varying two-centre potential and a superheavy quasi-molecule (or quasi-atom) may transiently be created. The study of such objects may be based on the observable effects following their excitations: x-ray or Auger electron emission, δ electron emission, emission of non-characteristic (quasi-molecular) x rays, positron emission (for $Z_1 + Z_2 \gtrsim 170$), etc. This field of research is very rapidly developing at present, and its current status is presented in several recent reviews [9.61–65]. The leading lines of research in this area are the studies of quasi-molecular x-ray radiation and the search for electron-positron pair formation in heavy ion collisions. Below, we shall give a brief account of the main results.

9.3.1 Quasi-Molecular X-Ray Radiation

In 1972, *Saris* and co-workers [9.66] discovered continuous x-ray emission in Ar-Ar collision which they attributed to transitions between molecular orbitals of the transient colliding complex. The MO x-ray continua were afterwards confirmed by many groups (see, e.g. [9.67, 68]). An example is shown in Fig. 9.12 for the Ni-Nb collisions [9.69]. The MO x-ray radiation originates from transitions into a $1s\sigma$ K-shell vacancy from higher occupied states. As can be seen from Fig. 9.12 this radiation is "non-characteristic", i.e. it does not originate from a filling of the K-vacancies in the separated atoms or the united atom. Its continuous character is due to the continuous change of the energy difference between the MO states involved with variation of the internuclear separation. X-ray continua in the energy range considered may also originate from the nucleus-nucleus bremsstrahlung, bremsstrahlung of the ejected electrons (δ rays), the tails of radiative electron capture (REC), etc. The relative contributions of these background x-ray continua depend on experimental conditions, but one can often find situations in which the MO x-ray continuous radiation dominates. The MO x-ray radiation can be experimentally distinguished in a clear manner through its characteristic directional anisotropy [9.70, 71].

The theory of quasi-molecular x-ray radiation has also been fairly well developed and reviewed elsewhere [9.72]. Calculations have been performed utilizing the quasi-classical approximation [9.73], a numerical Fourier transform analysis [9.74] and the static or stationary-phase approximation [9.75, 76]. The agreement is found to be satisfactory and illustrated in Fig. 9.13 [9.76].

9.3.2 Positron Emission

The Dirac equation for a single electron in the Coulomb field of a pointlike positive charge Z predicts that for $Z \geqq Z_{cr} = 137$ all the $j = 1/2$ states have

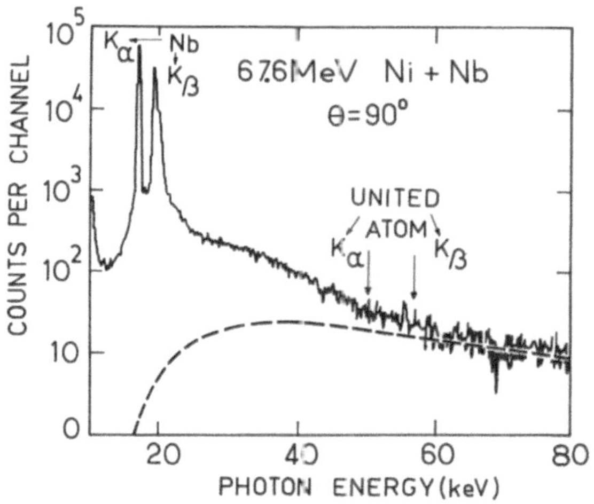

Fig. 9.12. Quasi-molecular x-ray continuum from Ni-Nb collisions [9.69]. The dashed curve represents the nucleus-nucleus bremsstrahlung background

▼ **Fig. 9.13.** Comparison of experimental and theoretical results for quasi-molecular x-ray radiation from 60 Mev Br-Br and Br-Zr collisions [9.76]. The dashed curve is the nucleus-nucleus bremsstrahlung

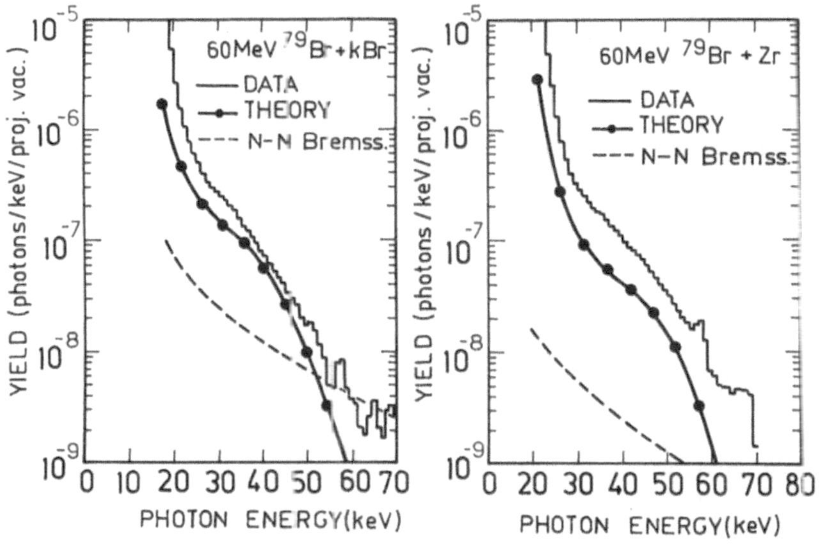

binding energies exceeding (at a corresponding $Z_{cr}^{(j)}$) twice the rest mass of the electron. For $Z \geqq Z_{cr}^{(j)}$ the electron "dives" into the negative energy continuum; its energy is spread and the state "autoionizes" by emission of an electron-positron (e-p) pair [9.77, 78]. The Dirac-Fock-Slater calculations taking into account the finite size of the nucleus have shown [9.79, 80] that for the $1s_{1/2}$, $2p_{1/2}$ and $2s_{1/2}$ states the values of $Z_{cr}^{(j)}$ are 173, 185 and 245, respectively (Fig. 9.14).

The critical charge Z_{cr} defines a boundary for positron creation: for $Z < Z_{cr}$ positrons may be created by different dynamic mechanisms (the

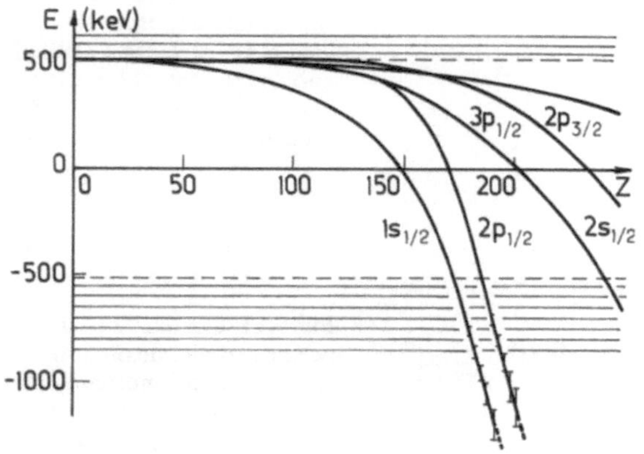

Fig. 9.14. Binding energies of the low-lying electron states in the field of a nucleus with a finite size [9.79, 80]

electron-positron pairs have a virtual character), while for $Z \geqq Z_{cr}$ positrons are created via the spontaneous decay of the (real) electron-positron pairs. The electrons from these processes are usually captured in the vacancies of the superheavy atomic system, whereas the positron emission may be observed.

Creation of the conditions close to the critical charge Z_{cr} (≈ 170) has been attempted recently with the use of energetic collisions of high-Z ions. When the combined charge of the collision partners, $Z_u = Z_1 + Z_2$, is close to, or above Z_{cr} one may expect positron emission due to dynamically induced e-p pair creation ($Z_u < Z_{cr}$), or due to spontaneous decay ($Z_u \geqq Z_{cr}$), provided the collision energy is sufficiently high to bring the nuclei close enough together.

For $Z_u \geqq Z_{cr}$ induced positron emission also exists due to the polarization of the e-p vacuum, and this emission is coherent with the one from the spontaneous e-p decay. Extrapolation of the dynamically induced positron emission from the $Z_u < Z_{cr}$ region into the $Z_u \geqq Z_{cr}$ region may be regarded as a background for the spontaneous positron emission in this latter region.

So far, positron emission in heavy ion collisions has been observed in Pb-Pb, Pb-U and U-U collisions at 5.9 MeV/amu laboratory energy [9.81, 82]. The positron production probability was found to decrease rapidly with increasing distance of closest approach, R_{min}, between the nuclei (i.e. with decreasing scattering angle in the centre-of-mass system). Although the above three systems cover both the $Z_u < Z_{cr}$ and the $Z_u > Z_{cr}$ cases ($Z_u = 164$, 174 and 184 for Pb-Pb, Pb-U and U-U systems, respectiely), no qualitative differences have been observed in the positron production probability, except its gradual increase with increasing Z_u. Such a behaviour of the data is in accord with the theoretical predictions for positron production by direct, Coulomb creation of e-p pairs ("shake-off" of the vacuum polarization cloud) [9.83]. The absence of a substantial increase in positron production for the overcritical systems, Pb-U and U-U, with respect to the subcritical one (Pb-Pb), indicates that no positrons from spontaneous e-p decay were observed in these experiments.

The theoretical studies of positron formation in heavy ion collisions is based on formulation of the collision problem in terms of coupled-channel equations for a single-electron two-Coulomb-centre systems, (Z_1, e, Z_2). The total wave function is expanded in terms of the solutions of the two-Coulomb-centre problem treated in the relativistic approximation [9.84, 85]. The dynamical problem for heavy ion collisions can be formulated in the semi-classical approximation, since for $Z_1 + Z_2 \sim 150$ the binding energy of a K electron is ~ 500 keV and collisions with an energy of ~ 10 MeV/amu can be considered slow. Thus, the form and the treatment of coupled-channel equations for the transition amplitudes are, generally speaking, the same as those presented in Sect. 6.2. However, all dynamical coupling elements for the positron creation processes in a superheavy quasi-molecular complex are large [9.83], and a perturbative approach to solving the coupled-channel equations is impossible in this case. Calculations of the positron production probabilities from different e-p creation mechanisms have been performed (see [9.83]) on the basis of a numerical treatment of coupled-channel equations for the experimentally studied collision systems, cited above. It was shown that, under the conditions of the above-mentioned experiments, the spontaneous positron production probability is negligible for these systems. Analogous calculations have shown that for the supercritical U-Cf system the spontaneous positron production would be approximately twice as large as the production by other dynamical mechanisms. Another possibility for enhancing the spontaneous positron creation is to enhance the nuclear contact time between colliding particles due to the "friction" of nuclei in deep inelastic collisions [9.86]. In such over-Coulomb-barrier collisions, the nuclear delay time may become of the order of the decay time of the e-p resonance and a sharp and intense positron peak may be expected. Such a phenomenon may occur for contact times $\sim 10^{-20}$ s [9.83].

10. Rate Coefficients of Elementary Processes

Different problems connected with the kinetics of elementary processes in plasmas (level populations, line intensities, ionization equilibrium, etc.) require knowledge of the rate coefficients of these processes. In this chapter, we evaluate the rate coefficients for the following processes: electron-impact excitation and ionization, photorecombination, dielectronic recombination and charge-transfer. Theoretical calculations are tabulated and compared with experimental data.

10.1 Energy (Velocity) Distribution Function

The rate coefficient $\langle v\,\sigma\rangle$ is determined by the effective cross section σ averaged over the energy (or velocity) distribution of the incident particles:

$$\langle v\,\sigma\rangle = \int_{\Delta E}^{\infty} v\,\sigma(E)\,\mathscr{F}(E)\,dE \quad [\mathrm{cm^3\,s^{-1}}]\,, \quad E = M\,v^2/2\,, \tag{10.1}$$

where $\mathscr{F}(E)$ is the energy distribution function, ΔE is the threshold energy for a given process, and M is the mass of an incident particle. The probability W of a two-particle collisional process is

$$W = N\langle v\,\sigma\rangle \quad [\mathrm{s^{-1}}], \tag{10.2}$$

where N is the density of incident particles. In the case of photoionization,

$$W = \int_{I}^{\infty} N_\omega\, c\,\sigma_\omega\, d\omega \quad [\mathrm{s^{-1}}], \tag{10.3}$$

where N_ω is the density of photons with a given frequency ω, c is the velocity of light, and σ_ω is the photoionization cross section for an ion with binding energy I.

The distribution function $\mathscr{F}(E)$ [or $\mathscr{F}(v)$] is normalized to unity:

$$\int_0^\infty \mathscr{F}(v)\,dv = \int_0^\infty \mathscr{F}(E)\,dE = 1\,. \tag{10.4}$$

The Maxwellian distribution function has the form:

$$\mathcal{F}(E) = \frac{2}{\pi^{1/2} kT} \left(\frac{E}{kT}\right)^{1/2} \mathrm{e}^{-E/T} \tag{10.5}$$

or

$$\mathcal{F}(v) = 4\pi \left(\frac{M}{2\pi T}\right)^{3/2} v^2 \, \mathrm{e}^{-Mv^2/2T} \tag{10.6}$$

where T is the plasma (or gas) temperature and k is the Boltzmann constant.

Maxwellian distributions (10.5, 6) are isotropic. In real plasmas the function $\mathcal{F}(v)$ may differ from a Maxwellian function. Generally, the distribution function is anisotropic and the velocity components do not follow a Maxwell distribution of the type (10.6). Here we will not discuss this question in detail, but only point out the main types of such differences. In the general case the distribution function may be obtained from the Boltzmann equation (see, for example [10.1]).

The presence of even a weak constant electric field in plasmas leads to a distortion of $\mathcal{F}(v)$ in the high velocity region and the appearance of a current of so-called non-thermal electrons [10.2]. Although the absolute number of non-thermal ("hot") electrons is relatively small compared to the number of electrons which follow a Maxwellian distribution, their contribution may be essential, because their distribution function decays with increasing energy much slower than the exponential factor in (10.6). This may be sufficient to affect the emission spectra, ionization equilibrium and other characteristics.

Another example of a non-Maxwellian distribution occurs in anisotropically heated plasmas which have an electron temperature gradient. In this case there is an increase of fast electrons in the cold-plasma region, and the distribution function $\mathcal{F}(v)$ is described by a two-temperature function [10.3].

In low-density hot plasmas anisotropic electron beams may occur with a power-law distribution function [10.4],

$$\mathcal{F}(E) = \text{const}\, E^{-\gamma} . \tag{10.7}$$

The presence of such electrons beams may be detected, for example, by the polarization of the shortwave radiation of the continuum or the spectral lines of the multiply charged ions [10.5].

All the examples mentioned above correspond to an increasing number of fast particles in the wing of a Maxwellian distribution function. In the absence of strong external perturbations (fields, particle beams, inhomogeneities) an opposite situation is, in principle, possible: due to inelastic collisions (excitation, ionization) the number of fast electrons may be less than the number of Maxwellian electrons. However, this situation is important only in plasmas with a low ionization stage, where elastic collisions restore the Maxwellian distribution and are much more effective than inelastic ones [10.6].

10.2 Electron-Impact Excitation

The excitation rate coefficient $\langle v\,\sigma_e\rangle$ in plasmas with a Maxwellian electron velocity distribution is determined by (10.1, 5):

$$\langle v\,\sigma_e\rangle = K\left(\frac{\Delta E}{\mathrm{Ry}}\right)^{1/2} \beta^{3/2}\,\mathrm{e}^{-\beta}\int_0^\infty (u+1)\,\frac{\sigma_e(u)}{\pi a_0^2}\,\mathrm{e}^{-\beta u}\,du\,, \tag{10.8}$$

$$u=\frac{E-\Delta E}{\Delta E}\,,\quad \beta=\frac{\Delta E}{kT}\,;\quad K=\frac{2\sqrt{\pi}\,\hbar\,a_0}{m}=2.18\times10^{-8}\,[\mathrm{cm^3\,s^{-1}}]\,, \tag{10.9}$$

where ΔE is the energy of the transition, σ_e is the electron-impact excitation cross section as a function of the electron energy u in threshold units, m is the mass of the electron, and T is the temperature.

It follows from (10.8), that the quantity

$$\langle v\,\sigma_e\rangle\,Z^3\,\mathrm{e}^{\beta}\sqrt{\theta}\,,\quad \theta=T/Z^2\,\mathrm{Ry}\,, \tag{10.10}$$

where Z is the spectroscopic symbol of the ion, has a weak dependence on Z and T for a $0-1$ transition along a given isoelectronic sequence. In other words, (10.10) gives the scaling law for the excitation rate coefficients.

According to the detailed balance principle, the excitation rate coefficient $\langle v\,\sigma_e\rangle_{01}$ and the de-excitation coefficient $\langle v\,\sigma_e\rangle_{10}$ are connected by the relation:

$$g_0\,\langle v\,\sigma_e\rangle_{01} = g_1\,\langle v\,\sigma_e\rangle_{10}\,\mathrm{e}^{-\beta}\,, \tag{10.11}$$

where g_0 and g_1 are the statistical weights of the states 0 and 1, respectively.

Generally speaking, the electron-impact excitation cross section of the ions consists of two parts: the 'direct' cross section, σ^{pot}, corresponding to potential scattering and an additional one, σ^{res}, due to resonance scattering, i.e.

$$\sigma_e = \sigma^{\mathrm{pot}} + \sigma^{\mathrm{res}}\,. \tag{10.12}$$

For the excitation rate coefficient we have

$$\langle v\,\sigma_e\rangle = \langle v\,\sigma^{\mathrm{pot}}\rangle + \langle v\,\sigma^{\mathrm{res}}\rangle\,. \tag{10.13}$$

As was discussed in Chap. 4, the additional part σ^{res} is connected with the possibility of electron attachment into a quasi-stationary state of the ion with subsequent decay. Although the energy range E of the incident electron, where resonance excitation takes place, is not large, the contribution of $\langle v\,\sigma^{\mathrm{res}}\rangle$ to the total excitation rate may be significant (see below).

Consider first the contribution from potential scattering, $\langle v\,\sigma^{\mathrm{pot}}\rangle$. At relatively small Z and orbital quantum number $\bar{l}$ of the incident electron the close-coupling effects usually play the principal role. In such cases the close-coupling method is used for calculations of σ^{pot} and a system of integro-differential equations (which includes as many excited ionic states as possible) is solved. Such calculations were performed for example, in [10.7] for excitation of H-like ions with $Z < 4$. For large $\bar{l}$ (but small Z) the close-coupling effects are not important and it is possible to use perturbation theory, for example, the distorted-wave method with exchange (DWE) [10.8].

For a large ionic charge Z the influence of the Coulomb field is predominant, and the DWE approximation gives the same results as the Coulomb-Born-Exchange method (CBE) [10.9, 10]. At large incident-electron energies $\tilde{E}_0 \gg \Delta E$, the CBE results coincide asymptotically with those given by the Born-Oppenheimer method. More information on original papers is given in [10.11].

The excitation cross section σ^{pot} for the transition $0-1$ in the CBE approximation and within the partial wave representation can be written in the form [10.12]

$$\sigma^{\mathrm{pot}}(0-1) = \sum_{\varkappa} \sigma_{\varkappa}(n_0 l_0 - n_1 l_1)\,, \tag{10.14}$$

$$\sigma_{\varkappa}(n_0 l_0 - n_1 l_1) = \frac{4\pi a_0^2}{(2l_0+1)\,\tilde{E}_0} \sum_{\bar{l}_0 \bar{l}_1} \Big[Q_{\varkappa} (R^{\mathrm{d}}_{\varkappa})^2 - Q_{\varkappa} R^{\mathrm{d}}_{\varkappa} \sum_{\varkappa''} R^{\mathrm{e}}_{\varkappa\varkappa''} + Q^{\mathrm{e}}_{\varkappa} \Big(\sum_{\varkappa''} R^{\mathrm{e}}_{\varkappa''\varkappa} \Big)^2 \Big]\,, \tag{10.15}$$

$$\left.\begin{aligned}
&\varkappa = \varkappa_{\min}, \varkappa_{\min}+2, \dots, \varkappa_{\max}\,; \quad \varkappa'' = \varkappa''_{\min}, \varkappa''_{\min}+2, \dots, \varkappa''_{\max}\,,\\
&\varkappa_{\min} = \max(|l_1 - l_0|, |\bar{l}_1 - \bar{l}_0|)\,; \quad \varkappa_{\max} = \min(l_0 + l_1, \bar{l}_0 + \bar{l}_1)\,,\\
&\varkappa''_{\min} = \max(|l_1 - \bar{l}_0|, |\bar{l}_1 - l_0|)\,; \quad \varkappa''_{\max} = \min(\bar{l}_0 + l_1, l_0 + \bar{l}_1)\,,
\end{aligned}\right\} \tag{10.16}$$

$$\tilde{E}_0 - \Delta E = \tilde{E}_1\,, \quad \Delta E > 0\,.$$

Here $\tilde{E}_0$ and $\tilde{E}_1$ are the energies of the incident and scattered electrons in Ry units, and l_0 and l_1 are the orbital quantum numbers of the ion electron in the initial and final states. The sum over $\bar{l}_0$, $\bar{l}_1$ means summation over partial waves of the incident and scattered electrons; R^{d} and R^{e} denote the direct and exchange radial integrals, which depend only on the quantum numbers $n_0\, l_0$, $n_1\, l_1$, of the ion:

$$R^{\mathrm{d}}_{\varkappa} = \left[\frac{(2l_0+1)(2l_1+1)(2\bar{l}_0+1)(2\bar{l}_1+1)}{2\varkappa+1}\right]^{1/2} \begin{pmatrix}\varkappa & l_0 & l_1\\ 0 & 0 & 0\end{pmatrix}\begin{pmatrix}\varkappa & \bar{l}_0 & \bar{l}_1\\ 0 & 0 & 0\end{pmatrix}$$

$$\times 2 \iint P_{n_1 l_1}(r')\, F_{\bar{E}_1 \bar{l}_1}(r'') \frac{r_<^{\varkappa}}{r_>^{\varkappa+1}} F_{\bar{E}_0 \bar{l}_0}(r'')\, P_{n_0 l_0}(r')\, dr'\, dr'' , \tag{10.17}$$

$$R^{\mathrm{e}}_{\varkappa''\varkappa} = (-1)^{\varkappa''+l_0+l_1} [(2\varkappa+1)(2l_0+1)(2l_1+1)(2\bar{l}_0+1)(2\bar{l}_1+1)]^{1/2}$$

$$\times \begin{pmatrix}\varkappa'' & l_0 & \bar{l}_1\\ 0 & 0 & 0\end{pmatrix}\begin{pmatrix}\varkappa'' & \bar{l}_0 & l_1\\ 0 & 0 & 0\end{pmatrix}\begin{Bmatrix}\varkappa'' & l_0 & \bar{l}_1\\ \varkappa & \bar{l}_0 & l_1\end{Bmatrix}$$

$$\times 2 \iint P_{n_1 l_1}(r')\, F_{\bar{E}_1 \bar{l}_1}(r'') \frac{r_<^{\varkappa}}{r_>^{\varkappa+1}} P_{n_0 l_0}(r'')\, F_{\bar{E}_0 \bar{l}_0}(r')\, dr'\, dr'' , \tag{10.18}$$

where $F_{El}(r)$ and $P_{nl}(r)$ are radial wave functions of a free and a bound electron. The functions $P_{nl}(r)$ are normalized to unity:

$$\int_0^{\infty} P^2_{nl}(r)\, dr = 1 . \tag{10.19}$$

As a rule, the radial integrals R^{d} are calculated with the Coulomb radial wave functions $F_{El}(r)$, and R^{e} with the orthogonalized wave functions (see [10.12] for details).

We note that the terms with direct matrix elements $R^{\mathrm{d}}_{\varkappa}$ in (10.14) are of the same parity $\varkappa$, while the sum over $\varkappa''$, i.e. the exchange part, consists of terms with both odd and even $\varkappa$ values, where the condition

$$\varkappa + 1 < l_0 + l_1 \tag{10.20}$$

must be satisfied.

The angular factors $Q_{\varkappa}$ and $Q^{\mathrm{e}}_{\varkappa}$ depend on the coupling scheme. For example, in the case of pure LS-coupling for the transition $l_0^q\, L_0\, S_0 - l_0^{q-1}\, [L_p\, S_p]\, l_1\, L_1\, S_1$, we have

$$Q_{\varkappa}(L_0\, S_0, L_1\, S_1) = \delta_{S_0 S_1}\, Q_{\varkappa}(L_0\, L_1) , \tag{10.21}$$

$$Q^{\mathrm{e}}_{\varkappa}(L_0\, S_0, L_1\, S_1) = \frac{2S_1+1}{2(2S_p+1)}\, Q_{\varkappa}(L_0\, L_1) , \tag{10.22}$$

$$Q_{\varkappa}(L_0\, L_1) = (2l_0+1)(2L_1+1) \begin{Bmatrix}\varkappa & L_0 & L_1\\ L_p & l_1 & l_0\end{Bmatrix}^2 |\, G^{L_0 S_0}_{L_p S_p}|^2\, q \tag{10.23}$$

where the G's are the fractional parentage coefficients.

For transitions between configurations $l_0^q - l_0^{q-1}\, l_1$, the summation over L_1, S_1, L_p, S_p gives

$$Q_\varkappa = Q_\varkappa^{\mathrm{e}} = q\,. \tag{10.24}$$

For the transition $\bar{a}_0 - \bar{a}_1$ in the intermediate coupling scheme, the angular factors $Q_\varkappa(\bar{a}_0, \bar{a}_1)$ and $Q_\varkappa^{\mathrm{e}}(\bar{a}_0, \bar{a}_1)$ are related to the 'amplitude' angular factors b in LS-coupling by

$$Q_\varkappa(\bar{a}_0, \bar{a}_1) = b_\varkappa^2(\bar{a}_0, \bar{a}_1)\,; \quad Q_\varkappa^{\mathrm{e}}(\bar{a}_0, \bar{a}_1) = \sum_{x_1 x_2} b^2_{\varkappa x_1 x_2}(\bar{a}_0, \bar{a}_1)\,, \tag{10.25}$$

$$b_\varkappa(\bar{a}_0, \bar{a}_1) = \sum_{L_0 S_0 L_1 S_1} (\bar{a}_0 \mid a_0)\, b_\varkappa(a_0, a_1)\, (a_1 \mid \bar{a}_1)\,, \tag{10.26}$$

$$b_{\varkappa x_1 x_2}(\bar{a}_0, \bar{a}_1) = \sum_{L_0 S_0 L_1 S_1} (\bar{a}_0 \mid a_0)\, b_{\varkappa x_1 x_2}(a_0, a_1)\, (a_1 \mid \bar{a}_1)\,, \tag{10.27}$$

where a_0, a_1 are the states in LS-coupling. For example, if $a_0 = l_0^q\, L_0\, S_0\, J_0$, $a_1 = l_0^{q-1}\,[L_p\, S_p]\, l_1\, L_1\, S_1\, J_1$, we have

$$b_\varkappa = \delta_{S_0 S_1} (-1)^{J_1 + S_1 + S_p} [(2 l_0 + 1)(2 J_1 + 1)(2 L_0 + 1)(2 L_1 + 1)]^{1/2} \\ \times \begin{Bmatrix} \varkappa & J_0 & J_1 \\ S_1 & L_1 & L_0 \end{Bmatrix} \begin{Bmatrix} \varkappa & L_0 & L_1 \\ L_p & l_1 & l_0 \end{Bmatrix} G^{L_0 S_0}_{L_p S_p} \sqrt{q}\,, \tag{10.28}$$

$$b_{\varkappa x_1 x_2} = (-1)^{S_p + L_p - S_1 + 1/2 + L_1} \frac{1}{\sqrt{2}} \\ \times [(2 l_0 + 1)(2 J + 1)(2 L_1 + 1)(2 S_1 + 1)(2 x_1 + 1)(2 x_2 + 1)]^{1/2} \\ \times \begin{Bmatrix} \varkappa & J_0 & x_1 \\ S_0 & L_1 & L_0 \end{Bmatrix} \begin{Bmatrix} x_1 & J_1 & x_2 \\ S_1 & S_0 & L_1 \end{Bmatrix} \begin{Bmatrix} x_2 & S_0 & S_1 \\ S_p & 1/2 & 1/2 \end{Bmatrix} \begin{Bmatrix} \varkappa & L_0 & L \\ L_p & l_1 & l_0 \end{Bmatrix} G^{L_0 S_0}_{L_p S_p} \sqrt{q}\,. \tag{10.29}$$

In the intermediate coupling the coefficients $(\bar{a} \mid a)$ are calculated numerically. In the case of some 'pure' coupling scheme ($j l$, jj, etc.) they can be expressed in terms of $3nj$-symbols [10.13].

Thus, the contribution to the excitation rate coefficient due to direct scattering $\langle v\, \sigma^{\mathrm{pot}} \rangle$ is given by (10.8, 14). Values of $\langle v\, \sigma^{\mathrm{pot}} \rangle$ in the CBE approximation are well described by the two fitting-parameter formulae [10.12]

– for transitions without change of spin, $\Delta S = 0$:

$$\langle v\, \sigma^{\mathrm{pot}} \rangle = 10^{-8} \left(\frac{\mathrm{Ry}}{\Delta E}\right)^{3/2} \left(\frac{E_1}{E_0}\right)^{3/2} \frac{Q}{2 l_0 + 1}\, \mathrm{e}^{-\beta} A\, \frac{\beta + 1}{\beta + \chi} \sqrt{\beta} \quad [\mathrm{cm^3\, s^{-1}}]\,, \\ \beta = \Delta E / kT\,, \tag{10.30}$$

– for transitions with change of spin, $\Delta S = 1$:

$$\langle v\, \sigma^{\mathrm{pot}} \rangle = 10^{-8} \left(\frac{\mathrm{Ry}}{\Delta E}\right)^{3/2} \left(\frac{E_1}{E_0}\right)^{3/2} \frac{Q^{\mathrm{e}}}{2 l_0 + 1}\, \mathrm{e}^{-\beta} A''\, \frac{\beta^{3/2}}{\beta + \chi''} \quad [\mathrm{cm^3\, s^{-1}}]\,, \\ \beta = \Delta E / kT\,, \tag{10.31}$$

where E_0 and E_1 are the energies of the bound electron in its initial and final state, respectively, A, c and A'', χ'' are fitting parameters, and Q is the angular factor. The accuracy of approximations (10.30, 31) is within 20% for $0.25 \leq \beta \leq 10$. The fitting parameters A, χ and A'', χ'' for some transitions in H-like and He-like ions with $Z > 3$ are given in [10.12] including transitions between closely spaced levels.

Consider now the part of the excitation rate coefficient $\langle v\,\sigma^{\rm res}\rangle$ arising due to resonance scattering. In Sect. 4.5 it was pointed out that the presence of resonances is connected with excitation of doubly-excited autoionizing states:

$$X_{Z+1}(\alpha_0) + e \rightarrow X_Z^{**}(\gamma) \begin{cases} X^*_{Z+1}(\alpha') + e\,, & (10.32\,a) \\ X_Z(\alpha) + \hbar\,\omega\,, & (10.32\,b) \end{cases}$$

Here autoionization decay (10.32 a) gives additional contributions to the cross sections and excitation rates, whereas radiative decay or dielectronic recombination (10.32 b) is a competing process.

Since resonance excitation exists in the immediate vicinity of the excitation threshold, the term $\langle v\,\sigma^{\rm res}\rangle$ may be comparable with $\langle v\,\sigma^{\rm pot}\rangle$ (or even larger) at relatively low electronic temperatures. The dielectronic recombination probability rises with increasing Z, and for large Z values the effect of resonance excitation becomes smaller. For this reason it is impossible to write down a simple scaling law describing $\langle v\,\sigma^{\rm res}\rangle$ as a function of Z (see Sect. 4.5).

We discuss now some results of numerical calculations. *Pradhan et al.* [10.14] published extensive tables of $\langle v\,\sigma\rangle$ values for He-like ions including the sum of both potential and resonance excitation terms. They did not take

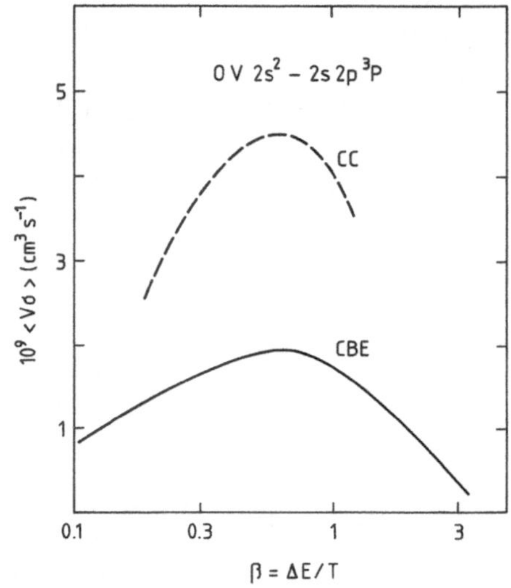

Fig. 10.1. Rate coefficients for the O $V\,2s^2\,{}^1S - 2s\,2p\,{}^3P$ excitation. (—): CBE result (10.14–18), (- - -) close-coupling result [10.16] taking resonance effects into account

into consideration the dielectronic recombination channel (10.32b), which can decrease significantly the resonance excitation contribution for $Z \gtrsim 10$; for $Z < 10$ radiative decay (10.32b) gives only small corrections to the process (10.32a). Therefore, the results in [10.14] are valid for light ions and give overestimated excitation rates for heavy ions. According to [10.14], the resonance part $\langle v\,\sigma^{\text{res}}\rangle$ is a few times larger than the potential (direct) part $\langle v\,\sigma^{\text{pot}}\rangle$ for the $1\,^1S - 2\,^3S$ transition in a helium-like iron ion, Fe XXV, at low temperatures. Refined calculations of *Pradhan* [10.15], in which dielectronic recombination (10.32b) is taken into consideration, give completely different results for this transition: $\langle v\,\sigma^{\text{res}}\rangle$ is much smaller than $\langle v\,\sigma^{\text{pot}}\rangle$.

At present, resonance contributions have been better investigated for light multiply charged ions ($Z < 10$) where radiative decay of resonances is less important. An example is given in Fig. 10.1, where potential and resonance excitations (neglecting dielectronic recombination) are shown for the $2s^2\,^1S - 2s\,2p\,^3P$ transition in a beryllium-like O V ion [10.16]. Here the main contribution to $\langle v\,\sigma^{\text{res}}\rangle$ is due to radiationless decay of f-resonances of the $2s\,2p\,(^1P^0)\,n\,l$ configurations. At low temperatures the resonance excitation rate is comparable with the direct excitation rate.

10.3 Electron-Impact Ionization

The ionization rate coefficient $\langle v\,\sigma_{\text{i}}\rangle$ in plasmas with a Maxwellian velocity distribution of electrons is given by an expression analoguous to the excitation rate (10.8), i.e.

$$\langle v\,\sigma_{\text{i}}\rangle = K\left(\frac{I_Z}{\text{Ry}}\right)^{1/2} \beta^{3/2}\,\text{e}^{-\beta} \int_{I_Z}^{\infty} (u+1)\,\frac{\sigma_{\text{i}}(u)}{\pi\,a_0^2}\,\text{e}^{-\beta u}\,du\,, \tag{10.33}$$

$$u = \frac{E - I_Z}{I_Z}\,,\quad \beta = \frac{I_Z}{kT}\,,\quad K = \frac{2\sqrt{\pi}\,\hbar\,a_0}{m} = 2.18\times 10^{-8}\quad [\text{cm}^3\,\text{s}^{-1}]\,, \tag{10.34}$$

where E is the incident electron energy, I_Z is the binding energy of the target ion, σ_{i} is the electron-impact ionization cross section as a function of energy u measured in threshold units and T is the plasma electron temperature.

According to the detailed balance principle the ionization rate coefficient $\langle v\,\sigma_{\text{i}}\rangle$ and the three-body recombination coefficient $\varkappa_{\text{r}}$ are connected by the relation

$$g_Z\,\langle v\,\sigma_{\text{i}}\rangle = 2\left(\frac{m\,T}{2\pi\,\hbar^2}\right)^{3/2} g_{Z+1}\,\text{e}^{-\beta}\,\varkappa_{\text{r}}\,, \tag{10.35}$$

$$\varkappa_{\text{r}} = \iint v_1\,v_2\,\sigma_{\text{r}}\,\mathcal{F}(E_1)\,\mathcal{F}(E_2)\,dE_1\,dE_2\quad [\text{cm}^6\,\text{s}^{-1}]\,, \tag{10.36}$$

where g_Z and g_{Z+1} are the statistical weights of the target X_Z and the resulting X_{Z+1} ions respectively.

The rate $\varkappa_r$ in (10.36) is averaged over the velocity distribution functions $\mathcal{F}(E)$ of two incident electrons.

Among semi-empirical formulae for $\langle v\,\sigma_i\rangle$ the Seaton formula [10.17],

$$\langle v\,\sigma_i\rangle = 4.3\times 10^{-8}\,q\left(\frac{\mathrm{Ry}}{I_Z}\right)^{3/2}\beta^{-1/2}\,\mathrm{e}^{-\beta}\quad [\mathrm{cm}^3\,\mathrm{s}^{-1}]\,, \tag{10.37}$$

or the Lotz formula [10.18],

$$\langle v\,\sigma_i\rangle = 6\times 10^{-8}\,q\left(\frac{\mathrm{Ry}}{I_Z}\right)^{3/2}\beta^{-1/2}\,\mathrm{e}^{-\beta}f(\beta)\quad [\mathrm{cm}^3\,\mathrm{s}^{-1}]\,, \tag{10.38}$$

$$f(\beta) = \beta\,\mathrm{e}^{\beta}\,|\,\mathrm{Ei}\,(-\beta)\,|\,,\quad \beta = I_Z/(k\,T)\,, \tag{10.39}$$

are often used. Here I_Z is the binding energy of shell $n\,l^q$ of the target ion, and Ei is the exponential integral, tabulated in [10.19] [see also (10.51)]. The Lotz formula is based on Coulomb-Born calculations for hydrogenic ions with $Z > 5$.

At present the Coulomb-Born-Exchange (CBE) approximation is the most reliable method for calculations of ionization cross sections. CBE ionization cross sections can be obtained using (10.14–20), where the wave function $P_{n_1 l_1}(r)$ of a bound electron in the final state is replaced by a continuous spectrum function, $P_{E_1 l_1}(r)$, with the asymptotic behaviour

$$P_{E_1 l_1}(r)\underset{r\to\infty}{\cong} k^{-1/2}\sin\left[k\,r + \frac{Z}{k}\ln\,(k\,r) - \frac{\pi\,l_1}{2} + \delta_{l_1}\right],\quad k^2 = E_1\,, \tag{10.40}$$

$$\int_0^\infty P_{E_1 l_1}(r)\,P_{E_1' l_1}(r)\,dr = \pi\,\delta\,(E_1 - E_1')\,, \tag{10.41}$$

where E_1 is the energy of the ejected electron and δ_{l_1} is the phase shift.

The CBE ionization cross section can be written in the form

$$\sigma_i^{\mathrm{CBE}} = q\sum_{l_1\varkappa}\int_0^{(E_0-I_z)/2}\sigma_\varkappa\,(n\,l,\,E_1\,l_1)\,dE_1\,, \tag{10.42}$$

where $\sigma_\varkappa$ is defined by (10.15), and E_0 is the energy of the incident electron. The ionization rate in the CBE approximation (10.42, 33) can be represented by

$$\langle v\,\sigma_i\rangle = q\times 10^{-8}\left(\frac{\mathrm{Ry}}{I_Z}\right)^{3/2}\mathrm{e}^{-\beta}\frac{A\sqrt{\beta}}{\beta+\chi}\,[\mathrm{cm}^3\,\mathrm{s}^{-1}]\,,\quad \beta = I_Z/kT \tag{10.43}$$

where A and χ are fitting parameters. The accuracy of the approximation (10.43) is within 6% for $\beta = 0.1-10$. In Table 10.1 the parameters A and χ are given for ionization of outer- and inner-shell electrons, for ions

Table 10.1. Fitting parameters A and χ (10.43) for ionization rates of outer and inner electrons of ions with $Z_n = 99$ [10.20]

Outer shells				Inner shells		
$n\,l^q$	Isoelectronic sequence	A	χ	nl^q	A	χ
$1\,s^q$	H, He	5.10	0.452	$1\,s^2$	4.84	0.397
$2\,s^q$	Li, Be	5.57	0.677	$2\,s^2$	5.44	0.632
$2\,p^q$	B – Ne	8.27	0.95	$2\,p^6$	8.50	0.925
$3\,s^q$	Na, Mg	4.79	0.746	$3\,s^2$	4.97	0.767
$3\,p^q$	Al – Ar	7.43	1.10	$3\,p^6$	7.47	1.10
$4\,s^q$	K, Ca	3.63	0.751			

belonging to isoelectronic sequences from H to Ca. These parameters were obtained on the basis of CBE results for ions with nuclear charge $Z_n = 99$ [10.20].

Using (10.43) and the properties of ionization cross sections, the scaling law for the ionization rate coefficients can be written in the form

$$e^{\beta}\, I_Z^{3/2} \langle v\, \sigma_i \rangle / q \sim Z^3\, e^{\beta} \langle v\, \sigma_i \rangle / q\,, \quad \beta = I_Z / kT\,, \tag{10.44}$$

i.e. in the representation (10.44) the rate coefficients for ionization of an electron with given quantum numbers $n\,l$ have a weak dependence on the factor β and the ion charge Z.

Experimental data on $\langle v\, \sigma_i \rangle$ are given mainly for laboratory plasma sources (θ-pinch [10.21–24], stellarator [10.25], tokamak [10.26]) and crossed-beam measurements [10.27–29]. Although beam experiments seem to be most capable of producing high accuracy and unambiguous details desirable for testing different theoretical approaches, they have been applied to studies of highly charged ions only recently, mainly because of the lack of suitable ion beam sources.

The method for determining the ionization rates in plasma experiments has been reviewed by *Kunze* [10.30]. It is based on a numerical model, which predicts the time evolution of spectral line intensities dI/dt with electron densities N, electron temperature T and the ionization rates $\langle v\sigma_i \rangle$, used as input parameters. The ionization rate coefficients are adjusted until the observed values dI/dt agree with the model ones. Although this method contains significant disadvantages (the use of values of N and T averaged in time and space, a Maxwellian electron velocity distributions and so on), it is the only method, at present, which has provided data for ions of charge greater than 6.

A comparison of the ionization rate coefficients $\langle v\, \sigma_i \rangle$ obtained by this method, with crossed-beam data and calculations in the CBE and DWE

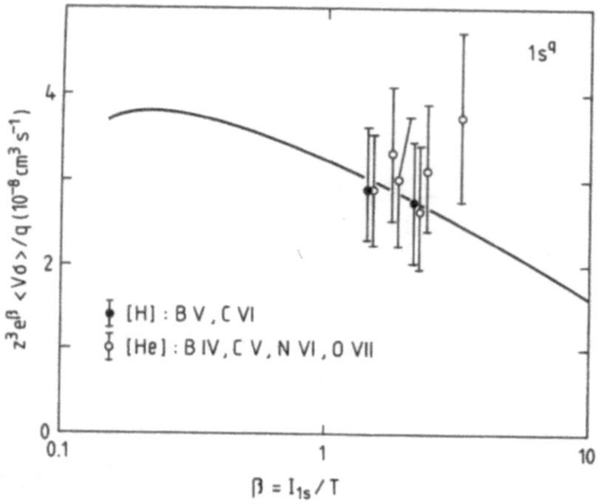

Fig. 10.2. Scaled rate coefficients for the ionization of $1\,s^q$ electrons. (—) CBE result for ions with nuclear charge $Z_n = 99$ [10.20]. Experiment: (⧳ ⦶) θ-pinch [10.23, 24]

▼ **Fig. 10.3.** Scaled rate coefficients for ionization of $2\,s^q$ electrons. (*S*) Seaton (10.16) and (*L*) Lotz (10.17) estimates, (*JM–CBE*) result for Fe *XXV* ions [10.31], (*SV–CBE*) result for ions with $Z_n = 99$ [10.20]. Experiment: θ-pinch (+, ○, ●, ×) [10.23, 24], (△, ▲) [10.21, 22]; stellarator (■) [10.25]; crossed beams (– – –) [10.27]

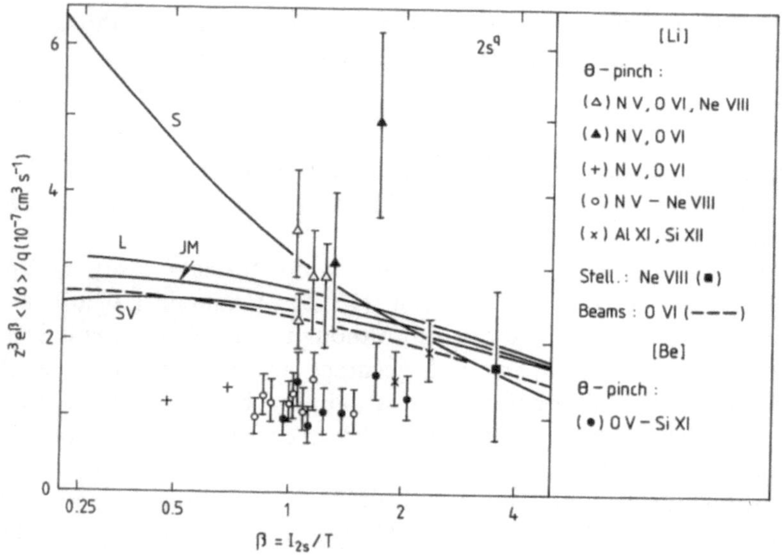

approximations, shows that the plasma spectroscopy method gives, as a rule, results which are 1.5–2 times underestimated. In Figs. 10.2–6 the scaled theoretical and experimental data $Z^3\,e^{\beta}\,\langle v\,\sigma_i\rangle/q$ as a function of $\beta = I_Z/kT$ are given, for ions belonging to isoelectronic sequences from H to Ar. Theoretical and experimental rate coefficients for the ionization of O^{5+} ions are compared in Fig. 10.7.

A comparison of different theoretical approximations with experimental data shows that the Lotz formula gives, as a rule, overestimated values of $\langle v\,\sigma_i\rangle$ especially for small β (large temperatures). A compilation of ratios of

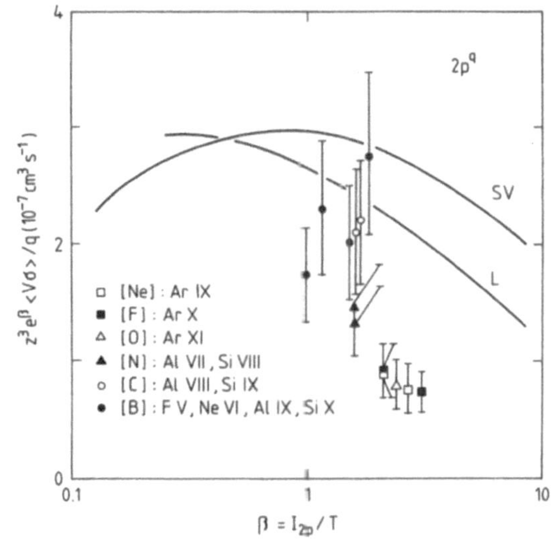

Fig. 10.4. Scaled rate coefficients for ionization of $2p^q$ electrons. (L) Lotz estimate (10.17), (SV–CBE) result for ions with $Z_n = 99$ [10.20]. Experiments: θ-pinch [10.24]

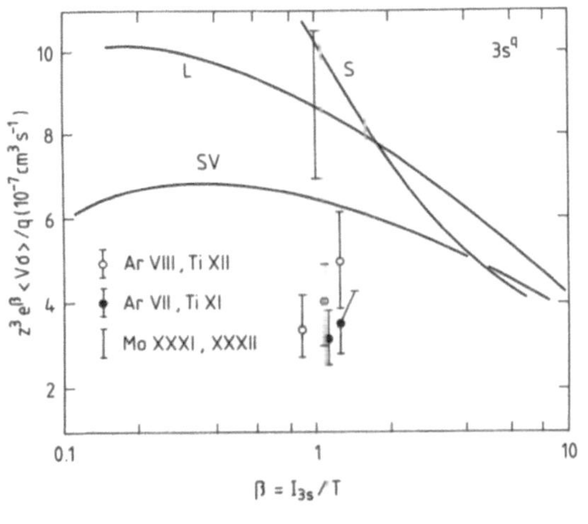

Fig. 10.5. Scaled rate coefficients for ionization of $3s^q$ electrons. (S) Seaton (10.16) and (L) Lotz (10.17) estimates, (SV–CBE) result for ions with $Z_n = 99$ [10.20]. Experiment: θ-pinch (ɸ, ɸ) [10.24]; (I) estimated from tokamak data [10.26]

experimentally observed rate coefficients to those estimated by the Lotz formula, showed [10.32] that, except for 7 out of 30 measurements, the ratio exp/Lotz was found to be between 0.5 and 1.0. This has given impetus to the impression that the Lotz formula gives rates that are too large. On the other hand, the crossed-beam data for ionization cross sections imply that the CBE results are more reliable at predicting the direct ionization process.

We note finally that calculations of ionization rate coefficients require one, generally speaking, to take into account the contributions from ionization and excitation-autoionization of inner shell electrons (Sect. 5.6).

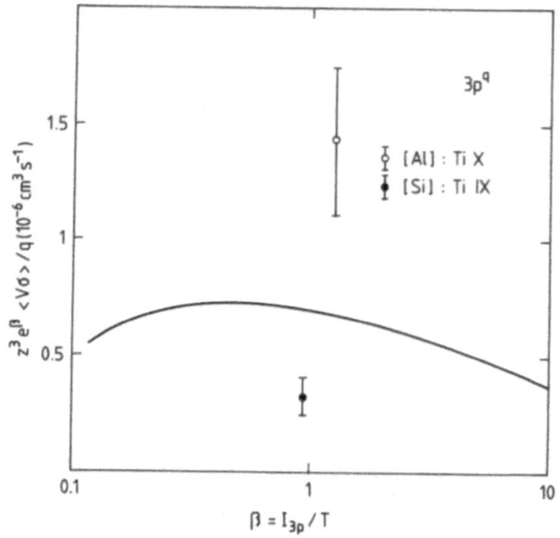

Fig. 10.6. Scaled ionization rate coefficients for $3p^q$ electrons. (—) CBE result for ions with $Z_n = 99$ [10.20]. Experiment: θ-pinch [10.24]

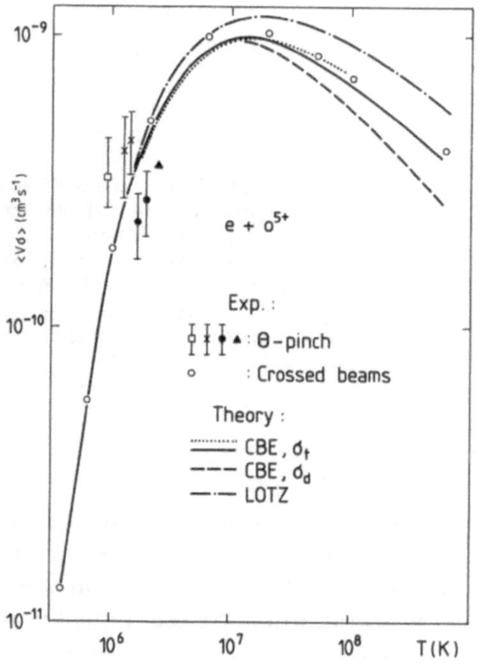

Fig. 10.7. Ionization rate coefficients for O^{5+} ions as a function of electron temperature. Theory: (···) CBE result [10.31] for the total cross section (ionization + autoionization); (—) CBE result for the total and (- - -) direct (ionization only) cross sections [10.20]; (–·–) Lotz formula (10.17). Experiment: θ-pinch (⧻) [10.21], (⊡) [10.22], (▲, ●) [10.23]; crossed beams (○) [10.27]

10.4 Photorecombination

The main properties of photorecombination processes

$$X_{Z+1} + e \rightarrow X_Z(n\,l) + \hbar\,\omega \tag{10.45}$$

are considered in Sect. 3.2. The photorecombination rate coefficient can be written in the form

$$\varkappa_\nu(n\,l) = \langle v\,\sigma_{r\nu}\rangle = K\left(\frac{I_{nl}}{\mathrm{Ry}}\right)^{1/2} \beta^{3/2}\int_0^\infty u\,\frac{\sigma_{r\nu}(u)}{\pi\,a_0^2}\,e^{-\beta u}\,du\,, \tag{10.46}$$

$$u = \frac{E}{I_{nl}}\,,\quad \beta = \frac{I_{nl}}{T}\,,\quad K = \frac{2\sqrt{\pi}\,\hbar\,a_0}{m} = 2.18\times 10^{-8}\quad [\mathrm{cm^3\,s^{-1}}]\,, \tag{10.47}$$

where I_{nl} is the binding energy of the ion X_Z in a given state $n\,l$, E is the energy of an incident electron, $\sigma_{r\nu}(u)$ is the photorecombination cross section of the ion X_Z, and T is the electron temperature.

It follows from (10.46) with the use of the classical Kramers formula for $\sigma_{r\nu}$ (3.47) that photorecombination predominates for levels with a binding energy $I_{nl} > T$; in this case $\varkappa_\nu(n\,l) \sim n^{-1}$. For higher levels with $I_{nl} < T$ the photorecombination rate is proportional to n^{-3}. For relatively large temperatures $T > I_{n_0 l_0}$ ($n_0\,l_0$ denotes the ground state of the ion X_Z), photorecombination leads mainly to electron capture into a group of levels with $n \leq n_0$ (the case $n < n_0$ takes place, for example, for Ca-like ions: $n_0\,l_0 = 4\,s$, $n\,l = 3\,d$).

The total rate of photorecombination, i.e. summed over all final states $n\,l$, is given by

$$\varkappa_\nu^t = \sum_{nl} \varkappa_\nu(n\,l)\,. \tag{10.48}$$

It is possible to estimate the contribution from highly excited states $n\,l$ in (10.46) also using the classical Kramers cross section σ_ν^{Kr} (3.47) for all levels above a certain level $\bar{n}$:

$$\begin{aligned}\sum_{n=\bar{n}} \varkappa_\nu^{\mathrm{Kr}}(n) &\equiv \sum_{l,\,n=\bar{n}}^{\infty} \langle v\,\sigma_{r\nu}^{\mathrm{Kr}}\rangle \\ &= K_1\,\bar{n}\,\beta_1^{1/2}\,Z\,[\ln(1.78\,\beta_1) + e^{\beta_1}(1+\beta_1/\bar{n})\,|\,\mathrm{Ei}(-\beta_1)\,|\,]\,,\end{aligned} \tag{10.49}$$

$$\beta_1 = \frac{Z^2\,\mathrm{Ry}}{\bar{n}^2\,T}\,,\quad K_1 = \frac{32\sqrt{\pi}\,a_0^2\,c}{3\sqrt{3}\cdot 137^4} = 2.60\times 10^{-14}\ \mathrm{cm^3\,s^{-1}}\,, \tag{10.50}$$

where c is the velocity of light and $\mathrm{Ei}(-x)$ is the exponential integral [10.19], which is well fitted by the formula

$$e^x \left| \mathrm{Ei}(-x) \right| \simeq \ln\left[1+\frac{0.562+1.4\,x}{x\,(1+1.4\,x)}\right] \tag{10.51}$$

for all x.

Thus, the total rate coefficient $\varkappa_\nu^{\mathrm{t}}$ for photorecombination into all states of the ion X_Z can be written in the form

$$\varkappa_\nu^{\mathrm{t}} = \sum_{n=n_0}^{\bar{n}-1} \sum_{l<n} \varkappa_\nu(n\,l) + \sum_{n=\bar{n}}^{\infty} \varkappa_\nu^{\mathrm{Kr}}(n)\,, \tag{10.52}$$

where $n_0\,l_0$ are the quantum numbers of the ground state and $\varkappa_\nu(n\,l)$ and $\varkappa_\nu^{\mathrm{Kr}}(n)$ are defined in (10.46) and (10.49) respectively.

Table 10.2. Fitting parameters A and χ (10.53) for the total photorecombination rates of oxygen ion: $O_{Z+1} \rightarrow O_Z$

Z	$1/8 \leqq \beta \leqq 8$		$10 \leqq \beta \leqq 800$	
	A	χ	A	χ
1	7.2	0.06	7.2	0.06
2	17.1	0.14	17.1	0.14
3	24.3	0.43	55	18
4	27.4	0.67	62.4	18.4
5	25.1	0.76	67.3	19.7
6	27.9	0.88	67.6	19.7
7	6.76	1.1	15.9	21.5
8	9.75	1.01	19.3	17.9

Table 10.3. Fitting parameters A and χ (10.53) for the total photorecombination rates of iron ions $Fe_{Z+1} \rightarrow Fe_Z$ in the interval $1/8 \leqq \beta \leqq 8$

Z	A	χ	Z	A	χ
9	34.1	0.69	18	23.6	1.17
10	38.0	0.69	19	25.6	1.15
11	41.7	0.71	20	27.3	1.14
12	49.4	0.74	21	29.0	1.15
13	52.1	0.77	22	30.5	1.16
14	54.9	0.79	23	30.0	1.05
15	48.3	0.82	24	31.1	0.96
16	51.3	0.82	25	6.56	0.92
17	20.2	1.22	26	8.39	0.73

Table 10.4. The total calculated photorecombination rates $\varkappa_\nu^t$ of oxygen ions $O_{Z+1} \to O_Z$ at temperature $T = 10^4$ K

Z	$10^{12}\,\varkappa_\nu^t\,[\mathrm{cm^3\,s^{-1}}]$ [10.32]	$10^{12}\,\varkappa_\nu^t\,[\mathrm{cm^3\,s^{-1}}]$ [10.33]
1	0.286	0.31
2	1.75	2.0
3	6.88	5.1
4	11.7	9.6
5	17.5	12
6	24.3	23
7	33.5	41
8	48.3	–

The value $\varkappa_\nu^t$ is well approximated by [10.12]

$$\varkappa_\nu^t = 10^{-14}\left(\frac{I_{n_0 l_0}}{\mathrm{Ry}}\right)^{1/2}\frac{A\,\beta^{3/2}}{\beta+\chi}\,[\mathrm{cm^3\,s^{-1}}]\,,\quad \beta = I_{n_0 l_0}/T\,, \tag{10.53}$$

where $I_{n_0 l_0}$ is the binding energy of the ion X_Z in its ground state and A and χ are fitting parameters. In Tables 10.2, 3 the parameters A and χ, obtained with the help of exact calculations (10.46), are given for oxygen and iron ions. These parameters were used in [10.33] for calculations of plasma ionization equilibrium. The interval $1/8 \le \beta \le 8$ corresponds to the temperature region where O_Z and Fe_Z ions have a maximum concentration. The parameters A and χ for oxygen ions in the interval $10 \le \beta \le 800$ were used in [10.33] to calculate ionization equilibrium of plasmas in a strong radiation field.

In Table 10.4 the total calculated photorecombination rates for all ionization stages of oxygen ions are given at a temperature $T = 10^4$ K.

10.5 Dielectronic Recombination

Calculation of the rate coefficients of dielectronic recombination

$$X_{Z+1}(\alpha_0) + e \to X_Z^{**}(\gamma) \to X_Z(\gamma') + \hbar\,\omega\,,\quad \gamma = \alpha\,n\,l\,,\ \gamma' = \alpha'\,n\,l\,, \tag{10.54}$$

is a more complicated problem compared to ionization and photorecombination processes due to its resonance character (see Sect. 3.3 for details). To calculate the rate coefficient $\varkappa_d$, one must take into account the alternative channel to radiative decay – the autoionization process

$$X_Z^{**}(\gamma) \to X_{Z+1}^{*}(\alpha') + e\,. \tag{10.55}$$

Radiative decay (10.54) goes, as a rule , to the ground state γ', whereas autoionization (10.55) goes to the ground and excited states α'. Moreover, the probability of autoionization decay to the excited states (when it is possible energetically) is much larger than to the ground state (see Sects. 2.5 and 3.4).

Calculations of $\varkappa_d$ rate coefficients including autoionization to the ground state give satisfactory results in the case of H- and He-like ions [10.35–37]. In the case of more complicated ions it is necessary to include additional channels of autoionization to excited states [10.38].

The rate of dielectronic recombination is well fitted by the formula [10.12]:

$$\varkappa_d = 10^{-13}\, B_d(T)\, \beta^{3/2}\, e^{-\beta\chi}\ [\mathrm{cm^3\,s^{-1}}]\,, \quad \beta = (Z+1)^2\,\mathrm{Ry}/T\,, \qquad (10.56)$$

$$\chi = E_{\alpha\alpha_0}/(Z+1)\,\mathrm{Ry}\,,$$

where $E_{\alpha\alpha_0} = E_\alpha - E_{\alpha_0}$ is the excitation energy and T is the electron temperature in Ry units. The function $B_d(T)$ can be calculated using the methods described in Sect. 3.3.2, which require, however, calculations of the radiative (A_r) and autoionization (A_a) probabilities for a large number of intermediate states γ. For various applications simple models for $\varkappa_d$ are usually used. In [10.39] the following formulae were proposed for $\varkappa_d$, summed over all quantum numbers $\alpha'\, n\, l$ of the final states taking into account all possible autoionization channels:

$$\varkappa_d(\alpha_0 - \alpha) \equiv \sum_{\alpha' n l} \varkappa_d(\alpha_0 - \alpha\, n\, l - \alpha'\, n\, l) \qquad (10.57)$$

$$= 10^{-13}\, B_d\, \beta^{3/2}\, e^{-\beta\chi_d}\,[\mathrm{cm^3\,s^{-1}}], \quad \beta = (Z+1)^2\,\mathrm{Ry}/T\,,$$

$$B_d = A\frac{Z}{n_1^4} f_{\alpha_0\alpha} \sum_{n \geqq n_1} \sum_{l<n} \frac{2l+1}{B + (n/n_s)^3}\,, \qquad (10.58)$$

$$A = 10^{13}\,\frac{4\,\pi^{3/2}}{137^3}\,\frac{a_0\,\hbar}{m}\left(\frac{Z}{Z+1}\right)^3 = 0.53\left(\frac{Z}{Z+1}\right)^3, \qquad (10.59)$$

$$n_1^2 = \frac{Z^2\,\mathrm{Ry}}{E_{\alpha\alpha_0}}\,, \quad n_s = 137\left(\frac{n_1^2\,\sigma(\alpha_0, \alpha\, l)}{\pi^2 a_0^2 (2\,l+1) f_{\alpha_0\alpha}}\right)^{1/3}, \qquad (10.60)$$

$$B = \sum_{\alpha'} \frac{E_{\alpha\alpha'}}{E_{\alpha\alpha_0}}\,\frac{g(\alpha')\,\sigma(\alpha', \alpha\, l)}{g(\alpha_0)\,\sigma(\alpha_0, \alpha\, l)}\,, \qquad (10.61)$$

where α_0 is the ground state of the recombining ion X_{Z+1}, $\alpha\, n\, l$ is the doubly excited state, α' is a set of quantum numbers belonging to the levels lying between α_0 and α, $g(\alpha')$ and $g(\alpha_0)$ are statistical weights of the corresponding states, $E_{\alpha\alpha_0}$ and $f_{\alpha_0\alpha}$ are the energy and the oscillator strength for the transition $\alpha_0 - \alpha$, $\sigma(\alpha_0, \alpha\, l)$ and $\sigma(\alpha', \alpha\, l)$ are the partial electron excitation cross sections of the ion X_Z at the threshold and $\sigma(\alpha_0, \alpha\, l) = \sum_{\bar{l}} \sigma(\alpha_0\, \bar{l}, \alpha\, l)$,

Table 10.5. Parameters B_d and χ_d (10.57) for the rate of dielectronic recombination of iron ions $Fe_{Z+1}(\alpha_0) \rightarrow Fe_Z(\alpha)$ $(1/8 \leqq \beta \leqq 8)$

Z	Transition $\alpha_0 - \alpha$	B_d	χ_d	Z	Transition $\alpha_0 - \alpha$	B_d	χ_d
9	$3p-4d$	0.86	0.054	17	$2p-3d$	49.3	0.14
	$-3d$	66.5	0.041		$2s-2p$	0.84	0.021
10	$3p-4d$	0.82	0.087	18	$2p-3d$	42.4	0.13
	$-3d$	43.6	0.034		$2s-2p$	1.3	0.017
11	$3p-4d$	1.32	0.061	19	$2p-3d$	32.4	0.124
	$-3d$	27.3	0.026		$2s-2p$	1.4	0.014
12	$3p-4d$	1.57	0.062	20	$2p-3d$	21.6	0.12
	$-3d$	14.2	0.021		$2s-2p$	1.3	0.011
13	$3p-4d$	1.33	0.055	21	$2p-3d$	10.5	0.11
	$-3d$	5.54	0.018		$2s-2p$	1.1	0.009
14	$3p-4d$	0.50	0.05	22	$2p-3d$	1.82	0.13
	$-3d$	11.0	0.01		$2s-2p$	0.78	0.007
15	$3p-4d$	0.59	0.048	23	$2p-3d$	1.30	0.14
	$-3d$	4.46	0.008		$2s-2p$	0.21	0.005
16	$2p-3d$	50.5	0.16	24	$1s-2p$	8.1	0.628
	$2s-2p$	5.1	0.17	25	$1s-2p$	4.0	0.598

Table 10.6. Parameters B_d and χ_d (10.57) for the rate of dielectric recombination of oxygen ions: $O_{Z+1}(\alpha_0) \rightarrow O_Z(\alpha)$ $(1/8 \leqq \beta \leqq 8)$

Z	Transition $\alpha_0 - \alpha$	B_d	χ_d	Z	Transition $\alpha_0 - \alpha$	B_d	χ_d
1	$2p-3d$	3.87	0.26	4	$2s-3p$	2.26	0.20
	$2s-2p$	42.9	0.27		$2s-2p$	30.6	0.058
2	$2p-3d$	7.94	0.30	5	$2s-3p$	1.79	0.16
	$2s-2p$	40.5	0.12		$2s-2p$	4.90	0.024
3	$2p-3d$	8.93	0.22	6	$1s-2p$	43.2	0.804
	$2s-2p$	32.7	0.072	7	$1s-2p$	31.1	0.69

where $\tilde{l}$ is the orbital angular momentum of the incident electron. Here the coefficient B_d is averaged over the states $n\,l$ and does not depend on the electron temperature T, and χ_d is the fitting parameter. In Tables 10.5, 6 the parameters B_d and χ_d for some transitions in iron and oxygen ions are given, calculated with the use of (10.57–61). The parameters χ_d were fitted using results of exact calculations (3.77–81) at a temperature $T = I_Z/4$ (I_Z is the binding energy of the ion X_Z in the ground state).

It is worth noting that the effect of autoionization decay on the rate coefficient $\varkappa_d$ is especially large for ions with a smaller nuclear and ion charge. For example, for the recombination

$$\begin{aligned} O\,IV(1s^2\,2s^2\,2p) + e &\rightarrow O\,III(1s^2\,2s^2\,3d,\, n\,l) \\ &\rightarrow O\,III(1s^2\,2s^2\,2p,\, n\,l) + \hbar\omega\,, \end{aligned} \tag{10.62}$$

the effect of the autoionization processes

$$\begin{aligned} \mathrm{O}\,III(3\,d, n\,l) &\rightarrow \mathrm{O}\,IV(2\,p) + \mathrm{e} \\ &\rightarrow \mathrm{O}\,IV(3\,s) + \mathrm{e} \\ &\rightarrow \mathrm{O}\,IV(3\,p) + \mathrm{e}\,, \end{aligned} \quad (10.63)$$

leads to a decrease of the rate coefficient $\varkappa_d(2\,p-3\,d)$ by a factor 3, whereas for the transition Fe *XXII* $(2\,p-3\,d)$ this effect practically does not change the $\varkappa_d$ value. This is connected with the fact that the number of autoionizing states which contribute to the rate $\varkappa_d$ decreases with increasing ion charge.

The influence of autoionization channels on the rate $\varkappa_d$ has been considered in [10.38,39]. However, the contribution of this effect in [10.38] seems to be greatly overestimated. In Fig. 10.8 the calculated temperature dependence of the dielectronic rate coefficient $\varkappa_d$ for the transition Fe *XVII* $(2\,p-3\,d)$ (10.57–62) with and without the contribution of autoionization is given in comparison with the *Burgess* formula [10.40] (3.82–84) and the results of *Jacobs* et al. [10.38]. It is seen that the Burgess formula, which does not describe the individual properties of the recombining ion X_{Z+1}, gives results that are ~ 1.5 times overestimated. This is consistent with the well-established degree of accuracy of this formula [10.41].

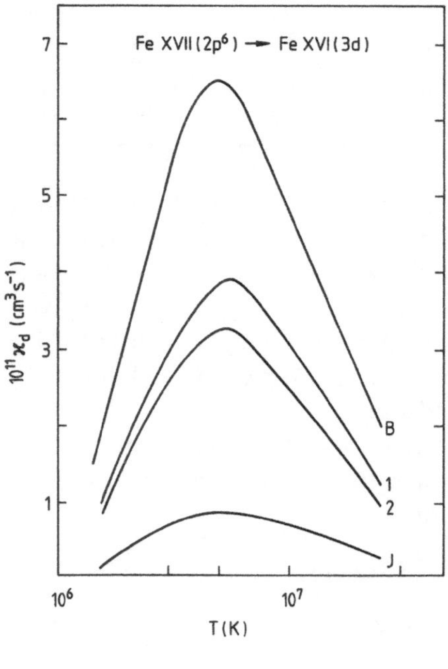

Fig. 10.8. Dielectronic recombination rates for the Fe *XVII* $(2\,p^6) \rightarrow$ Fe *XVI* $(2\,p^6\,3\,d)$ transition as a function of electron temperature. (*B*) Burgess formula (3.82–84), (*1* and *2*)-calculated without and with autoionization decay considerations [10.39], (*J*) calculated including autoionization decay [10.38]

We note finally that the total rate of dielectronic recombination $\varkappa_d^t$ of the ion $X_{Z+1}(\alpha_0)$ is given by

$$\varkappa_d^t = \sum_{\alpha} \varkappa_d(\alpha_0 - \alpha). \tag{10.64}$$

10.6 Charge Transfer

The main properties of charge-transfer processes arising in ion-atom collisions

$$X^{Z+} + A \to X^{(Z-m)+} + A^{m+}, \quad m \geqq 1, \tag{10.65}$$

were considered in Chap. 7. Under certain conditions, such reactions may proceed more effectively than recombination (radiative and dielectronic) and ionization (electron-impact and photoionization) processes [10.42–44].

In this section we will consider the rate coefficient for one-electron capture ($m = 1$), which has been investigated in detail theoretically and experimentally. The rate coefficient for charge transfer $\langle v\,\sigma_c \rangle$ is determined by the general formulae (10.1, 6) for a Maxwellian distribution function of the incident ions X^{Z+}, or by the product $v\,\sigma_c$ in collisions of monokinetic ion beams with gaseous or solid targets.

To date the experimental rates of charge-exchange processes have been measured for ions X^{Z+} ($Z < 10$) colliding with neutral gases. For example, capture-rate coefficients $v\,\sigma_c$ in the collisions $Ne^{8+}(1s\,2p\,^3P_1) + A$; A = He, Ne, Ar, Xe, CH_4, were obtained by observing the quenching of the K_α x-ray lines $1s\,2p\,^{1,3}P_1 - 1s^2\,^1S_0$ in He-like neon ions [10.45, 46]. Low-velocity Ne^{8+} ions ($v = 8 \times 10^5$ cm s^{-1}) were produced by the impact of 1.4 MeV/amu uranium ions on a neutral neon-gas target using the Universal Linear Accelerator (UNILAC) at GSI (Darmstadt). The corresponding values are given in Table 10.7

Table 10.7. Experimental capture-rate coefficients $v\,\sigma_c$ for the reaction $Ne^{8+}(1s\,2p\,^3P_1) + A$ at velocity $v = 8 \times 10^5$ cm s^{-1} [10.45]

Target A	Target binding energy [eV]	$v\,\sigma_c$ [10^{-9} cm^3 s^{-1}]
He	24.49	1.0 ± 0.3
Ne	21.55	2.9 ± 0.9
Ar	15.76	8.2 ± 2.5
Xe	12.40	23 ± 7
CH_4	12.60	8.0 ± 2.4

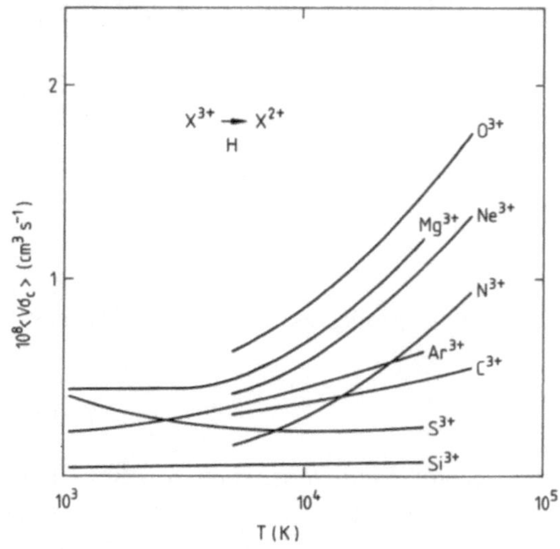

Fig. 10.9–10. Charge-exchange rate coefficients for the $X^{3+,4+}$ + H reactions: quantum calculations [10.49, 50]

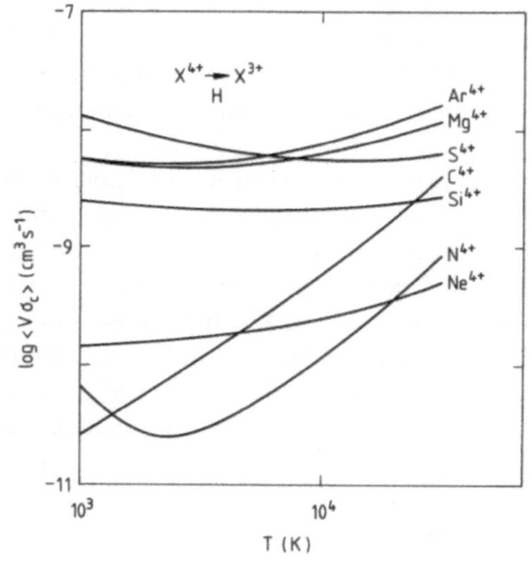

Fig. 10.10.

The rate coefficients $\langle v\,\sigma_c \rangle$ for a number of charge-transfer reactions of astrophysical importance, X^{Z+} + H, He $(Z = 1 - 4)$ at low temperatures are presented in [10.47–50]. Values of $\langle v\,\sigma_c \rangle$ in [10.49, 50] were calculated using configuration interaction wave functions in a close-coupled formulation of the scattering. The corresponding rate coefficients for charge transfer in hydrogen are given in Figs. 10.9, 10.

References

Chapter 1

1.1 I. S. Bowen, R. A. Millikan: Phys. Rev. **25,** 591 (1925)

1.2 B. Edlén: "Atomic Spectra", in *Spectroscopy I,* ed. by S. Flügge, Handbuch der Physik, Vol. **27** (Springer, Berlin, Göttingen, Heidelberg 1964) p. 80

1.3 B. Edlén: Phys. Scr. **T3,** 5 (1983)

1.4 C. E. Moore: *Atomic Energy Levels,* Vols. 1–3 (Natl. Bur. Stand. Washington, D.C. 1949, 1950, 1952)

1.5 B. Edlén, F. Tyren: Nature, **143,** 940 (1939)

1.6 V. V. Korneev, V. V. Krutov, S. L. Mandel'stam, I. A. Zhitnik: Sol. Phys. **63,** 319, 329 (1979); **68,** 381 (1980)

1.7 G. E. Brueckner: Philos. Trans. R. Soc. London, Ser. A **281,** 443 (1976)

1.8 A. K. Duprée: Adv. At. Mol. Phys. **14,** 393 (1978)

1.9 H. W. Drawin: Phys. Rep. **34,** 125 (1978); Phys. Scr. **24,** 622 (1981)

1.10 M. F. A. Harrison: Phys. Rep. **37,** 59 (1978)

1.11 J. T. Hogan: Phys. Rep. **37,** 83 (1978)

1.12 See, e.g.: Phys. Rep. **37,** (1978); Phys. Scr. **23** Vol. 2 (1981)

1.13 M. R. C. McDowell, A. M. Ferendeci (eds.): *Atomic and Molecular Processes in Controlled Thermonuclear Fusion* (Plenum, New York 1980)

1.14 C. J. Joachain, D. E. Post (eds.): *Atomic and Molecular Physics of Controlled Thermonuclear Fusion* (Plenum, New York 1983)

1.15 B. Grasemann (ed.): *Atomic Inner Shell Processes,* Vols. 1, 2 (Academic, New York 1975)

1.16 I. A. Sellin, D. J. Pegg (eds.): *Beam Foil Spectroscopy: Heavy Ion Atomic Physics* (Plenum, New York 1975)

1.17 S. Bashkin (ed.): *Beam-Foil Spectroscopy,* Topics Current Phys., Vol. 1 (Springer, Berlin, Heidelberg 1976)

1.18 I. A. Sellin (ed.): *Structure and Collisions of Ions and Atoms,* Topics Current Phys. Vol. 5 (Springer, Berlin, Heidelberg 1978) Chap. 7

1.19 Ya. B. Zel'dovich, V. S. Popov: Usp. Fiz. Nauk **105,** 403 (1972) [English transl.: Sov. Phys.-Usp. **14,** 673 (1972)]

1.20 J. Reinhardt, W. Greiner: Rep. Prog. Phys. **40,** 219 (1977)

1.21 L. P. Presnyakov, V. P. Shevelko: Plis'ma Zh. Eksp. Theor. Fiz. **13,** 286 (1971) [English transl.: JETP Lett. **13,** 203 (1971)]

1.22 A. V. Vinogradov, I. I. Sobel'man: Zh. Eksp. Theor. Fiz. **63,** 2113 (1972) [English transl.: Sov. Phys.-JETP **36,** 1115 (1973)]

1.23 F. V. Bunkin, V. I. Derzhiev, S. I. Yakovlenko: Kvantovaya Electron. **8,** 1621 (1981) [English transl.: Sov. j. Quantum Electron. **11,** 981 (1981)]

Chapter 2

2.1 E. U. Condon, G. H. Shortley: *The Theory of Atomic Spectra* (Cambridge Univ. Press, Cambridge 1935)

2.2 I. I. Sobel'man: *Atomic Spectra and Radiative Transitions,* Springer Ser. Chem Phys., Vol. 1 (Springer, Berlin Heidelberg 1979)

2.3 B. Edlén, F. Tyren: Nature **143,** 940 (1939)
2.4 A. H. Gabriel: In *Highlights of Astronomy,* ed. by C. de Jager (Reidel, Dordrecht 1971) pp. 486–494
2.5 L. P. Presnyakov: Usp. Fiz. Nauk **119,** 49 (1976) [English transl.: Sov. Phys. – Usp. **19,** 387 (1976)]
2.6 V. A. Boiko, A. Ya. Faenov, S. A. Pikuz: J. Quant. Spectrosc. Radiat. Transfer **19,** 11 (1978)
2.7 V. A. Boiko, A. V. Vinogradov, S. A. Pikuz, I. Yu. Skobelev, A. Ya. Faenov: *X-Ray Spectroscopy of Laser Produced Plasmas* (Plenum, New York 1984), in press
2.8 U. Feldman: Phys. Scr. **24,** 681 (1981)
2.9 N. J. Peacock: In Proc. XIII Int. Conf. Physics Ionized Gases, Invited Lectures (Physical Society of G.D.R., Berlin 1977) p. 383
2.10 L. P. Presnyakov, A. M. Urnov: J. Phys. (Paris), Colloq. Suppl. n7 **40,** C7–C279 (1979)
2.11 R. D. Cowan: *The Theory of Atomic Structure and Spectra* (University of California Press, Berkeley 1981)
2.12 Yu. I. Grineva, V. I. Karev, V. V. Korneev, V. V. Krutov, S. L. Mandel'stam, L. A. Vainshtein, B. N. Vasiljev, I. A. Zhitnik: In Proc. 16th Cosmic Space Research Meeting (Seattle, USA, 1971) p. 243
2.13 Yu. I. Grineva, V. I. Karev, V. V. Korneev, V. V. Krutov, S. L. Mandel'stam, L. A. Vainshtein, B. N. Vasiljev, I. A. Zhitnik: Sol. Phys. **29,** 441 (1973)
2.14 G. A. Doschek, U. Feldman, R. W. Kreplin, L. Cohen: Astrophys. J. **239,** 725 (1980)
2.15 U. Feldman, G. A. Doschek, R. W. Kreplin: Astrophys. J. **238,** 265 (1980)
2.16 A. H. Gabriel, C. Jordan: Mon. Not. R. Astron. Soc. **145,** 241 (1969)
2.17 A. H. Gabriel: Mon. Not. R. Astron. Soc. **160,** 99 (1972)
2.18 C. P. Bhalla, A. H. Gabriel, L. P. Presnyakov: Mon. Not. R. Astron. Soc. **172,** 359 (1975)
2.19 F. Bely-Dubau, A. H. Gabriel, S. Volonte: Mon. Not. R. Astron. Soc. **186,** 405 (1979)
2.20 J. Dubau, A. H. Gabriel, M. Loulerque, L. Steenman-Clark: Mon. Not. R. Astron. Soc. **195,** 705 (1981)
2.21 E. Ya. Goltz, I. A. Zhitnik, E. Ya. Kononov, S. L. Mandel'stam, Yu. V. Sidelnikov: Dokl. Akad. Nauk SSSR **220,** 560 (1975) [English transl.: Sov. Phys. – Dokl. **20,** 49 (1975)]
2.22 J. L. Schwob, B. S. Fraenkel: Phys. Lett. **40 A,** 81 (1972)
2.23 J. J. Turechek, H. – J. Kunze: Z. Phys. A **273,** 111 (1975)
2.24 R. Beier, H.-J. Kunze: Z. Phys. A **285,** 347 (1978)
2.25 R. G. Burkhalter, C. M. Dozier, D. J. Nagel: Phys. Rev. A **15,** 700 (1977)
2.26 P. G. Burhalter, R. Schmieder, C. M. Dozier, R. D. Cowan: Phys. Rev. A **18,** 718 (1978)
2.27 M. Bitter, S. von Goeler, R. Horton, M. Goldman, K. W. Hill, N. R. Sauthoff, W. Strodick: Phys. Rev. Lett. **43,** 129 (1979)
2.28 J. L. Schwob, M. Klapisch, L. Schwertzer, C. Breton, C. de Michelis, M. Mattioli: Phys. Lett. **26,** 544 (1977)
2.29 H. Gould, R. Marrus, R. W. Schmieder: Phys. Rev. Lett. **31,** 501 (1973)
2.30 I. A. Armour, S. Bashkin, N. A. Jelley, R. O'Brien, J. D. Silver, E. A. Trabert: J. Phys. (Paris), Collog. Suppl. n2 **40,** C1–C211 (1979)
2.31 I. A. Sellin: "Extensions of Beam Foil Spectroscopy", in *Structure and Collisions of Ions and Atoms,* ed. by I. A. Sellin, Topics Curr. Phys., Vol. 5 (Springer, Berlin, Heidelberg 1978) p. 273
2.32 A. M. Ermolaev, M. Jones: J Phys. B **7,** 199 (1974)
2.33 L. N. Ivanov, E. P. Ivanova, U. I. Safronova: J. Quant. Spectrosc. Radiab. Transfer **15,** 553 (1975)
2.34 D. Layser, J. Bahcall: Ann. Phys. (N.Y.) **17,** 177 (1962)
2.35 L. A. Vainshtein, U. I. Safronova: At. Data Nucl. Data Tables **21,** 49 (1978)
2.36 C. D. Lin, W. R. Johnson, A. Dalgarno: Phys. Rev. A **15,** 154 (1977)
2.37 R. H. Garstang: Publ. Astron. Soc. Pac. **81,** 482 (1969)

2.38 R. J. Fortner, R. H. Howell, D. L. Matthews: J. Phys. B **13**, 3545 (1980)
2.39 S. Morita, J. Fujita: J. Phys. Soc. Jpn. **52**, 2673 (1983)
2.40 U. I. Safronova: Phys. Scri. **23**, 241 (1981)
2.41 I. L. Beigman, U. I. Safronova: Sov. Phys. – JETP **60**, 2045 (1971)
2.42 J. P. Buchet, M. C. Buchet, M. C. Buchet-Poulizac, A. Demis, J. Desesquelles, M. Druetta, J. P. Grandin, X. Husson: Phys. Rev. A **23**, 3354 (1981)
2.43 B. N. Chichkov, M. A. Mazing, A. P. Shevelko, A. M. Urnov: Phys. Lett. **83 A**, 412 (1981)
2.44 F. Folkmann, H. F. Beyer, R. Mann, K.-H. Schartner: Nucl. Instrum. Methods **181**, 99 (1981)
2.45 B. Edlén: J. Phys. Chem. Ref. Data **3**, 825 (1974)
2.46 J. Reader, Ch. Corliss, W. Wiese, W. A. Martin: *Wavelengths and Transition Probabilities for Atoms and Ions,* NSRDS **68** (Natl. Bur. Stand., Washington, D.C. 1980)
2.47 H. A. Bethe, E. E. Salpeter: *Quantum Mechanics of One- and Two-Electron Atoms* (Springer, Berlin, Göttingen, Heidelberg 1957)
2.48 U. I. Safronova, A. F. Shestakov: Preprint n7 (Inst. of Spectroscopy of the USSR Acad. Sci. 1976)
2.49 G. W. Ericson: J. Phys. Chem. Ref. Data. **6**, 325 (1977)
2.50 P. J. Mohr: Ann. Phys. **88**, 25 (1974); Phys. Rev. Lett. **34**, 1050 (1975); Phys. Rev. A **26**, 2338 (1982)
2.51 S. R. Lundeen, F. M. Pipkin: 2nd Int. Conf. Precise Meas. of Fundamental Constants, NBS, Gaithersburg, Maryland, Book of Abstr. pp. 52–53 (1981)
2.52 I. A. Sellin: Adv. Atom. Mol. Phys. **12**, 215 (1976)
2.53 P. Pellegrin, Y. Mazzi, L. Falffy, A. Prieels: Phys. Rev. Lett. **49**, 1762 (1982)
2.54 E. T. Nelson, F. M. Pipkin ibid. pp. 92–93
2.55 R. C. Elton: Astrophys. J. **148**, 573 (1967)
2.56 G. W. F. Drake, A. Dalgarno: Astrophys. J. **157**, 459 (1968)
2.57 G. W. F. Drake: Astrophys. J. **158**, 1199 (1969)
2.58 G. W. F. Drake, G. A. Viktor, A. Dalgarno: Phys. Rev. **180**, 25 (1969)
2.59 K. T. Cheng, Y. K. Kim, J. P. Desclaux: At. Data Nucl. Data Tables 24, **111** (1979)
2.60 U. I. Safronova, T. G. Lisina: At. Data Nucl. Data Tables **24**, 49 (1979)
2.61 U. I. Safronova, V. S. Senashenko: Phys. Scr. **25**, 37 (1982)
2.62 P. Mohr: At. Data Nucl. Data Tables **29**, 453 (1983)
2.63 B. Edlén: "Atomic Spectra", in *Spectroscopy I,* ed. by S. Flügge, Handbuch der Physik, Vol. 27 (Springer, Berlin, Göttingen, Heidelberg 1964) pp. 80–220
2.64 A. Dalgarno: Adv. Atom. Mol. Phys. **11**, 281 (1962)
2.65 A. F. Shestakov, S. V. Kchristenko: Opt. Spektrosk. **36**, 635 (1974) [English transl.: Sov. Phys. – Opt. Spectrosc. **36**, 635 (1974)]
2.66 F. Weinhold: Proc. Roy. Soc. A **327**, 209 (1972)
2.67 S. A. Zapryagaev, N. L. Manakov, V. G. Palchikov: Opt. Spektrosk. **46**, 214 (1978) [English transl.: Sov. Phys. – Opt. Spectrosc. **46**, 119 (1979)]
2.68 R. F. Stewart: Mol. Phys. **30**, 787 (1975); **30**, 1283 (1975)
2.69 E. Markewicz, R. P. McEachran, A. D. Stauffer: J. Phys. B **14**, 949 (1981)
2.70 S. Fraga, J. Karwowski, K. M. S. Saxena: *Handbook of Atomic Data* (Elsevier, Amsterdam 1976)
2.71 L. Curtis: Phys. Scr. **21**, 162 (1980)
2.72 A. Dalgarno, A. E. Kingston: Proc. Phys. Soc. **73**, 455 (1959)
2.73 A. Lindgaard, L. J. Curtis, I. Martinson, S. E. Nielsen: Phys. Scr. **21**, 47 (1980)
2.74 S. O. Kastner, M. L. Wolf: J. Opt. Soc. Am. **69**, 1279 (1979)
2.75 V. P. Shevelko, A. V. Vinogradov: Phys. Scr. **19**, 275 (1979)
2.76 B. Edlén: Phys. Scr. **17**, 565 (1978)
2.77 P. Vogel: Nucl. Instrum. Methods **110**, 241 (1973)
2.78 V. P. Shevelko: Kratk. Soobshch. Fiz. **4**, 47 (1982) [English transl.: Sov. Phys. – Short Commun. Phys. **4**, 25 (1982)]
2.79 R. R. Teachout, R. T. Pack: At. Data Nucl. Data Tables **3**, 195 (1971)

2.80 S. I. Vetchinkin, S. V. Kchristenko: Chem. Phys. Lett. **1**, 437 (1967)
2.81 S. V. Kchristenko, S. I. Vetchinkin: Opt. Spektrosk. **26**, 310 (1969) [English transl.: Sov. Phys. – Opt. Spectrosc. **26**, 186 (1969)]
2.82 N. Natori, M. Matsuzawa, T. Watanabe: J. Phys. Soc. Jpn. **30**, 518 (1971)
2.83 V. G. Palchikov: Izv. Vyssh. Uchebn. Zaved., Fiz. **6**, 105 (1980)
2.84 J. P. Briand, J. P. Mossé, P. Indelicato, et al.: Phys. Rev. A **28**, 1413 (1983)
2.85 H. F. Beyer, R. D. Deslattes, F. Folkman, R. E. La Villa: *Determination of the 1s Lamb shift in one-electron Argon recoil ions,* Preprint GSI-84-55 (Darmstadt 1984)
2.86 R. D. Deslattes, H. F. Beyer, F. Folkman: J. Phys. B **17**, L 35 (1984)

Chapter 3

3.1 A. Sommerfeld: Ann. Phys. **11**, 257 (1931); *Atombau und Spektrallinien,* Vol. 2 (Vieweg, Braunschweig 1951)
3.2 H. A. Bethe, S. E. Salpeter: Quantum Mechanics of One- and Two-Electron Atoms (Springer, Berlin, Göttingen, Heidelberg 1957) Sect. 59
3.3 A. Nordsieck: Phys. Rev. **93**, 785 (1954)
3.4 M. Abramowitz, I. S. Stegun (eds.) *Handbook of Mathematical Functions* (Dover, New York 1965)
3.5 V. B. Berestetskiy, E. M. Lifshitz, L. P. Pitaevskiy: *Relativistic Quantum Theory,* Part I (Nauka, Moscow 1968)
3.6 J. M. Berger: Phys. Rev. **105**, 35 (1957)
3.7 L. D. Landau, E. M. Lifschitz: *Klassische Feldtheorie* (Akademie Verlag, Berlin 1971); *The Classical Theory of fields,* 3rd ed. (Pergamon, Oxford 1971)
3.8 I. I. Sobel'man: *Atomic Spectra and Radiative Transitions,* Springer Ser. Chem. Phys., Vol. 1 (Springer, Berlin, Heidelberg 1979)
3.9 V. Gervids, V. I. Kogan: Zh. Eksp. Teor. Fiz. Pis'ma **22**, 511 (1975) [English transl.: Sov. Phys. – JETP Lett. **22**, 308 (1975)]
3.10 H. A. Bethe, W. Heitler: Proc. Roy. Soc. London A **146**, 83, (1934)
3.11 F. Sauter: Ann. Phys. **20**, 404 (1934)
3.12 H. A. Bethe, L. Maximon: Phys. Rev. **93**, 768 (1954)
3.13 R. L. Gluckstern, M. N. Hill, G. Breit: Phys. Rev. **90**, 1026 (1953)
3.14 H. W. Koch, J. W. Motz: Rev. Mod. Phys. **31**, 920 (1959)
3.15 H. McMaster: Rev. Mod. Phys. **33**, 8 (1961)
3.16 I. L. Beigman, L. A. Vainshtein, B. N. Chichkov: Zh. Eksp. Teor. Fiz. **80**, 964 (1981). English transl.: Sov. Phys. – JETP **53**, 541 (1981)
3.17 I. I. Sobel'man, L. A. Vainshtein, E. A. Yukov: *Excitation of Atoms and Broadening of Spectral Lines,* Springer Ser. Chem. Phys., Vol. 7 (Springer, Berlin, Heidelberg 1981)
3.18 D. W. Missavage, S. T. Manson, G. R. Duam: Phys. Rev. A **15**, 1001 (1977)
3.19 R. F. Reilman, S. T. Manson: Phys. Rev. A **18**, 2124 (1978)
3.20 W. D. Barfield: J. Phys. B **13**, 931 (1980)
3.21 T. N. Chang, T. Olsen: Phys. Rev. A **24**, 1091 (1981)
3.22 W. Ong, S. T. Manson, H. K. Tseng, R. H. Pratt: Phys. Lett. **69 A**, 319 (1979)
3.23 A. Burgess: Astrophys. J. **139**, 776 (1964)
3.24 I. L. Beigman, L. A. Vainshtein, R. Sunyaev: Usp. Fiz. Nauk **95**, 267 (1968) [English transl.: Sov. Phys. – Usp. **11**, 411 (1968)]
3.25 J. N. Gau, Y. Hahn, J. A. Retter: J. Quant. Spectrosc. Radiat. Transfer **23**, 131 (1980)
3.26 L. A. Vainshtein: Proc. P. N. Lebedev Inst. **119**, 3 (1980)
3.27 J. Dubau, S. Volonté: Rep. Prog. Phys. **43**, 199 (1980)
3.28 R. L. Brooks, R. U. Datla, H. R. Griem: Phys. Rev. Lett A **41**, 107 (1978)
3.29 R. L. Brooks, R. U. Datla, A. D. Krumbein, H. R. Griem: Phys. Rev. A **21**, 1387 (1980)

3.30 C. Breton, C. DeMichelis, M. Finkenthal, M. Mattioli: Phys. Rev. Lett. **41**, 110 (1978)
3.31 F. Bely-Dubau, M. Bitter, J. Dubau, P. Faucher, A. H. Gabriel, K. W. Hill, S. von Goeler, N. Sauthoff, S. Volonté: Phys. Lett. **93A**, 189 (1983)
3.32 B. N. Chichkov, M. A. Mazing, A. P. Shevelko, A. M. Urnov: Phys. Lett. **83A**, 412 (1981)
3.33 J. B. A. Mitchell, C. T. Ng, J. L. Forand, D. P. Levac, R. E. Mitchell, A. Sen, D. B. Mico, J. Wm. McGowan: Phys. Rev. Lett. **50**, 335 (1983)
3.34 D. S. Belic, G. H. Dunn, T. J. Morgan, D. W. Mueller, C. Timmer: Phys. Rev. Lett. **50**, 339 (1983)
3.35 K. La Gattuta, Y. Hahn: J. Phys. B **15**, 2101 (1982)
3.36 U. Fano: Phys. Rev. **124**, 1866 (1968)
3.37 O. Bely: J. Phys. B **1**, 23 (1968)
3.38 E. H. S. Burhop: *The Auger Effect and other Radiationless Transitions* (Cambridge Univ. Press, Cambridge 1952)
3.39 A. Temkin (ed.): *Autoionization* (Mono Book Co., Baltimore 1966)
3.40 M. J. Seaton: Proc. Roy. Soc. **77**, 184 (1961)

Chapter 4

4.1 A. M. Lane, R. G. Thomas: Rev. Mod. Phys. **30**, 257 (1958)
4.2 H. Feshbach: Ann. Phys **5**, 357 (1958)
4.3 L. D. Landau, E. M. Lifshitz: *Quantum Mechanics* (Pergamon, New York 1975)
4.4 N. F. Mott, H. S. W. Massey: *Theory of Atomic Collisions*, 2nd ed. (Clarendon Press, Oxford 1965)
4.5 A. I. Baz': Zh. Eksp. Theor. Fiz. **36**, 1762 (1959) [English transl.: Sov. Phys. – JETP **36**, 1256 (1959)]
4.6 L. Fonda, R. Newton: Ann. Phys. **7**, 133 (1959)
4.7 R. Newton, L. Fonda: Ann. Phys. **9**, 416 (1960)
4.8 L. Fonda: Nuovo Cimento, Suppl. **20**, 116 (1961)
4.9 M. Gailitis: Zh. Eksp. Teor. Fiz. **44**, 1974 (1963) [English transl.: Sov. Phys – JETP **17**, 1328 (1963)]
4.10 M. J. Seaton: Proc. Phys. Soc. **88**, 801 (1966); **88**, 815 (1966)
4.11 M. J. Seaton: J. Phys. B **2**, 5 (1969)
4.12 W. Eissner, M. J. Seaton: J. Phys. B **5**, 2187 (1972)
4.13 M. D. Hershkowitz, M. J. Seaton: J. Phys. B **6**, 1176 (1973)
4.14 L. P. Presnyakov, A. M. Urnov: J. Phys. B **8**, 1280 (1975)
4.15 L. P. Presnyakov, A. M. Urnov: Zh. Eksp. Teor. Fiz. **68**, 61 (1975) [English transl.: Sov. Phys. – JETP **41**, 31 (1975)]
4.16 I. I. Sobel'man, L. A. Vainshtein, E. A. Yukov: *Excitation of Atoms and Broadening of Spectral Lines*, Springer Ser. Chem. Phys., Vol. 7 (Springer, Berlin, Heidelberg 1981)
4.17 M. H. Hull, G. Breit: "Coulomb Wave Functions", in *Nuclear Reactions II: Theory*, ed. by S. Flügge, Handbuch der Physik, Vol. XLI (I) (Springer, Berlin, Göttingen, Heidelberg 1959) p. 408
4.18 C. E. Froberg: Rev. Mod. Phys. **27**, 399 (1955)
4.19 Bateman Manuscript Project, ed. by A. Erdelyi: *Higher Transcendental Functions* (McGraw-Hill, New York 1953)
4.20 M. Abramowitz, I. S. Stegun (eds.): *Handbook of Mathematical Functions* (Dover, New York 1965)
4.21 A. I. Baz', Ya. B. Zeldovich, A. M. Perelomov: *Scattering, Reactions and Decay in Non-relativistic Quantum Mechanics* (Nauka: Moscow 1971) (in Russian)
4.22 P. C. W. Davies, M. J. Seaton: J. Phys. B **2**, 341 (1969)
4.23 V. I. Ochkur: Zh. Eksp. Teor. Fiz. **45**, 735 (1963) [English transl.: Sov. Phys. – JETP **18**, 435 (1963)]

4.24 I. L. Beigman, L. A. Vainshtein: Zh. Eksp. Teor. Fiz. **52,** 185 (1967) [English transl.: Sov. Phys. – JETP, **25,** 119 (1967)]
4.25 L.A. Vainshtein: Zh. Eksp. Teor. Fiz. **67,** 63 (1974) [English transl.: Sov. Phys. – JETP **40,** 36 (1974)]
4.26 L. P. Presnyakov, A. M. Urnov, B. N. Chichkov: 12th Int. Conf. Phys. Elect. At. Collisions, Abstracts, p. 480 (Gathlinburg, Tenn. 1981)
4.27 A. K. Pradhan, D. W. Norcross, D. G. Hummer: Phys. Rev. A **23,** 619 (1981)
4.28 A. K. Pradhan: Phys. Rev. Lett. **47,** 79 (1981)
4.29 K. A. Berrington, P. G. Burke, P. L. Dufton, A. E. Kingston, A. L. Sinfailam: J. Phys. B **12,** L275 (1979)
4.30 P. O. Taylor, D. Gregory, G. H. Dunn, R. A. Phaneuf, D. H. Crandall: Phys. Rev. Lett. **39,** 1256 (1977)
4.31 D. Gregory, G. H. Dunn, R. A. Phaneuf, D. H. Crandall: Phys. Rev. A **20,** 410 (1979)
4.32 P. O. Taylor, R. A. Phaneuf, G. H. Dunn: Phys. Rev. A **22,** 435 (1980)
4.33 W.-L. van Wyngaarden, R. J. W. Henry: J. Phys. B **9,** 146 (1976)

Chapter 5

5.1 J. J. Thomson: Philos. Mag. **33,** 449 (1912)
5.2 M. R. H. Rudge: Rev. Mod. Phys. **40,** 564 (1968)
5.3 O. Bely, H. van Regemorter: Ann. Rev. Astron. Astrophys. **8,** 329 (1970)
5.4 D. H. Crandall: Phys. Scr. **23,** 153 (1981)
5.5 V. A. Bazylev, M. I. Chibisov: Usp. Fiz. Nauk **133,** 617 (1981) [English transl: Sov. Phys.-Usp. **33,** 372 (1981)]
5.6 R. K. Peterkop: *Theory of Ionization of Atoms by Electron Impact* (Associated University Press, Colorado 1977)
5.7 R. K. Peterkop: Zh. Eksp. Teor. Fiz **43,** 616 (1962) [English transl.: Sov. Phys. – JETP, **16,** 442 (1963)]
5.8 H. S. W. Massey, J. H. Moore: Proc. Roy. Soc. **142,** 143 (1933)
5.9 G. H. Wannier: Phys. Rev. **90,** 817 (1953)
5.10 R. Peterkop: J. Phys. B **4,** 513 (1971)
5.11 A. R. P. Rau: Phys. Rev. A **4,** 207 (1971)
5.12 T. A. Roth: Phys. Rev. A **5,** 476 (1972)
5.13 Y. Itikawa, T. Kato: *Empirical Formulas for Ionization Cross Section of Atomic Ions for Electron Collisions,* Report IPPJ-AM-17, Inst. of Plasma Physics, Nagoya University (1981)
5.14 W. Lotz: Z. Phys. **216,** 241 (1968); **220,** 466 (1969)
5.15 A. Burgess, I. C. Percival: Adv. At. Mol. Phys. **4,** 109 (1968)
5.16 R. Fox, W. Hickman, T. Kjeldaas: Phys. Rev. **89,** 555 (1953)
5.17 L. Goldberg, A. K. Duprée, J. W. Allen: Ann. Astrophys. (Paris) **28,** 589 (1965)
5.18 B. Peart, K. T. Dolder: J. Phys. B **1,** 872 (1968)
5.19 D. H. Sampson, L. B. Golden: J. Phys. B **12,** 785 (1979)
5.20 D. H. Crandall, R. A. Falk, D. S. Belić, G. H. Dunn: Phys. Rev. A **25,** 143 (1982)
5.21 G. H. Dunn: IEEE Trans. Nucl. Sci. NS-**23,** 929 (1976)
5.22 E. D. Donets, V. P. Ovsyannikov: Zh. Eksp. Teor. Fiz. **80,** 916 (1981) [English transl.: Sov. Phys. – JETP **53,** 466 (1981)]
5.23 M. R. H. Rudge, S. B. Shwartz: Proc. Phys. Soc. **88,** 563 (1966)
5.24 E. D. Donets, V. P. Ovsyannikov: *Ionization of the Positive Ions of N, O, Ne and Ar by Electron Impact,* Rpt. P7-10780, Joint Institute for Nuclear Research, Dubna (1977)
5.25 D. L. Moores: J. Phys. B **11,** L403 (1978)
5.26 L. B. Golden, D. H. Sampson: J. Phys. B **10,** 2229 (1977); B **11,** 541 (1978); B **14,** 903 (1981)

5.27 A. Müller, E. Salzborn, R. Frodl, R. Becker, H. Klein, H. Winter: J. Phys. B **13**, 1877 (1980)

5.28 A. Salop: Phys. Rev. A **14**, 2095 (1976)

5.29 V. P. Shevelko, L. A. Vainshtein, A. M. Urnov, A. Müller: Mon. Not. R. Astron. Soc., Short Commun. **203**, 45 (1983)

5.30 S. M. Younger: Phys. Rev. A **22**, 111 (1980)

5.31 D. C. Griffin, C. Bottcher, M. S. Pindzola: Phys. Rev. A **25**, 154 (1982)

5.32 A. Müller, R. Frodl: Phys. Rev. Lett. **44**, 29 (1980)

5.33 C. Achenbach, A. Müller, E. Salzborn, R. Becker: Phys. Rev. Lett. **50**, 2070 (1983)

5.34 V. P. Shevelko, L. A. Vainshtein: *Electron Impact Ionization of Highly Charged Ions*, Rpt. No. 78, P. N. Lebedev Inst., Moscow (1983)

Chapter 6

6.1 I. V. Komarov, L. I. Ponomarev, S. Yu. Slavyanov: *Spheroidal and Coulomb Spheroidal Functions* (Nauka, Moscow 1976)

6.2 L. D. Landau, E. M. Lifshitz: *Quantum Mechanics* (Pergamon, Oxford 1975)

6.3 S. S. Gershtein, V. D. Krivchenkov: Zh. Eksp. Teor. Fiz. **40**, 1942 (1961) [English transl.: Sov. Phys. – JETP **13**, 1044 (1961)]

6.4 R. J. Damburg, R. Kh. Propin: J Phys. B **1**, 681 (1968)

6.5 I. V. Komarov, S. Yu. Slavyanov: J. Phys. B **1**, 1066 (1968)

6.6 J. D. Power: Phil. Trans. Roy. Soc. London A **274**, 663 (1973)

6.7 J. von Neumann, E. Wigner: Z. Phys. **30**, 467 (1929)

6.8 S. P. Alliluev, A. V. Matveenko: Zh. Eksp. Teor. Fiz. **51**, 1873 (1966) [English transl.: Sov. Phys. – JETP **24**, 1260 (1967)]

6.9 C. A. Coulson, A. Joseph: Int. J. Quant. Chem. **1**, 337 (1967)

6.10 E. E. Nikitin, S. Ya. Umanskii: *Non-adiabatic Transitions in Slow Atomic Collisions* (Atomizdat, Moscow 1979); *Theory of Slow Atomic Collisions*, Springer Ser. Chem. Phys., Vol. 30 (Springer, Berlin, Heidelberg 1984)

6.11 E. A. Solov'ev: Zh. Eksp. Teor. Fiz. **81**, 1681 (1981) [English transl.: Sov. Phys – JETP **54**, 893 (1981)]

6.12 R. K. Janev, C. J. Joachain, N. N. Nedeljković: Phys. Rev. A **26**, 116 (1982)

6.13 I. V. Komarov, E. A. Solov'ev: Teoret. Mat. Fiz. **40**, 130 (1979) (in Russian)

6.14 R. E. Olson, A. Salop: Phys. Rev. A **14**, 579 (1976)

6.15 L. I. Ponomarev: Zh. Eksp. Teor. Fiz **55**, 1836 (1968) [English transl.: Sov. Phys. – JETP **28**, 971 (1969)]

6.16 M. I. Chibisov: Zh. Eksp. Teor. Fiz. **76**, 1898 (1979) [English transl.: Sov. Phys. – JETP **49**, 962 (1979)]

6.17 E. L. Duman, L. I. Men'shikov: Zh. Eksp. Teor. Fiz. **77**, 858 (1979) [English transl.: Sov. Phys. – JETP **50**, 433 (1979)]

6.18 M. I. Chibisov: Zh. Tekh. Fiz. **51**, 470 (1981) [English transl.: Sov. Phys. – Tech. Phys. **26**, 284 (1981)]

6.19 W. G. Baber, H. R. Hassé: Proc. Cambridge Philos. Soc. **31**, 564 (1935)

6.20 G. Jaffé: Z. Phys. **87**, 535 (1934)

6.21 K. Helfrich, H. Hartmann: Theor. Chim. Acta. **16**, 263 (1970)

6.22 M. M. Madsen, J. M. Peek: At. Data **2**, 171 (1971)

6.23 L. I. Ponomarev, T. P. Puzynina: JINR-P4-5040 preprint, Joint Institute for Nuclear Research, Dubna (1970); USSR comp. Math. math. Phys. **8**, (No. 6) 94 (1968)

6.24 J. D. Power: *Quantum Chemistry Program Exchange*, Vol. X, No. 233 (Indiana University Chemistry Department, 1974)

6.25 R. S. Mulliken: Phys. Rev. **32**, 186 (1928)

6.26 F. Hund: Z. Phys. **77**, 12 (1932)

6.27 J. Eichler, U. Wille: Phys. Rev. A **11**, 1973 (1975)

6.28 P. P. Szydlik, A. E. S. Green: Phys. Rev. A **9**, 1885 (1974)

6.29 C. Zener: Proc. Roy. Soc. London A **137**, 696 (1932)

6.30 H. Hellman, J. K. Syrkin: Acta Physicochim. URSS **2,** 433 (1935)
6.31 W. Lichten: Phys. Rev. **131,** 229 (1963); **139,** 27 (1965); **164,** 132 (1967)
6.32 U. Fano, W. Lichten: Phys. Rev. Lett. **14,** 627 (1965)
6.33 M. Barat, W. Lichten: Phys. Rev. A **6,** 211 (1972)
6.34 J. Eichler, U. Wille, B. Fastrup, K. Taulbjerg: Phys. Rev. A **14,** 707 (1976)
6.35 H. Rosenthal: Phys. Rev. A **4,** 1030 (1971)
6.36 V. Sidis: "Some Aspects of the Molecular Approach to Atomic Collisions", in *The Physics of Electronic and Atomic Collisions,* ed. by J. S. Risley, R. Geballe (University of Washington Press, Seattle 1975) p. 295
6.37 J. P. Gauyacq: "Diabatic Molecular Orbitals in Atomic Collisions. (Outer Shell Excitation)", in *Electronic and Atomic Collisions* ed. by G. Watel (North-Holland, Amsterdam 1978) p. 431
6.38 W. Lichten: J. Phys. Chem. **84,** 2102 (1980)
6.39 N. F. Mott, H. S. W. Massey: *The Theory of Atomic Collisions,* 3rd ed. (Oxford University Press, Oxford 1965)
6.40 M. R. C. McDowell, J. P. Coleman: *Introduction to the Theory of Ion-Atom Collisions* (North-Holland, Amerdam 1970)
6.41 B. H. Bransden: *Atomic collision Theory,* 2nd ed. (Benjamin, New York 1982)
6.42 T. A. Green: Proc. Phys. Soc. **86,** 1017 (1965)
6.43 D. R. Bates, R. McCarroll: Proc. Roy Soc. London A **245,** 754 (1958)
6.44 S. B. Schneiderman, A. Russek: Phys. Rev. **181,** 311 (1969)
6.45 W. R. Thorson, H. Levy: Phys. Rev. **181,** 240 (1969)
6.46 W. R. Thorson, I. B. Delos: Phys. Rev. A **18,** 117, 156 (1978)
6.47 K. Taulbjerg, J. S. Waaben, B. Fastrup: Phys. Rev. A **12,** 2235 (1975)
6.48 M. E. Riley, T. A. Green: Phys. Rev. A **4,** 619 (1971)
6.49 D. S. F. Crothers, J. G. Hughes: Proc. Roy. Soc. London A **357,** 345 (1978)
6.50 T. A. Green: Phys. Rev. A **23,** 519, 532 (1981)
6.51 L. F. Errea, L. Méndez, A. Riera: J. Phys. B **15,** 101 (1982)
6.52 J. Waaben, K. Taulbjerg: J. Phys. B **14,** 1815 (1981)
6.53 N. F. Mott: Proc. Cambridge Philos. Soc. **27,** 553 (1933)
6.54 R. D. Piacentini, A. Salin: J. Phys. B **7,** 1666 (1974)
6.55 R. K. Janev: Adv. At. Mol. Phys. **12,** 1 (1976)
6.56 F. T. Smith: Phys. Rev. **179,** 111 (1969)
6.57 H. Gabriel, K. Taulbjerg: Phys. Rev. A **10,** 741 (1974)
6.58 J. B. Delos: Rev. Mod. Phys. **53,** 287 (1981)
6.59 I. A. Poluektov, L. P. Presnyakov: Zh. Eksp. Teor. Fiz. **54,** 120 (1968) [English transl.: Sov. Phys. – JETP **27,** 67 (1968)]
6.60 I. A. Poluektov, L. P. Presnyakov: Proc. Lebedev Phys. Inst. (Moscow) **51,** 73 (1970) (in Russian)
6.61 B. H. Bransden: Rep. Prog. Phys. **35,** 949 (1972)
6.62 D. F. Gallaher, L. Wilets: Phys. Rev. **169,** 139 (1968)
6.63 I. M. Cheshire, D. F. Gallaher, A. J. Taylor: J. Phys. B **3,** 813 (1970)
6.64 W. Fritsch, C. D. Lin: J. Phys. B **15,** 1255 (1982)
6.65 J. F. Reading, A. L. Ford, G. L. Swafford, A. Fitchard: Phys. Rev. A **20,** 130 (1979)
6.66 R. K. Janev, L. P..Presnyakov: J. Phys. B **13,** 4233 (1980)
6.67 J. F. Reading, A. L. Ford, R. L. Becker: J. Phys. B **14,** 1995 (1981)
6.68 L. D. Landau: Phys. Z. Sowjetunion **2,** 46 (1932)
6.69 C. Zener: Proc. Roy. Soc. London, A **137,** 696 (1932)
6.70 E. C. G. Stueckelberg: Helv. Phys. Acta **5,** 369 (1932)
6.71 N. Rosen, C. Zener: Phys. Rev. **40,** 502 (1932)
6.72 Yu. N. Demkov: Zh. Eksp. Teor. Fiz. **45,** 159 (1963) [English transl.: Sov. Phys. – JETP **18,** 138 (1964)]
6.73 E. E. Nikitin: Optika Spektrosk. **13,** 761 (1962); Discuss. Faraday Soc. **33,** 41 (1962)
6.74 M. S. Child: *Molecular Collision Theory* (Academic, New York 1974)
6.75 L. A. Vainshtein, L. P. Presnyakow, I. I. Sobel'man: Zh. Eksp. Teor. Fiz. **43,** 518 (1962) [English transl.: Sov. Phys. – JETP **18,** 1383 (1964)]

6.76 D. S. F. Crothers: J. Phys. B **6**, 1418 (1973)
6.77 R. Aaron, R. D. Amado, B. W. Lee; Phys. Rev. **121**, 319 (1961)
6.78 K. Dettmann, G. Leibfried: Phys. Rev. **148**, 1271 (1966)
6.79 R. Shakeshaft, L. Spruch: Rev. Med. Phys. **51**, 369 (1979)
6.80 K. R. Greider, L. R. Dodd: Phys. Rev. **146**, (1966)
6.81 D. R. Bates: Proc. Phys. Soc. **73**, 227 (1969)
6.82 D. R. Bates: Proc. Roy. Soc. London A **247**, 294 (1968)
6.83 A. M. Dykhne: Zh. Eksp. Teor. Fiz **41**, 1324 (1961) [English transl.: Sov. Phys. – JETP **14**, 941 (1962)
6.84 I. M. Cheshire: Proc. Phys. Soc. **84**, 89 (1964)
6.85 T. Pradhan: Phys. Rev. **105**, 1250 (1957)
6.86 M. R. C. McDowell: Proc. Roy. Soc. (London) A **264**, 277 (1961)
6.87 D. Basu, S. C. Mukherjee, D. P. Sural: Phys. Rep. **42**, 145 (1978)
6.88 Dž. Belkić, R. Gayet, A. Salin: Phys. Rep. **56**, 279 (1979)
6.89 Dž. S. Belkić: J. Phys. B **10**, 3491 (1977)
6.90 J. P. Coleman: "The Impulse Approximation and Related Methods in the Theory of Atomic Collisions", in *Case Studies in Atomic Collision Physics,* ed. by E. W. McDaniel, M. R. C. McDowell, Vol. **1** (North-Holland, Amsterdam 1969) p. 101
6.91 J. J. Thomson: Philos. Mag. **23**, 449 (1912)
6.92 A. Burgess, I. C. Percival: Adv. At. Mol. Phys. **4**, 109 (1968)
6.93 L. Vriens: "Binary-Encounter and Classical Collisions Theories", in *Case Studies in Atomic Collision Physics,* ed. by E. W. McDaniel, M. R. C. McDowell, Vol. **1** (North-Holland, Amsterdam 1969) p. 335
6.94 D. R. Bates, A. E. Kingston: Adv. At. Mol. Phys. **6**, 269 (1970)
6.95 V. Fock: Z. Phys. **98**, 145 (1935)
6.96 I. C. Percival, D. Richards: Adv. At. Mol. Phys. **11**, 1 (1975)
6.97 M. Gryzinski: Phys. Rev. **115**, 374 (1959); **138**, 305, 322, 336 (1965)
6.98 L. H. Thomas: Proc. Roy. Soc. London A **114**, 561 (1927)
6.99 J. Hirschfelder, H. Eyring, B. Topley: J. Chem. Phys. **4**, 170 (1936)
6.100 R. Abrines, I. C. Percival: Proc. Phys. Soc. **88**, 861, 873, (1966)
6.101 M. Carplus, R. N. Porter, R. D. Sharma: J. Chem. Phys. **43**, 3259 (1936)

Chapter 7

7.1 L. P. Presnyakov: "Charge Exchange of Highly Charged Ions in Neutral Atoms", in *"Electronic and Atomic Collisions"*, ed by G. Watel (North-Holland, Amsterdam 1977) p. 407
7.2 R. E. Olson: "Electron Capture between Multiply-Charged Ions and One-Electron Targets", in *Electronic and Atomic Collisions,* ed by N. Oda and K. Takayanagi (North-Holland, Amsterdam 1980) p. 391
7.3 E. Salzborn, A. Müller: "Electron Capture by Multi-Charged Ions Colliding with Multi-Electron Targets", in *Electronic and Atomic Collisions,* ed. by N. Oda, K. Takayanagi (North-Holland, Amsterdam 1980) p. 407
7.4 R. K. Janev, L. P. Presnyakov: Phys. Rep. **70**, 1 (1981)
7.5 H. B. Gilbody: Phys. Scr. **24**, 712 (1981)
7.6 P. T. Greenland: Phys. Rep. **81**, 131 (1982)
7.7 B. H. Bransden, R. K. Janev: Adv. At. Mol. Phys. **19**, 1 (1983)
7.8 M. R. C. McDowell, A. M. Ferendeci (eds.): *Atomic and Molecular Processes in Controlled Thermonuclear Fusion* (Plenum, New York 1980)
7.9 C. J. Joachain, D. E. Post (eds.): *Atomic and Molecular Physics of Controlled Thermonuclear Fusion* (Plenum, New York 1983)
7.10 T. P. Grozdanov, R. K. Janev: Phys. Rev. A **17**, 880 (1978)
7.11 A. R. Radzig, B. M. Smirnov: *Handbook on Atomic and Molecular Physics Data* (Atomizdat, Moscow 1980) (in Russian); *Reference Data on Atoms, Molecules and Ions,* Springer Ser. Chem. Phys., Vol. 31 (Springer, Berlin, Heidelberg 1985)

7.12 M. I. Chibisov: Zh. Eksp. Theor. Fiz. Pis'ma **24**, 56 (1976) [English transl.: JETP Lett. **24**, 46 (1976)]
7.13 E. L. Duman, B. M. Smirnov: Fizika plazmy **4**, 1161 (1978) [English transl.: Sov. J. Plasma Phys. **4**, 650 (1978)]
7.14 R. E. Olson, A. Salop: Phys. Rev. A. **14**, 579 (1976)
7.15 T. P. Grozdanov, R. K. Janev: Phys. Lett. **66**A, 191 (1978)
7.16 S. S. Gershtein: Zh. Eksp. Teor. Fiz. **43**, 706 (1962) [English transl.: Sov. Phys. – JETP **16**, 501 (1962)]
7.17 A. Salop, R. E. Olson: Phys. Rev. A **13**, 1312 (1976)
7.18 V. A. Abramov, F. F. Baryshnikov, V. S. Lisitsa: Zh. Eksp. Teor. Fiz. **74**, 897 (1978) [English transl.: Sov. Phys. – JETP **47**, 469 (1979)]
7.19 R. K. Janev, D. S. Belić, B. H. Bransden: Phys. Rev. A **28**, 1293 (1983)
7.20 R. K. Janev, B. S. Belić: Phys. Scr. **T3**, 246 (1983)
7.21 I. V. Komarov, N. F. Truskova: Preprint JINR No. P4-11445, Joint Inst. for Nuclear Research, Dubna (1978)
7.22 J. I. Casaubon, R. D. Piacentini, A. Salin: J. Phys. B **14**, L 297 (1981)
7.23 C. Harel, A. Salin: J. Phys. B **10**, 3511 (1977)
7.24 A. Salop, R. E. Olson: Phys. Rev. A **16**, 1811 (1977)
7.25 A. Salop, R. E. Olson: Phys. Rev. A **19**, 1921 (1979)
7.26 A. Salop, R. E. Olson: Phys. Lett. **71**A, 407 (1979)
7.27 T. G. Winter, G. J. Hatton: Phys. Rev. A **21**, 793 (1980)
7.28 M. Kimura, W. Thorson: Phys. Rev. A **24**, 3019 (1981)
7.29 D. S. F. Crothers, N. R. Todd: J. Phys. B **14**, 2233, 2251 (1981)
7.30 T. A. Green, E. J. Shipsey, J. C. Browne: Phys. Rev. A **25**, 1364 (1982)
7.31 E. J. Shipsey, T. A. Green, J. C. Browne: Phys. Rev. A **27**, 821 (1983)
7.32 R. A. Phaneuf, I. Alvarez, F. W. Meyer, D. H. Crandall: Phys. Rev. A **26**, 1892 (1982)
7.33 H. Ryufuku, T. Watanabe: Phys. Rev. A **18**, 2005 (1978); **19**, 1538 (1979)
7.34 L. P. Presnyakov, D. B. Uskov, R. K. Janev: Zh. Eksp. Teor. Fiz. **83**, 933 (1982) [English transl.: Sov. Phys. – JETP **83**, 525 (1982)]
7.35 S. E. Butler, T. G. Heil, A. Dalgarno: Astrophys. J. **241**, 422 (1980)
7.36 A. Dalgarno, T. G. Heil, S. E. Butler: Astrophys. J. **245**, 793 (1981)
7.37 R. E. Olson, E. J. Shipsey, J. C. Browne: J. Phys. B **11**, 699 (1978)
7.38 R. McCarroll, P. Valiron: Astron. Astrophys. **53**, 83 (1976)
7.39 E. J. Chipsey, J. C. Browne, R. E. Olson: J. Phys. B **14**, 869 (1981)
7.40 V. A. Basylev, N. K. Zhevago: Zh. Eksp. Teor. Fiz. **69**, 853 (1975) [English transl.: Sov. Phys. – JETP **42**, 436 (1975)]
7.41 A. Z. Devdariani, V. N. Ostrovskii, Yu. N. Sebyakin: Zh. Eksp. Teor. Fiz. **71**, 909 (1976) [English transl.: Sov. Phys. – JETP **44**, 477 (1976)]
7.42 L. P. Presnyakov, D. B. Uskov, R. K. Janev: Phys. Lett. **84**A, 243 (1981)
7.43 B. H. Bransden, C. T. Noble, J. Chandler: J. Phys. B **16**, 4191 (1983)
7.44 B. H. Bransden, C. J. Noble: J. Phys. B **15**, 451 (1982)
7.45 W. Fritsch: J. Phys. B **15**, L 389 (1982)
7.46 W. Fritsch, C. D. Lin: J. Phys. B **15**, L 281 (1982); Phys. Scr. **T3**, 241 (1983)
7.47 A. L. Ford, J. F. Reading, R. L. Becker: J Phys. B **15**, 3257 (1982)
7.48 M. B. Shah, T. V. Goffe, H. B. Gilbody: J. Phys. B **11**, L 233 (1978)
7.49 W. Seim, A. Müller, I. Wirkner-Bott, E. Salzborn: J. Phys. B **14**, 3475 (1981)
7.50 T. G. Winter: Phys. Rev. A **25**, 697 (1982)
7.51 L. P. Presnyakov, A. D. Ulantsev: Kvantovaya Elektronika **1**, 2377 (1974) [English transl.: Sov. J. Quantum Electron. **4**, 1320 (1975)]
7.52 H. Ryufuku, T. Watanabe: Phys. Rev. A **20**, 1828 (1979)
7.53 H. Ryufuku: Phys. Rev. A **25**, 720 (1982)
7.54 D. H. Crandall, R. A. Phaneuf, F. W. Meyer: Phys. Rev. A **22**, 379 (1980)
7.55 L. D. Gardner, J. E. Bayfield, P. M. Coch, H. J. Kim, P. H. Stelson: Phys. Rev. A **16**, 1415 (1977)

7.56 F. W. Meyer, R. A. Phaneuf, H. J. Kim, P. Hvelplund, P. H. Stelson: Phys. Rev. A **19**, 515 (1979)
7.57 K. H. Berkner, W. G. Graham, R. V. Pyle, A. S. Schlachter, J. W. Stearns, R. E. Olson: J. Phys. B **11**, 875 (1978)
7.58 K. H. Berkner, W. G. Graham, R. V. Pyle, A. S. Schlachter, J. W. Stearns: Phys. Lett. **62** A, 407 (1977)
7.59 R. E. Olson, A. Salop: Phys. Rev. A **16**, 531 (1977)
7.60 A. Salop: J. Phys. B **12**, 919 (1979)
7.61 R. E. Olson: Phys. Rev. A **24**, 1726 (1981)
7.62 N. Bohr, J. Lindhard: K. Dan. Vidensk. Selsk. Mat. Fys. Medd. **28**, 1 (1954)
7.63 I. V. Komarov, E. A. Solov'ev: "Quasi-classical Approximation for the Charge Exchange Reactions of Highly Charged Ions with Hydrogen Atoms", in *Problems in the Theory of Atomic Collisions,* Vol. 2, ed. by G. F. Drukarev (Leningrad State Univ. Press, Leningrad 1980) p. 234 (in Russian)
7.64 T. P. Grozdanov: J. Phys. B **13**, 3835 (1980)
7.65 H. Knudsen, H. K. Haugen, P. Hvelplund: Phys. Rev. A **23**, 597 (1981)
7.66 R. A. Mapleton: *Theory of Charge Exchange* (Wiley, New York 1972)
7.67 Dž. Belkić, R. Gayet, A. Salin: Phys. Rep. **56**, 279 (1979)
7.68 J. R. Oppenheimer: Phys. Rev. **31**, 349 (1928)
7.69 H. C. Brinkman, H. A. Kramers: Proc. K. Ned. Acad. Sci. (Amsterdam) **33**, 973 (1930)
7.70 A. V. Vinogradov, V. P. Shevelko: Zh. Eksp. Teor. Fiz. **59**, 593 (1970) [English transl.: Sov. Phys. – JETP **23**, 323 (1971)]
7.71 V. P. Shevelko: Z. Phys. A **287**, 19 (1978)
7.72 R. May: Phys. Rev. **136** A, 669 (1964)
7.73 D. S. F. Crothers, N. R. Todd: J. Phys. B **13**, 2277 (1980)
7.74 A. Müller, V. P. Shevelko: Zh. Tekh. Fiz. **50**, 985 (1980) [Englisch transl.: Sov. Phys. – Tech. Phys. **25**, 905 (1980)]
7.75 K. Omidvar: Phys. Rev. **153**, 121 (1967)
7.76 V. P. Shevelko: Fizika (Zagreb) **13**, 185 (1981)
7.77 L. P. Presnyakov, V. P. Shevelko: Zh. Tekh. Fiz. **43**, 1714 (1973) [English transl.: Sov. Phys. – Tech. Phys **18**, 1079 (1974)]
7.78 R. K. Janev, L. P. Presnyakov, V. P. Shevelko: Phys. Lett. **76** A, 121 (1980)
7.79 F. W. Martin, J. T. MacDonald: Phys. Rev. A **4**, 1974 (1971)
7.80 P. Hvelplund, J. Heinemeir, E. Horsdal Pedersen, F. R. Simpson: J. Phys. B **9**, 491 (1976)
7.81 S. M. Ferguson, J. R. MacDonald, T. Chiao, L. D. Ellsworth, S. A. Savoy: Phys. Rev. A **8**, 2417 (1973)
7.82 J. R. MacDonald, S. M. Ferguson, T. Chiao, L. D. Ellsworth, S. A. Savoy: Phys. Rev. A **5**, 1188 (1972)
7.83 K. H. Berkner, W. G. Graham, R. V. Pyle, A. S. Schlachter, J. W. Stearns: Proc. X Int. Conf. Phys. Electron Atom Collision (Abstract of Papers), Paris, 1977, p. 542
7.84 W. Erb: Thesis, GSI-Rpt. p-7-78, Darmstadt 1978
7.85 H. D. Betz: Rev. Mod. Phys. **44**, 465 (1972)
7.86 M. Rødbro, E. Horsdal Pederson: Proc. X Int. Conf. Phys. Electron Atom Collisions (Abstract of Papers), Paris, 1977, p. 48
7.87 J. R. MacDonald, C. L. Cocke, W. W. Edison: Phys. Rev. Lett. **32**, 648 (1974)
7.88 F. T. Chan, J. Eichler: Phys. Rev. Lett. **42**, 58 (1979)
7.89 J. Eichler, F. T. Chan: Phys. Rev. A **20**, 104 (1979)
7.90 D. P. Dewangan: J. Phys. B **10**, 1053 (1977)
7.91 J. Eichler, A. Tsuji, T. Ishihara: Phys. Rev. A 23, 2833 (1981)
7.92 T. S. Ho, D. Umberger, R. L. Day, M. Lieber, F. T. Chan: Phys. Rev. A **24**, 705 (1981)
7.93 R. Shakeshaft, L. Spruch: Rev. Mod. Phys. **51**, 369 (1979)
7.94 J. S. Briggs, L. J. Dubé: J. Phys. B **13**, 771 (1980)

7.95 L. J. Dubé, J. S. Briggs: J. Phys. B **14,** 4595 (1981)
7.96 R. M. Drisko: Thesis, Carnegie Institute of Technology (1955, unpublished)
7.97 R. Shakeshaft: J. Phys. B **7,** 1059 (1974)
7.98 L. H. Thomas: Proc. Roy. Soc. London A **114,** 561 (1927)
7.99 P. J. Kramer: Phys. Rev A **6,** 2125 (1972)
7.100 J. E. Miraglia, R. D. Piacentini, R. D. Rivarola, A. Salin: J. Phys. B **14,** L 197 (1981)
7.101 K. Dettmann, G. Leibfried: Z. Phys. **210,** 43 (1968); **218,** 1 (1969)
7.102 R. Shakeshaft: Phys. Rev. A **17,** 1011 (1978)
7.103 Dž. Belkić, R. McCarroll: J. Phys. B **10,** 1933 (1977)
7.104 D. S. F. Crothers: J. Phys. B **14,** 1035 (1981)
7.105 J. S. Briggs: J. Phys. B **10,** 3075 (1977)
7.106 J. H. Macek, R. Shakeshaft: Phys. Rev. A **22,** 1441 (1980)
7.107 J. Macek, K. Taulbjerg: Phys. Rev. Lett. **46,** 170 (1981)
7.108 J. Macek, S. Alston: Phys. Rev. A **26,** 250 (1982)
7.109 G. B. Crooks, M. E. Rudd: Phys. Rev. Lett. **25,** 1599 (1970)
7.110 A. Salin: J. Phys. B **2,** 631, 1255 (1969)
7.111 J. Macek: Phys. Rev. A **1,** 235 (1970)
7.112 K. Dettmann, K. G. Harrison, M. W. Lucas: J. Phys. B **7,** 269 (1974)
7.113 R. Shakeshaft, L. Spruch: J. Phys. B **11,** L 621 (1978)
7.114. R. K. Janev, B. H. Bransden, J. W. Gallagher: J. Phys. Chem. Ref. Data **12,** 829 (1983)
7.115 J. W. Gallagher, B. H. Bransden, R. K. Janev: J. phys. Chem. Ref. Data **12,** 873 (1983)
7.116 R. K. Janev, B. H. Bransden: "Charge Exchange between Highly Charged Ions and Atomic Hydrogen: A Critical Analysis of Theoretical Data", International Atomic Energy Agency Report, INDC (NDC) – 135/GA (1982)
7.117 H. B. Gilbody: Phys. Scr. **23,** 143 (1981)
7.118 R. K. Janev, P. Hvelplund: Comm. At. Mol. Phys. **11,** 75 (1981)
7.119 S. Bliman, S. Dousson, R. Geller, B. Jaquot, D. van Houtte: 12th Int. Conf. Phys. Electron. Atom. Collisions, Gatlingburg, Tenn. USA, 1981, Abstracts of Contr. Papers, p. 704
7.120 B. A. Huber, H. J. Kahlert: J. Phys. B **13,** L 159 (1980)
7.121 H. Knudsen, H. K. Haugen, P. Hvelplund: Phys. Rev. A **23,** 597 (1981)
7.122 G. M. Raisbeck, H. J. Crowford, P. J. Lindstrom, D. E. Greiner, F. S. Beiser, H. H. Heckman: Proc. X Int. Conf. Phys. Electron Atom Collisions, Paris 1977, Abstract of Papers, p. 854
7.123 A. Müller, C. Achenbach, E. Salzborn: Phys. Lett. **70** A, 410 (1979)
7.124 H. Schrey, B. A. Huber: J. Phys. B **14,** 3197 (1981)
7.125 R. C. Isler: Phys. Rev. Lett. **38,** 1359 (1977); R. H. Dixon, R. C. Elton: Phys. Rev. Lett. **38,** 1072 (1977)
7.126 V. V. Afrosimov, Yu. S. Gordeev, A. N. Zinov'ev, A. A. Korotkov: Pis'ma Zh. Eksp. Teor. Fiz. **28,** 540 (1978) [English transl. Sov. Phys. JETP Lett. **28,** 500 (1978)]
7.127 F. J. de Heer: "Collisions between Multiply Charged Ions and Atoms Resulting in Radiation and Autoionization", in *Electronic and Atomic Collisions,* ed. by N. Oda, K. Takayanagi (North-Holland, Amsterdam 1980) p. 427
7.128 M. N. Panov: "Electron Capture into Different Excited States of Multiply Charged Ions" in *Electronic and Atomic Collisions,* ed. by N. Oda, K. Takayanagi (North-Holland, Amsterdam 1980) p. 437
7.129 R. K. Janev: Comm. At. Mol. Phys. **12,** 277 (1983)
7.130 R. K. Janev: Phys. Scr. **T3,** 208 (1983)
7.131 V. A. Abramov, F. F. Baryshnikov, V. S. Lisitsa: Pis'ma Zh. Eksp. Teor. Fiz. **27,** 494 (1978) [English transl.: Sov. Phys. JETP Lett. **27,** 464 (1978)]
7.132 P. Hvelplund, E. Samsø, L. H. Andersen, H. K. Haugen, H. Knudsen: Phys. Scr. **T3,** 176 (1983)
7.133 V. V. Afrosimov, A. A. Basalaev, Yu. S. Gordeev, E. D. Donets, A. N. Zinov'ev, S. Yu. Ovchinnikov, M. N. Panov: Pis'ma Zh. Eksp. Teor. Fiz. **34,** 332 (1981) [English transl.: Sov. Phys. – JETP Lett. **34,** 316 (1981)]

7.134 R. W. McCullough, M. Lennon, F. G. Wilkie, H. B. Gilbody: J. Phys. B **16**, L 173 (1983)
7.135 K. Kadota, D. Dijkkamp, R. Van der Woude, Pan Guang Yan, F. J. de Heer: Phys. Lett. **88** A, 135 (1982)
7.136 H. Ryufuku, K. Sasaki, T. Watanabe: Phys. Rev. A **21**, 745 (1980)
7.137 V. V. Afrosimov, A. A. Basalaev, E. D. Donets, M. N. Panov: Pis'ma Zh. Eksp. Teor. Fiz. **31**, 635 (1980) [English transl.: Sov. Phys. – JETP Lett. **31**, 575 (1980)]
7.138 S. Bliman, J. J. Bonnet, S. Dousson, D. Hitz, B. Jacquot: Phys. Scr. **T3**, 63 (1983)
7.139 T. Iwai, Y. Kaneko, M. Kimura, N. Kobayashi, S. Ohtani, K. Okuno, S. Takagi, H. Tawara, S. Tsurubuchi: Phys. Rev. A **26**, 105 (1982)
7.140 H. J. Kim, P. Hvelplund, F. W. Meyer, R. A. Phaneuf, P. H. Stelson, C. Bottcher: Phys. Rev. Lett. **40**, 1635 (1978)
7.141 H. B. Gilbody: "Collisions between Positive Ions", in *Physics of Electronic and Atomic Collisions*, ed. by S. Datz (North-Holland, Amsterdam 1982) p. 223
7.142 V. M. Galitskii, E. E. Nikitin, B. M. Smirnov: *Teoriya Stolknoveniya Atomnykh Chastits* (Atomic Collision Theory) (Nauka, Moscow 1981)
7.143 R. K. Janev, D. S. Belić: Phys. Lett. **89** A, 190 (1982)
7.144 V. P. Zhdanov: Zh. Tekh. Fiz. **46**, 204 (1976) [English transl.: Sov. Phys. Tech. Phys. **21**, 117 (1976)]
7.145 A. Jognaux, F. Brouillard, S. Szucz: J. Phys. B **11**, L 669 (1978)
7.146 V. A. Bazylev, M. I. Chibisov: Fizika plazmy **5**, 584 (1979) [English transl.: Sov. J. Plasma Phys. **5**, 327 (1979)]
7.147 R. K. Janev, D. S. Belić: J. Phys. B **15**, 3479 (1982)
7.148 B. H. Bransden, C. J. Noble: J. phys. B **14**, 1849 (1981)
7.149 T. G. Winter, G. J. Hatton, N. F. Lane: Phys. Rev. A **22**, 930 (1980)
7.150 W. Fritsch, C. D. Lin: J Phys. B **15**, 1255 (1982)
7.151 B. Peart, R. Grey, K. T. Dolder: J. Phys. B **10**, 2675 (1977)
7.152 G. C. Angel, E. C. Sewell, K. F. Dunn, H. B. Gilbody: J. Phys. B **11**, L 297 (1978)
7.153 T. P. Grozdanov, R. K. Janev: J. Phys. B **13**, L 69 (1980)
7.154 S. Bliman, S. Dousson, R. Geller, B. Jacquot, D. Van Houtte: J. Phys. (Paris) **42**, 399 (1981)
7.155 S. Bliman, J. Aubert, R. Geller, B. Jacquot, D. Van Houtte: Phys. Rev. A **23**, 1703 (1981)
7.156 D. H. Crandall, R. A. Phaneuf, F. W. Meyer: Phys. Rev. A **19**, 504 (1979)
7.157 B. A. Huber: „Zum Elektronentransfer zwischen mehrfach geladenen Ionen und Atomen oder Molekülen", Ph. D. Thesis Ruhr-Universität, Bochum (1981). See also H. J. Kahlert, B. A. Huber, K. Wieseman: J. Phys. B **16**, 449 (1983)
7.158 T. V. Goffe, M. B. Shah, H. B. Gilbody: J. Phys. B **12**, 3763 (1979)
7.159 T. P. Grozdanov, R. K. Janev: J. Phys. B **13**, 3431 (1980)
7.160 I. V. Komarov, R. K. Janev: Zh. Eksp. Teor. Fiz. **51**, 1712 (1966) [Englisch transl.: Sov. Phys. – JETP **24**, 1259 (1977)]
7.161 M. I. Chibisov: Zh. Eksp. Teor. Fiz. **70**, 1687 (1976) [English transl.: Sov. Phys. – JETP **43**, 879 (1976)]
7.162 V. V. Afrosimov, G. A. Leiko, Yu. A. Mamaev, M. N. Panov: Zh. Eksp. Teor. Fiz. **67**, 1330 (1974) [English transl.: Sov. Phys. – JETP **40**, 781 (1974)]
7.163 M. B. Shah, H. B. Gilbody: J. Phys. B **7**, 256 (1974)
7.164 J. E. Bayfield, G. A. Khayrallah: Phys. Rev. A **11**, 920 (1975)
7.165 D. H. Crandall, R. E. Olson, E. J. Shipsey, J. C. Browne: Phys. Rev. Lett. **36**, 858 (1976)
7.166 D. H. Grandall: Phys. Rev. Rev. A **16**, 958 (1977)
7.167 A. Müller, E. Salzborn: Phys. Lett. **59** A, 19 (1976)
7.168 C. Harel, A. Salin: Proc. XII Int. Conf. Phys. Electron. Atom Collisions, Gatlinburg, 1981, p. 575
7.169 P. H. Woerlee, T. M. El. Sherbini, F. J. de Heer, F.W. Saris: J. Phys. B**12**, L 235 (1979)
7.170 G. N. Ogurtsov, V. M. Mikoushkin, I. P. Flaks: Proc. XI Int. Conf. Phys. Electron. Atom Collisions, Kyoto, 1979, p. 650
7.171 A. Müller, E. Salzborn: Phys. Lett. **62** A, 391 (1977)

7.172 S. Bliman, N. Chan-Tung, S. Dousson, B. Jacquot, D. van Houtte: Phys. Rev. A **21,** 1856 (1980)
7.173 R. Gayet, R. D. Rivarola, A. Salin: J. Phys. B **14,** 2421 (1981)
7.174 I. S. Dmitriev, Yu. A. Tashaev, V. S. Nikolaev, Ya. A. Teplova, B. M. Popov: Zh. Eksp. Teor. Fitz. **73,** 1684 (1977) [English transl.: Sov. Phys. – JETP **46,** 884 (1977)]
7.175 I. S. Dmitriev, V. S. Nikolaev, L. N. Fateeva, Ya. A. Teplova: Zh. Eksp. Teor. Fiz. **43,** 361 (1962) [English transl.: Sov. Phys. – JETP **16,** 259 (1963)]
7.176 H. Damsgaard, H. K. Haugen, P. Hvelplund, H. Knudsen: Phys. Rev. A **27,** 112 (1983)
7.177 H. Klinger, A. Müller, E. Salzborn: J. Phys. B **8,** 235 (1975)
7.178 A. Müller, H. Klinger, E. Salzborn: Phys. Lett. **55** A, 11 (1975)
7.179 K. Okuno, T. Koizumu, Y. Kaneko: Proc. XI Int. Conf. Phys. Electron Atom Collision, Kyoto 1979, p. 594
7.180 M. I. Chibisov: Zh. Tekh. Fiz. **51,** 463 (1981) [English transl.: Sov. Phys. – Tech. Phys. **26,** 280 (1981)]

Chapter 8

8.1 B. H. Bransden: *Atomic Collision Theory,* 2nd ed. (Benjamin, New York 1982)
8.2 M. R. C. McDowell, J. P. Coleman: *Introduction to the Theory Ion-Atom Collisions* (North-Holland, Amsterdam 1970)
8.3 R. K. Janev, L. P. Presnyakov: Phys. Rep. **70,** 1 (1981)
8.4 B. H. Bransden: "The Theory of Charge Exchange and Ionization by Heavy Particles", in *Atomic and Molecular Physics of Controlled Thermonuclear Fusion,* ed. by C. J. Joachain, D. E. Post (Plenum, New York 1983) p. 245
8.5 R. K. Janev: Phys. Scr. **23,** 180 (1981)
8.6 H. B. Gilbody: Phys. Scr. **24,** 712 (1981)
8.7 F. J. de Heer: "Experiments on Electron Capture and Ionization by Ions", in *Atomic and Molecular Processes in Controlled Thermonuclear Fusion,* ed. by M. R. C. McDowell, A. M. Ferendeci (Plenum, New York 1979) p. 351
8.8 H. Winter, E. Bloemen, F. J. de Heer: J. Phys. B **10,** L 311 (1977)
8.9 D. S. F. Crothers, J. G. Hughes: Phys. Rev. Lett. **43,** 1584 (1979)
8.10 M. Kimura, W. R. Thorson: Phys. Rev. **24,** 1780 (1981)
8.11 I. M. Cheshire, D. F. Gallaher, A. J. Taylor: J. Phys. B **3,** 813 (1970)
8.12 W. Fritch, C. D. Lin: Phys. Rev. A **26,** 762 (1982)
8.13 J. F. Reading, A. L. Ford, R. L. Becker: J. Phys. B **14,** 1995 (1981)
8.14 D. F. Gallaher, L. Wilets: Phys. Rev. **169,** 139 (1968)
8.15 R. Shakeshaft: Phys. Rev. A **14,** 1626 (1976)
8.16 D. R. Bates: Proc. Phys. Soc. **73,** 277 (1959)
8.17 L. P. Presnyakov: Proc. P. N. Lebedev Phys. Inst. **30,** 256 (1964) (in Russian)
8.18 R. K. Janev, L. P. Presnyakov: J. Phys. B **13,** 4233 (1980)
8.19 M. J. Seaton: Planet. Space Sci. **12,** 55 (1964)
8.20 L. Landau, E. Lifshitz: *Quantum Mechanics,* 3rd. ed. (Pergamon, Oxford 1977)
8.21 H. Ryufuku: Phys. Rev. A **25,** 720 (1982)
8.22 J. T. Park, J. E. Aldag, J. E. George, J. L. Peacher: Phys. Rev. A **14,** 608 (1976)
8.23 H. A. Bethe, E. E. Salpeter: *Quantum Mechanics of One- and Two-Electron Atoms* (Springer, Berlin, Göttingen, Heidelberg 1957)
8.24 M. J. Seaton: Proc. Phys. Soc. **79,** 1105 (1962)
8.25 L. P. Presnyakov, A. D. Ulantsev: Short Comm. Phys. **8,** 42 (1973) (in Russian)
8.26 K. Alder, A. Bohr, T. Huus, B. Mottelson, A. Winter: Rev. Mod. Phys. **28,** 432 (1956)
8.27 J. C. Briggs: Rep. Prog. Phys. **39,** 217 (1976)
8.28 E. Fitchard, A. L. Ford, J. F. Reading: Phys. Rev. A **16,** 1325 (1977)
8.29 E. L. Duman, L. I. Men'shikov, B. M. Smirnov: Zh. Eksp. Teor. Fiz. **76,** 512 (1979) [English transl.: Sov. Phys. – JETP **49,** 260 (1979)]

8.30 J. M. Hansteen, O. M. Johnsen, L. Kocbach: Atom. Data Nucl. Data Tables **15,** 305 (1975)
8.31 L. Kocbach, J. M. Hansteen, R. Gundersen: Nucl. Instrum. Methods **169,** 281 (1980)
8.32 W. Magnus: Commun. Pure Appl. Math. **7,** 649 (1954)
8.33 J. Eichler: Phys. Rev. A **15,** 1856 (1977)
8.34 A. Salop, J. Eichler: J. Phys. B **12,** 257 (1979)
8.35 E. Merzbacher, W. H. Lewis: In *Corpuscles and Radiation in Matter II,* ed. by S. Flügge, Handbuch der Physik, Vol. **34** (Springer, Berlin, Göttingen, Heidelberg 1958) p. 166
8.36 H. van Regemorter: Astrophys. J. **132,** 906 (1962)
8.37 R. K. Janev: Phys. Lett. **83** A, 5 (1981)
8.38 M. B. Shah, H. B. Gilbody: J. Phys. B **14,** 2361, 2931 (1981); **15,** 413 (1982)
8.39 P. Hvelplund, H. K. Haugen, K. Knudsen: Phys. Rev. A **22,** 1930 (1980)
8.40 H. K. Haugen, L. Andersen, P. Hvelplund, H. Knudsen: Phys. Rev. A **25,** 1950 (1982)
8.41 R. K. Janev, P. Hvelpund: Comm. Atom. Mol. Phys. **11,** 75 (1981)
8.42 R. Shakeshaft: Phys. Rev. A **18,** 1930 (1978)
8.43 R. E. Olson, A. Salop: Phys. Rev. A **16,** 531 (1977)
8.44 R. E. Olson: Phys. Rev. A **18,** 2464 (1978)
8.45 J. D. Garcia, E. Gerjuoy, J. E. Welker: Phys. Rev. **165,** 66 (1968)
8.46 T. P. Grozdanov: (unpublished, 1982)
8.47 D. R. Bates, A. E. Kingston: Adv. At. Mol. Phys. **6,** 269 (1970)
8.48 J. J. Thomson: Philos. Mag. **23,** 449 (1912)
8.49 N. Bohr: K. Dan. Vidensk. Selsk. Mat. Fys. Medd. **18,** No. 8 (1948)
8.50 E. Gerjuoy, B. K. Thomas: Rep. Prog. Phys. **37,** 1345 (1974)
8.51 J. E. Golden, J. H. McGuire: Phys. Rev. A **12,** 80 (1975)
8.52 J. H. McGuire: Phys. Rev. A **26,** 143 (1982)
8.53 Dž. Belkić: J. Phys. B **11,** 3529 (1978)
8.54 Dž. Belkić: J. Phys. B **13,** L 589 (1980)
8.55 D. R. Bates, G. Griffing: Proc. Phys. Soc. A **66,** 961 (1958)
8.56 W. L. Fite, R. J. Stebbings, D. G. Hummer, R. T. Brackman: Phys. Rev. **119,** 663 (1960)
8.57 H. B. Gilbody: "Collisions between Positive Ions", in *Physics of Electronic and Atomic Collisions,* ed. by S. Datz (North-Holland, Amster 1982) p. 223
8.58 K. H. Berkner, W. G. Graham, R. V. Pyle, A. S. Schlachter, J. W. Stearns: Phys. Rev. A **23,** 2891 (1981)
8.59 A. S. Schlachter, K. H. Berkner, W. G. Graham, J. W. Stearns, J. A. Tanis: Phys. Rev. A **24,** 1110 (1981)
8.60 M. B. Shah, H. B. Gilbody: J. Phys. B **15,** 3441 (1982)
8.61 K. H. Berkner, W. G. Graham, R. V. Pyle, A. S. Schlachter, J. W. Stearns, R. E. Olson: J. Phys. B **11,** 875 (1978)
8.62 J. H. McGuire, H. Weaver: Phys. Rev. A **16,** 41 (1977)
8.63 R. E. Olson: J. Phys. B **12,** 1842 (1979)
8.64 K. H. Berkner, W. G. Graham, R. V. Pyle, A. S. Schlachter, J. W. Stearns: Proc. X Int. Conf. Phys. Electron Atom Collisions (Abstract of papers), Paris (1977) p. 542
8.65 H. Damsgaard, H. K. Haugen, P. Hvelplund, H. Knudsen: Phys. Rev. A **27,** 112 (1983)
8.66 H. Knudsen, P. Hvelplund, L. H. Andersen, S. Bjørnelund, M. Frost, H. K. Haugen, E. Samsø: Phys. Scr. **T3,** 101 (1983)
8.67 L. V. Keldlysh: Zh. Eksp. Teor. Fiz. **47,** 1945 (1964) [English transl.: Sov. Phys. – JETP **20,** 1307 (1965)]
8.68 E. L. Duman, L. I. Men'shikov, B. M. Smirnov: Zh. Eksp. Teor. Fiz. **76,** 516 (1979) [English transl.: Sov. Phys. – JETP **49,** 260 (1979)]
8.69 I. V. Komarov, E. A. Solov'ev: "Quasi-classical Approximation for the Charge Exchange Reactions of Highly Charged Ions with Hydrogen Atom", in *Problem in the*

Theory of Atomic Collisions, Vol. **2**, ed. by G. F. Drukarev (Leningrad State Univ. Press, Leningrad 1980) p. 234 (in Russian)
8.70 A. Salop, R. E. Olson: Phys. Lett. **71** A, 407 (1979)
8.71 R. E. Olson: Phys. Rev. A **18**, 2464 (1978)
8.72 R. R. Janev: Phys. Rev. A **28**, 1810 (1983)
8.73 R. E. Olson, K. H. Berkner, W. G. Graham, R. V. Pyle, A. S. Schlachter, J. W. Stearns: Phys. Rev. Lett. **41**, 163 (1978)
8.74 H. Knudsen, H. K. Haugen, P. Hvelplund: Phys. Rev. A **23**, 597 (1981)
8.75 I. S. Dmitriev, V. S. Nikolaev: Zh. Eksp. Teor. Fiz. **44**, 660 (1963) [English transl.: Sov. Phys. – JETP **17**, 447 (1963)]
8.76 I. S. Dmitriev, Ya. M. Zhileikin, V. S. Nikolaev: Zh. Eksp. Teor. Fiz. **49**, 500 (1965) [English transl.: Sov. Phys. – JETP **22**, 352 (1966)]
8.77 G. H. Gillespie: Phys. Rev. A **18**, 1967 (1978)
8.78 H. Knudsen, L. H. Andersen, H. K. Haugen, P. Hvelplund: Phys. Scr. **26**, 132 (1982)
8.79 T. R. Dillingham, J. R. MacDonald, P. Richard: Phys. Rev. A **24**, 1237 (1981)
8.80 C. F. Barnett, H. K. Reynolds: Phys. Rev. **109**, 355 (1958)
8.81 I. S. Dmitriev, Yu. A. Tashaev, V. S. Nikolaev, Ya. A. Teplova, B. M. Popov: Zh. Eksp. Teor. Fiz. **73**, 1684 (1977) [English transl.: Sov. Phys. – JETP **46**, 884 (1977)
8.82 I. S. Dmitriev, N. F. Vorobiev, V. P. Zaikov, Zh. M. Konovalova, V. S. Nikolaev, Ya. A. Teplova, Yu. A. Fainberg: J. Phys. B **15**, L 351 (1982)

Chapter 9

9.1 B. Crasemann (ed.): *Atomic Inner-Shell Processes,* Vols. **1**, **2** (Academic, New York 1975)
9.2 I. Sellin (ed.) *Structure and Collisions of Ions and Atoms,* Topics Curr. Phys., Vol. **5** (Springer, Berlin Heidelberg, New York 1978)
9.3 A. Niehaus: Comments At. Mol. Phys. **9**, 153 (1980)
9.4 R. K. Janev, L. P. Presnyakov: Phys. Rep. **70**, 1 (1981)
9.5 B. M. Smirnov: *Asymptotic Methods in the Theory of Atomic Collisions* (Nauka, Moscow 1973)
9.6 T. P. Grozdanov, R. K. Janev, A. A. Zembekov: (unpublished, 1981)
9.7 T. P. Grozdanov, R. K. Janev: J. Phys. B **13**, 3431 (1980)
9.8 H. Winter, E. Bloemen, F. J. de Heer: J. Phys. B **10**, L 311 (1977)
9.9 P. H. Woerlee, T. M. El Sherbini, F. J. de Heer, F. W. Saris: J. Phys. B **12**, L 235 (1979)
9.10 L. K. Kishinevskii, E. S. Parilis: Zh. Eskp. Teor. Fiz. **55**, 1932 (1968) [English transl.: Sov. Phys. – JETP **28**, 1020 (1969)]
9.11 I. P. Flaks, G. N. Ogurtsov, N. V. Fedorenko: Zh. Eksp. Teor. Fiz. **41**, 1438 (1961) [English transl.: Sov. Phys. – JETP **14**, 781 (1962)]
9.12 U. Fano: Phys. Rev. **124**, 1866 (1961)
9.13 V. A. Bazylev, N. K. Zhevago: Zh. Eksp. Teor. Fiz. **69**, 853 (1975) [English transl.: Sov. Phys. – JETP **42**, 436 (1976)]
9.14 V. A. Bazylev, N. K. Zhevago, M. I. Chibisov: Zh. Eksp. Teor. Fiz. **71**, 1285 (1976) [English transl.: Sov. Phys. – JETP **44**, 671 (1976)]
9.15 A. Z. Devdariani, V. N. Ostrovskii, Yu. N. Sebyakin: Zh. Eksp. Teor. Fiz. **71**, 909 (1976) [English transl.: Sov. Phys. – JETP **44**, 477 (1976)]; **73**, 412, (1977) [**46**, 215 (1977)]
9.16 V. N. Ostrovskii: Zh. Eksp. Teor. Fiz. **72**, 2079 (1977) [English transl.: Sov. Phys. – JETP **45**, 1092 (1977)]
9.17 A. Z. Devdariani, V. N. Ostrovskii, Yu. N. Sebyakin: Zh. Eksp. Teor. Fiz. **76**, 529 (1979) [English transl.: Sov. Phys. – JETP **49**, 266 (1979)]
9.18 H. Winter, Th. M. El Sherbini, F. J. de Heer, A. Salop: Phys. Lett. **68** A, 211 (1978)
9.19 C. L. Cocke, R. DuBois, T. J. Gray, E. Justiniano, C. Can: Phys. Rev. Lett. **46**, 1671 (1981)

9.20 E. Justiniano, C. L. Cocke, T. J. Gray, R. DuBois, C. Can: Phys. Rev. A **24**, 2953 (1981)
9.21 H. Damsgaard, H. K. Haugen, P. Hvelplund, H. Knudsen: Phys. Rev. A **27**, 112 (1983)
9.22 W. Groh, A. Müller, C. Achenbach, A. S. Schlachter, E. Salzborn: Phys. Lett. **85**A, 77 (1981)
9.23 W. Groh, A. S. Schlachter, A. Müller, E. Salzborn: J. Phys. B **15**, L 207 (1982)
9.24 E. Salzborn, W. Groh, A. Müller, A. S. Schlachter: Phys. Scr. **T3**, 148 (1983)
9.25 J. D. Garcia, R. J. Fortner, T. M. Kavanagh: Rev. Mod. Phys. **45**, 111 (1973)
9.26 Q. C. Kessel, B. Fastrup: In *Case Studies in Atomic Physics,* Vol. **3**, ed. by M. R. C. McDowell, E. W. McDaniel (North-Holland, Amsterdam 1973) p. 137
9.27 J. S. Briggs: Rep. Prog. Phys. **39**, 217 (1976)
9.28 B. Fastrup: "Experimental Studies of Inner-Shell Excitation in Slow Ion-Atom Collisions", in *The Physics of Electronic and Atomic Collisions,* ed. by J. S. Risley, R. Geballe (Univ. of Washington Press, Seattle 1975) p. 361
9.29 W. Meyerhof, K. Taulbjerg: Ann. Rev. Nucl. Sci. **27**, 279 (1977)
9.30 N. Stolterfoht: "Introduction to Inner Shell Processes in Heavy Ion-Atom Collisions", in *The Physics of Ionized Gases* (Invited Lectures of SPIG-78), ed. by R. K. Janev (Inst. of Physics, Beograd 1978) p. 93
9.31 E. Merzbacher, H. W. Lewis: In *Corpuscles and Radiation in Matter II,* ed. by S. Flügge, Handbuch der Physik, Vol. **34**, (Springer, Berlin, Göttingen, Heidelberg 1958) p. 166
9.32 J. Bang, J. M. Hansteen: K. Dan. Vidensk. Selsk. Mat. Fys. Medd. **31**, No. 13 (1959)
9.33 J. H. McGuire, P. Richard: Phys. Rev. A **1374** (1973)
9.34 U. Fano, W. Lichten: Phys. Rev. Lett. **14**, 627 (1965)
9.35 W. Lichten: Phys. Rev. **164**, 131 (1967)
9.36 V. V. Afrosimov, N. V. Fedorenko: Sov. Phys. – Tech. Phys. **2**, 2378 (1957)
9.37 G. H. Morgan, E. Everhart: Phys. Rev. **128**, 667 (1962)
9.38 J. S. Briggs, J. H. Macek: J. Phys. B **5**, 579 (1972); **6**, 982 (1973)
9.39 J. H. Macek, J. S. Briggs: J. Phys. B **6**, 841 (1973)
9.40 B. Fastrup, E. Bøving, G. A. Larsen, P. Dahl: J. Phys. B **7**, L 206 (1974)
9.41 D. R. Bates, D. Williams: Proc. Phys. Soc. **83**, 425 (1964)
9.42 D. R. Bates, D. Spreavak: J. Phys. B **3**, 1483 (1970)
9.43 K. Taulbjerg, J. S. Briggs, J. Vaaben: J. Phys. B **8**, 1351 (1976)
9.44 R. D. Piacentini, A. Salin: J. Phys. B **7**, L 311 (1974); **7**, 1666 (1974)
9.45 K. Taulbjerg, J. S. Briggs: J. Phys. B **8**, 1895 (1975)
9.46 Yu. N. Demkov, C. V. Kunasz, V. N. Ostrovskii: Phys. Rev. A **18**, 2097 (1978)
9.47 W. E. Meyerhof: Phys. Rev. Lett. **31**, 1341 (1973)
9.48 E. Bøving: J. Phys. B **10**, L 63 (1977)
9.49 W. E. Meyerhof, A. Ruetschi, Ch. Stoller, M. Stöckli, W. Wölfli: Phys. Rev A **20**, 154 (1979)
9.50 J. S. Greenberg, P. Vincent, W. Lichten: Phys. Rev. A **16**, 964 (1977)
9.51 V. V. Afrosimov, Yu. S. Gordeev, A. N. Zinov'ev, D. H. Rasulov, A. P. Shergin: Pis'ma Zh. Eksp. Teor. Fiz. **21**, 535 (1975) [English transl.: Sov. Phys. – JETP Lett. **21**, 302 1975)]
9.52 W. Wölfli, Ch. Stoller, G. Bonani, M. Suter, M. Stöckli: Phys. Rev. Lett. **35**, 656 (1976)
9.53 T. P. Hoogkamer, P. Woerlee, F. W. Saris, M. Gavrila: J. Phys. B **9**, L 145 (1976)
9.54 A. R. Knudson, K. W. Hill, P. G. Burkhalter, D. J. Nagel: Phys. Rev. Lett. **37**, 679 (1976)
9.55 T. Åberg, K. A. Jamison, P. Richard: Phys. Rev. Lett. **37**, 63 (1976)
9.56 S. V. Khristenko: Phys. Lett **59** A, 202 (1976)
9.57 M. Ya. Amusia, I. S. Lee, A. N. Zinoviev: Phys. Lett. **60** A, 300 (1977)
9.58 U. I. Safronova, V. S. Senashenko: J. Phys. B **10**, L 271 (1977)
9.59 L. M. Kishinevskii, V. I. Matveev, E. S. Parilis: Pis'ma Zh. Tekhn. Fiz. **2**, 710 (1976) [English transl.: Sov. Phys. – Tech. Phys. Lett. **6**, 609 (1976)]

9.60 L. N. Ivanov, U. I. Safronova, V. S. Senashenko, D. S. Viktorov: J. Phys. B **11,** L 175 (1978)
9.61 D. Liesen, P. Armbruster, H.-H. Behncke, F. Bosch, S. Hagmann, P. H. Mokler, H. Schmidt-Bocking, R. Schuch: "Intermediate, Super-heavy Atomic Collission Systems – Experiments at the UNILAC", in *Electronic and Atomic Collisions,* ed. by N. Oda, K. Takayangi (North-Holland, Amsterdam 1980) p. 337
9.62 J. S. Greenberg: "Radiative Processes in Super-heavy Quasiatoms – A Search for Position Emission" in *Electronic and Atomic Collisions,* ed. by N. Oda, K. Takayanagi (North-Holland, Amerstam 1980) p. 351
9.63 J. Reinhardt, W. Greiner: Rep. Prog. Phys. **40,** 219 (1977)
9.64 Ya. B. Zel'dovich, V. S. Popov: Sov. Phys. Usp. **14,** 673 (1972)
9.65 W. Betz, G. Heiligenthal, J. Reinhardt, R. K. Smith, W. Greiner: "Important Problems in Future Heavy Ion Atomic Physics", in *The Physics of Electronic and Atomic Collisions,* ed. by J. S. Risley, R. Geballe (U. Washington Press, Seattle 1975) p. 531
9.66 F. W. Saris, W. F. van der Weg, H. Tawara, R. Laubert: Phys. Rev. Lett. **28,** 717 (1972)
9.67 F. W. Saris, Th. P. Hoogkamer: in *Atomic Physics* Vol. **5,** ed. by R. Prior, H. Shugart (Plenum, New York 1977) p. 509
9.68 P. Mokler, F. Folkman: In *Structure and Collisions of Ions and Atoms,* ed. by I. Sellin, Topics Curr. Phys., Vol. 5 (Springer, Berlin, Heidelberg 1978) Chap. 6
9.69 P. Vincent, C. K. Davis, J. C. Greenberg: Phys. Rev. **18,** 1978 (1978)
9.70 J. C. Greenberg, C. K. Davies, P. Vincent: Phys. Rev. Lett. **33,** 473 (1974)
9.71 G. Kraft, P. H. Mokler, H. J. Stein: Phys. Rev. Lett. **33,** 476 (1974)
9.72 B. Müller: "Radiative Processes in Transient Quasi-Molecules" in *The Physics of Electronic and Atomic Collisions,* ed. by J. S. Risley, R. Geballe (Univ. of Washington Press, Seattle 1975) p. 481
9.73 W. Lichten: Phys. Rev. A **9,** 1458 (1974)
9.74 J. S. Briggs, J. Macek: J. Phys. B **7,** 1312 (1974)
9.75 B. Müller, W. Greiner: Phys. Rev. Lett. **33,** 469 (1974)
9.76 W. E. Meyerhof, T. K. Saylor, S. M. Lazarus, A. Little, B. B. Triplett, L. F. Chase, A. R. Anholt: Phys. Rev. Lett. **32,** 1279 (1974)
9.77 I. Pomeranchuk, J. Smorodinsky: J. Phys. USSR **9,** 97 (1945)
9.78 S. S. Gershtein, Ya. B. Zel'dovitch: Sov. Phys. – JETP **30,** 358 (1970)
9.79 W. Pieper, W. Greiner: Z. Phys. **218,** 327 (1969)
9.80 B. Müller, J. Rafelski, W. Greiner: Z. Phys. **257,** 82, 163 (1972)
9.81 H. Backe, L. Handschug, F. Hessberger, E. Kankeleit, L. Richter, F. Weik, R. Willwater, H. Bokemeyer, Y. Nakayama, J. S. Greenberg: Phys. Rev. Lett. **40,** 1443 (1978)
9.82 C. Kozhusharov, P. Kienle, E. Berdermann, H. Bokemeyer, J. S. Greenberg, Y. Nakayama, P. Vincent, H. Backe, L. Handschug, E. Krankeleit: Phys. Rev. Lett. **42,** 376 (1979)
9.83 J. Reinhardt, W. Greiner, B. Müller: "Positrons and Other Messengers from Superheavy Quasiatoms: A Theoretical Survey", in *Electronic and Atomic collisions,* ed. by N. Oda, K. Takayanagi (North-Holland, Amsterdam 1980) p. 369
9.84 B. Müller, W. Greiner: Z. Naturforsch. **31** a, 1 (1976)
9.85 J. Rafelski, B. Müller: Phys. Rev. Lett. **36,** 517 (1976)
9.86 J. Rafelski, B. Müller, W. Greiner: Z. Phys. A **285,** 49 (1978)

Chapter 10

10.1 R. H. Fowler: *Statistische Mechanik* (Akademische Verlagsgesellschaft, Leipzig 1931)
10.2 A. V. Gurevitch: Zh. Eksp. Teor. Fiz. **39,** 1296 (1960) [English transl.: Sov. Phys. – JETP **12,** 798 (1960)]

10.3 A. V. Gurevitch, Ya. A. Istomin: Zh. Eksp. Teor. Fiz. **77,** 939 (1979) [English transl.: Sov. Phys. – JETP **50,** 541 (1979)]
10.4 S. I. Syrovatski, B. V. Somov: Usp. Fiz. Nauk. **120,** 217 (1976) [English transl.: Sov. Phys. – Usp. **20,** 207 (1976)
10.5 A. S. Shlyaptseva, A. M. Urnov, A.V. Vinogradov: *On Diagnostics of Suprathermal Electrons in High-Temperature Plasmas,* Rpt. No. 193, P. N. Lebedev Inst., Moscow (1981)
10.6 L. M. Biberman, V. S. Vorobjev, I. T. Yakubov: Usp. Fiz. Nauk, **128,** 233 (1977) [English transl.: Sov. Phys. – Usp. **28,** 202 (1977)]
10.7 M. A. Hayes, M. J. Seaton: J. Phys. B **10,** L 573 (1977)
10.8 M. R. C. McDowell, L. A. Morgan, V. P. Myerscough, T. Scott: J. Phys. B **10,** 2727 (1977)
10,9 A. Burgess, D. G. Hummer, J. Tully: Phil. Trans. Roy. Soc. A **266,** 225 (1975)
10.10 L. A. Vainshtein: Zh. Eksp. Teor. Fiz. **40,** 32 (1975) [English transl.: Sov. Phys. – JETP **13,** 18 (1975)]
10.11 J. Dubau, M. Loulerque: Phys. Scr. **23,** 136 (1981)
10.12 I. I. Sobel'man, L. A. Vainshtein, E. A. Yukov: *Excitation of Atoms and Broadening of Spectral Lines,* Springer Ser. Chem. Phyas., Vol. 7 (Springer, Berlin, Heidelberg 1981)
10.13 I. I. Sobel'man: *Atomic Spectra and Radiative Transitions,* Springer Ser. Chem. Phys., Vol. 1 (Springer, Berlin, Heidelberg 1979)
10.14 A. K. Pradhan, D. W. Norcross, D. G. Hummer: Astrophys. J. **246,** 1031 (1981)
10.15 A. K. Pradhan: Phys. Rev. Lett. **47,** 494 (1981)
10.16 P. L. Dufton, K. A. Berrington, P. G. Burke, A. E. Kingston: Astron. Astrophys. **62,** 111 (1978)
10.17 M. J. Seaton: Planet. Space Sci. **12,** 55 (1964)
10.18 W. Lotz: Z. Phys. **232,** 101 (1970)
10.19 M. Abramowitz, I. A. Stegun: *Handbook of Mathematical Functions* (Dover, New York 1965)
10.20 V. P. Shevelko, L. A. Vainshtein: *Electron Impact Ionization of Highly Charged Ions,* Rpt. No. 78, P. N. Lebedev Inst., Moscow (1983); V. P. Shevelko, A. M Urnov, L. A. Vainshtein, A. Müller: Mon. Not. R. Astron. Soc., Short Commun. **203,** 45 (1983)
10.21 L. A. Jones, E. Källne, D. B. Thomson: J. Phys. B **10,** 187 (1977)
10.22 W. L. Rowan, J. R. Roberts: Phys. Rev A **19,** 90 (1979)
10.23 P. Greve, M. Kato, H.-J. Kunze, R. S. Hornady: Phys. Rev. A **24,** 429 (1981)
10.24 P. Greve: Thesis, Bochum University (1981)
10.25 E. Hinnov: J. Opt. Soc. Am. **56,** 1179 (1966); **57,** 1392 (1968)
10.26 C. Breton, C. de Michelis, M. Finkelthal, M. Mattioli: Phys. Rev. Lett. **41,** 110 (1978)
10.27 D. H. Crandall, R. A. Phaneuf, D. G. Gregory: *Electron Impact Ionization of Multicharged Ions,* Rpt. No. ORNL/TM-7020, Oak Ridge, Tenn. (1979)
10.28 D. H. Crandall: Phys. Scr. **23,** 153 (1981)
10.29 A. Müller, R. Frodl: Phys. Rev. Lett. **44,** 29 (1980)
10.30 H.-J. Kunze: Space Sci. Rev. **13,** 565 (1972)
10.31 H. Jacubowitz, D. L. Moores: J. Phys. B **14,** 3733 (1981)
10.32 L. A. Jones: Bull. Am. Phys. Soc. **24,** 775 (1979)
10.33 I. L. Beigman, A. M. Urnov, B. N. Chichkov: *Ionization Equilibrium in Plasmas,* Rpt. No. 7, P. N. Lebedev Inst., Moscow (1981)
10.34 J. Bergeron, S. Colin-Souffrin: Astron. Astrophys. **36,** 27 (1974)
10.35 C. Jordan: Mon. Not. R. Astron. Soc. **142,** 501 (1969)
10.36 M. Landini, B. L. Monsignori-Fossi: Astron. Astrophys. Suppl. **7,** 291 (1972)
10.37 H. Nussbaumer, P. J. Storey: Astron. Astrophys. **44,** 321 (1975)
10.38 V. L. Jacobs, J. Davis, P. C. Kepple, M. Blaha: Astrophys. J. **211,** 605 (1977)
10.39 I. L. Beigman, L. A. Vainshtein, B. N. Chichkov: Zh. Eksp. Teor. Fiz. **80,** 964 (1981) [English transl.: Sov. Phys. – JETP **53,** 541 (1981)]
10.40 A. Burgess: Astrophys. J. **139,** 776 (1964)

10.41 F. Bely-Dubau, M. Bitter, J. Dubau, P. Faucher, A. H. Gabriel, K. W. Hill, S. von Goeller, N. Sauthoff, S. Volonté: Phys. Rev. Lett. **93** A, 189 (1983)
10.42 R. A. Hulse, D. E. Post, D. R. Mikkelsen: J. Phys. B **13**, 3895 (1980)
10.43 H. B. Gilbody: Phys. Scr. **23**, 143 (1981)
10.44 T. G. Heil, S. E. Butler, A. Dalgarno: Phys. Rev. A **23**, 1100 (1981)
10.45 H. F. Beyer, R. Mann, F. Folkmann: J. Phys. B **14**, L 377 (1981)
10.46 H. F. Beyer, R. Mann, F. Folkmann: *Electron Capture by Slow* Ne^{8+} *Recoil Ions,* Preprint GSI-81-20 a (Darmstadt 1981)
10.47 R. McCarroll, P. Valiron: Astron. Astrophys. **53**, 83 (1976)
10.48 W. D. Watson: "Gas Space Reactions in Astrophysics", in Ann. Rev. Astron. Astrophys. **16**, 445 (1978)
10.49 S. E. Butler, T. G. Heil, A. Dalgarno: *Charge Transfer of Multiply-Charged Ions with Hydrogen and Helium,* Preprint No. 1281 a, Center for Astrophysics, Cambridge, MA (1981)
10.50 S. E. Butler, A. Dalgarno: Preprint No. 1281 b, Center or Astrophysics, Cambridge, MA (1981)

Subject Index